SOLID MATTER

SOLID MATTER

Editor

Skender Avakian

scitus
academics

Solid Matter

Edited by **Skender Avakian**

This Book is Printed on Acid-Free Paper. ♾

ISBN: 9781681178578
Library of Congress Control Number: 2017948884

SCITUS Academics LLC,
616, Corporate Way, Suite 2, 4766,
Valley Cottage, NY 10989

www.scitusacademics.com

Printed in United States of America.

Preface

Matter is the "stuff" of the universe, the atoms, molecules and ions that make up all physical substances. Solid is one of the three fundamental states of matter. It is characterized by structural rigidity and resistance to changes of shape or volume. In a solid, these particles are packed closely together and are not free to move about within the substance. Molecular motion for the particles in a solid is confined to very small vibrations of the atoms around their fixed positions; therefore, solids have a fixed shape that is difficult to change. Solids also have a definite volume; that is, they keep their size no matter how you try to change them. Unlike a liquid, a solid object does not flow to take on the shape of its container, nor does it expand to fill the entire volume available to it like a gas does. Even in solids there is a small space between the atoms. Depending on how tight the atoms are packed determines the density of matter. This means that a one inch block of wood is not as dense as a one inch block of gold. There is more space between the atoms of the wood than the atoms of the gold. The atoms in a solid are tightly bound to each other, either in a regular geometric lattice or irregularly (an amorphous solid such as common window glass).

This volume 'Solid Matter' covers the research in condensed matter and materials science. Emphasis is placed on key aspects of solid state synthesis, structure-property relationships, theory and functionalities, in relation with experiments. It encompasses all aspects of the field from the more classical problems of structural analysis to mechanics of solids continually interacting with other media and including fracture, flow, wave propagation, heat transfer, thermal effects in solids, optimum design methods, model analysis, structural topology and numerical techniques. This book will be interest to research workers, students and engineers in the field.

Table of Contents

CHAPTER 1

Solid-Phase Aromatic Nitration with $Mg(NO_3)_2$ on Silica Gel

Tomoko Matsumoto[1], Ayaka Yamauchi[2], Jun Ishikawa[2], Guan-Hong Jin[2], Jin Matsumoto[2], Yoshiyuki Fueda[3], Masahide Yasuda[2*]

[1]*Center for Collaborative Research and Community Cooperation, University of Miyazaki, Miyazaki, Japan*

[2]*Department of Applied Chemistry, Faculty of Engineering, University of Miyazaki, Miyazaki, Japan*

[3]*Fuji Silysia Chemical Ltd., Miyazaki, Japan*

ABSTRACT

Nitroaromatics are usually prepared using a mixed acid of nitric acid with strong acids. However, the use of strong acids caused dangerous work-up and the disposal of large amounts of acid-waste. Therefore, much effort has been made on the improvement of nitration process without strong acids. We examined solid-phase aromatic nitration with $Mg(NO_3)_2$ on silica gel in order to establish the nitration process without strong acids. The nitration of 1,2- and 1,3-, 1,4-dimethoxybenzenes and 4-methylanisole with $Mg(NO_3)_2$ proceeded by heating on silica gel at 150°C for 4 - 5 h to produce the nitroaromatics. The nitration of 1,3,5-trimethoxybenzene produced the nitrated dimer, 2,4,6,2', 4',6'-hexamethoxy-3-nitrobiphenyl, which was not isolated in other solid-phase nitration. In the cases of naphthalene derivatives, the α-nitrated compounds were obtained. In the cases of p-cresol and 2-naphthol, the esterification occurred at the hydroxyl group to give 4-tolyl nitrate and 2-naphthyl nitrate, respectively. It is synthetic interest to note that nitrate esters were isolated in solid phase. Thus

$Mg(NO_3)_2$-SiO_2composite was mild reagent for solid-phase nitration. Acidity of $Mg(NO_3)_2$-SiO_2composite was determined to be pH 0.96 by the measurement of absorption spectra on a micro spectrophotometer using meso-tetra(p-cyanophenyl)porphyrin as a pH-indicator. $Mg(NO_3)_2$-SiO_2 composite made acidic conditions. Therefore, it was suggested that $Mg(NO_3)_2$ reacted with proton on silica gel to form the NO_2^+. Thus, electron-rich aromatic hydrocarbons led the efficient nitration through electrophilic attack of NO_2^+. After the nitration, acidic $Mg(NO_3)_2$-SiO_2 composite could be turned into neutrality by exposing wet conditions and disposed safely since the composite did not involve harmful elements. Thus the solid-phase nitration using $Mg(NO_3)_2$-SiO_2 composite will provide safety and environmentally conscious chemical process.

Keywords: Aromatic Nitration, Silica Gel, $Mg(NO_3)_2$, Solid State

1. INTRODUCTION

Nitroaromatics are important chemicals which are applicable to dyes, explosives, pharmaceuticals, and the intermediates to prepare amines. Industrial synthesis of nitroaromatics has been achieved using a mixed acid of nitric acid with strong acids (e.g. sulfuric acid). However, the use of strong acids caused dangerous work-up and the disposal of large amounts of acid-waste. Therefore, much effort has been made on the improvement of nitration process without strong acids [1]. Preliminarily in order to avoid the risk of sulfuric acid, sulfuric acid was supported on silica gel to apply to the aromatic nitration with nitric acid (HNO_3) [2]. Recently, silica gel which is the most commonly desiccant [3] has been used as a dehydration agent instead of sulfuric acid together with nitration reagents such as $Bi(NO_3)_3$ [4] , HNO_3 [5] , $Ce(NH_4)(NO_3)_5$ [6] , and $AcONO_2$ [7] in solid state. Also, montmorillonite and charcoal were used for solid-state aromatic nitration with $Bi(NO_3)_3$[8] and $Zn(NO_3)_2$ [9]. Also, $Bi(NO_3)_3$ [10] and $NaNO_2$ [11] were used as reagents for indirect nitration in solution phase. Thus solid-phase nitration using nitrate salts such as $Bi(NO_3)_3$, $Al(NO_3)_3 \cdot 9H_2O$ [12] [13] and $Zn(NO_3)_2$ on silica gel has been reported so far.

More than a decade ago, we started to develop cobalt-free humidity indicator for silica gel desiccant using Mg salts and porphyrins [14]. In those days, the desiccant ability of silica gel has been checked by color change of the $CoCl_2$ adsorbed on silica gel ($CoCl_2$-SiO_2, silica gel blue). However, considerable caution had to be paid to the $CoCl_2$-SiO_2 because $CoCl_2$ was determined to be carcinogenic to humans by International Agency for Research on Cancer [15]. Therefore, it was required to use the humidity indicators instead of $CoCl_2$-SiO_2. On the other hand, Gordeeva and co-workers have reported that acidic conditions were made by the reaction of SiO_2 with $CaCl_2$ under dry conditions [16]. In order to develop new type of humidity indicator, we mixed $MgCl_2$-SiO_2 with pH-sensitive tetraphenylporphyrin and dried under heating to prepare a porphyrin-$MgCl_2$- SiO_2(Indicator F™, Fuji Silysia), which caused color change from green under dry conditions to pink under wet conditions [17]. During the investigations, we measured the pH of the composites of Mg salts ($MgCl_2$, $MgSO_4$, $Mg(NO_3)_2$) with SiO_2 using several kinds of tetraarylporphyrins with different basicity [18]. In the case of the combination of $Mg(NO_3)_2$-SiO_2 and tetra(p-methoxyphenyl) porphyrin, the porphyrin was nitrated. This observation led us to use $Mg(NO_3)_2$ for the solid-phase aromatic nitration since there was no report on the nitration using $Mg(NO_3)_2$.

Here, we investigated the solid-phase nitration of aromatic hydrocarbons (1) with a composite of $Mg(NO_3)_2$ with silica gel ($Mg(NO_3)_2$-SiO_2) in order to establish the nitration process without strong acids.

2. EXPERIMENT

2.1. Instrument

^1H NMR (400 MHz) and ^{13}C NMR (100 MHz) spectra were taken with a Bruker AV 400M spectrometer for $CDCl_3$ solution using $SiMe_4$ as an internal standard. High-resolution mass spectra (HRMS) were measured on a Thermo Scientific Q Exactive mass spectrometer equipped with an electrospray ionization source. Almost HRMS spectra were measured at positive mode except for the cases of 4f and 4j which were measured in negative mode. Microscopic spectrophotometry was performed on a confocal laser scanning microscope (CLSM; Olympus FV? 300, Japan) equipped with a spectrophotometer (STFL 250, Seki Technotron, Japan) linked to the CLSM by an optical fiber. Using a 10 folds magnification lens, the measurable area was restricted to the inside of a circle with a 8.56 μm-diameter [18]. Microscopic absorption spectra were taken using a back-light as the light source.

2.2. Nitration of Aromatic Hydrocarbons (1) with $Mg(NO_3)_2$ on Silica Gel

General procedure of solid-phase nitration was performed as follows. A MeOH solution (10 mL) containing 1,4-dimethoxybenzene (1a; 3.62 mmol, 500 mg) and $Mg(NO_3)_2 \cdot 6H_2O$ (504.9 mg, 1.97 mmol, Wako Chemicals, Japan) was added slowly to silica gel (2.92 g, 48.7 mmol, Fuji Silysia A type, the average diameter = 79 μm) in a flask. After standing for 30 min to adsorb 1a and $Mg(NO_3)_2$ on silica gel, the solvent was removed by evaporation. The resulting composite of 1a, $Mg(NO_3)_2$, and silica gel was heated at a given temperature under N_2 atmosphere for 4-10 h under vigorous magnetic steering. The reacted composite was set on a silica gel column (Fuji Silysia BW 300, 50 mL) and was subjected to chromatography. Starting material (1a) and the nitrated products (2a) were isolated by elution with hexane and $CHCl_3$, respectively. The products were listed in Scheme 1. The 2a-2c, 2e-2i, 3d, 4f, and 4j had the following spectral data.

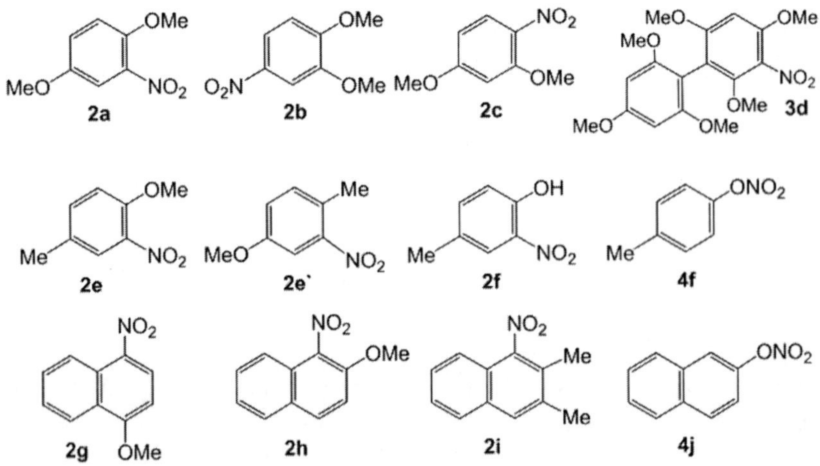

Scheme 1. Products (2a-2i, 3d, and 4f, 4j) of solid-phase nitration. Reaction conditions: 1 (3.62 mmol), $Mg(NO_3)_2$ (1.97 mmol), silica gel (2.92 g) under heating at 150°C.

1,4-Dimethoxy-2-nitrobenzene (2a). Yellow solid. M.p. 69°C - 70°C (lit. 72°C - 73°C [19]). ^1H NMR δ = 3.82 (s, 3H), 3.92 (s, 3H), 7.04 (d, J = 9.2 Hz, 1H), 7.16 (dd, J = 9.2, 3.1 Hz, 1H), 7.39 (d, J = 3.08 Hz, 1H). ^{13}C NMR δ = 55.03, 56.09, 108.96, 114.12, 119.91, 138.52, 146.37, 151.86. HRMS: m/z calcd for $C_8H_{10}NO_8$: ($[M + H]^+$) 184.0604, found 184.0605.

1,2-Dimethoxy-4-nitrobenzene (2b). Yellow solid. M.p. 99°C (lit 97.4°C [20]). ^1H NMR δ = 3.90 (3, 3H), 3.92 (s, 3H), 6.85 (d, J = 8.9 Hz, 1H), 7.69 (d, J = 2.6 Hz, 1H), 7.86 (dd, J = 8.9, 2.6 Hz, 1H). ^{13}C NMR δ = 56.47, 56.33, 106.45, 109.86, 117.81, 141.50, 148.87, 154.52. HRMS: m/z calcd for $C_8H_{10}NO_8$: ($[M + H]^+$) 184.0604, found 184.0602.

2,4-Dimethoxy-1-nitrobenzene (2c). Brown solid. M.p. 70°C (lit 72°C - 76°C [21]). ^1H NMR δ = 3.89 (s, 3H), 3.95 (s, 3H), 6.51 (d, J = 8.9, 2.5 Hz, 1H),

6.54 (d, J = 2.5 Hz, 1H), 8.01 (d, J = 8.9, 2.5 Hz, 1H). ^{13}C NMR δ = 55.92, 56.48, 99.67, 104.70, 128.53, 129.87, 155.70, 164.80. HRMS: m/z calcd for $C_8H_{10}NO_8$: ([M + H]$^+$) 184.0604, found 184.0602.

4-Methyl-2-nitroanisole (2e). M.p. 128°C. ^{1}H NMR δ = 2.34 (s, 3H), 3.93 (s, 3H), 6.98 (d, J = 8.5 Hz, 1H), 7.34 (dd, J = 8.6, 2.2 Hz, 1H), 7.65 (d, J = 1.8 Hz, 1H). HRMS: m/z calcd for $C_8H_9NO_3$: ([M + H]$^+$) 168.0655, found 168.0654.

4-Methyl-3-nitroanisole (2e'). ^{1}H NMR δ = 2.44 (s, 3H), 3.89 (s, 3H), 7.01 (d, J = 8.8 Hz, 1H), 7.84 (d, J = 8.8, 1H), 8.13 (s, 1H).

4-Methyl-2-nitrophenol (2f). Oil, ^{1}H NMR δ = 2.77 (s, 3H), 6.98 (d, J = 8.6 Hz, 1H), 7.32 (d, J = 8.8 and 2.0 Hz, 1H), 7.83 (d, J = 2.0 Hz, 1H).

1-Methoxy-4-nitronaphthalene (2g). Yellow solid. M.p. 80.0°C (lit 81°C - 83°C [22]), ^{1}H NMR δ = 4.11 (s, 3H), 6.82 (d, J = 8.7 Hz, 1H), 7.59 (dd, J = 8.2, 6.9 Hz, 1H), 7.74 (dd, J = 8.5, 6.9 Hz, 1H), 8.37 (d, J = 8.5 Hz, 1H), 8.40 (d, J = 8.7 Hz, 1H), 8.78 (d, J = 8.7 Hz, 1H). ^{13}C NMR δ = 56.29, 101.90, 122.77, 123.50, 125.62, 126.58, 126.88, 127.20, 129.46, 130.07, 160.60. HRMS: m/z calcd for $C_{11}H_9NO_3$: ([M + H]$^+$) 204.0655, found 204.0654.

2-Methoxy-1-nitronaphthalene (2h). Green solid. M.p. 126°C, ^{1}H NMR δ = 4.00 (s, 3H), 7.31 (d, J = 9.2 Hz, 1H), 7.43 (dd, J = 8.2, 6.8 Hz, 1H), 7.58 (dd, J = 8.6, 6.8 Hz, 1H), 7.66 (d, J = 8.6 Hz, 1H), 7.82 (d, J = 8.2 Hz, 1H), 7.92 (d, J = 9.2 Hz, 1H). ^{13}C NMR δ = 57.05, 113.06, 120.39, 125.15, 125.66, 128.07, 128.20 129.14, 132.21, 136.03, 148.62. HRMS: m/z calc. for [M + H], $C_{11}H_9NO_3^+$: 204.0655; Found: 204.0655.

2,3-Dimethyl-1-nitronaphthalene (2i). Yellow solid. M.p. 86.0°C - 88.0°C, ^{1}H NMR δ = 2.33 (s, 3H), 2.43 (s, 3H), 7.45-7.60 (m, 2H), 7.58 (d, J = 7.6 Hz), 7.68 (s, 1H), 7.75 (d, J = 7.2 Hz, 1H). ^{13}C NMR δ = 14.84, 20.47, 120.95, 126.66, 127.19, 127.50, 128.02, 129.76, 132.03, 133.70, 135.23, 148.51. HRMS: m/z calcd for $C_{12}H_{12}NO_2$: ([M + H]$^+$) 202.0863, found 202.0862.

2,4,6,2',4',6'-Hexamethoxy-3-nitrobiphenyl (3d). Yellow solid. M.p. 69.0°C - 70.0°C, ^{1}H NMR δ = 3.46 (s, 3H), 3.72 (s, 6H), 3.76 (s, 3H), 3.92 (s, 3H), 3.86 (s, 3H), 6.21 (s, 2H), 6.34 (s, 1H). ^{13}C NMR δ = 55.31, 55.95, 56.32, 56.36, 61.70, 90.87, 91.54, 102.73, 110.34, 130.87, 151.94, 152.52, 159.07, 160.24, 161.54. HRMS: m/z calcd for $C_{18}H_{22}NO_8$: ([M + H]$^+$) 380.1340, found 380.1337.

4-Tolyl nitrate (4f). Oil, ^{1}H NMR δ = 2.24 (s, 3H), 6.72 (d, J = 8.4 Hz, 2H), 7.00 (dd, J = 8.4 Hz, 1H). ^{13}C NMR δ = 19.46, 114.12, 128.93, 129.04, 152.08. HRMS: m/z calcd for $C_7H_7NO_3$: ([M − H]$^-$) 152.0353, found 152.0353.

2-Naphtyl nitrate (4j). Black solid. M.p. 150.0°C, ^{1}H NMR δ = 7.09 (dd, J = 8.8, 2.5 Hz, 1H), 7.14 (d, J = 2.5 Hz, 1H), 7.31 (dd, J = 8.1, 6.9 Hz, 1H), 7.42 (dd, J = 8.1, 6.9 Hz, 1H), 7.66 (d, J = 8.1, 1H), 7.73 (d, J = 8.8 Hz, 1H), 7.75 (d, J

= 8.1 Hz, 1H). ^{13}C NMR δ = 109.48, 117.78, 123.59, 126.37, 126.52, 127.77, 129.84, 131.46, 134.60, 153.42. HRMS: m/z calcd for $C_{10}H_7NO_3$: ($[M - H]^-$) 188.0353, found 188.0353.

2.3. Preparation of Meso-Tetra(p-Cyanophenyl)Porphyrin

As pH-indicator, meso-tetra(p-cyanophenyl)porphyrin (H_2tcp) was prepared according to the reported method as follows [23]. $BF_3 \cdot OEt_2$ (0.1 mL) were added to $CHCl_3$ solution (500 mL) of p-cyanobenzaldehyde (1048 mg; 8.0 mmol). A $CHCl_3$ solution (300 mL) of pyrrole (0.56 mL; 8.0 mmol) was then added. After the solution turned from pale orange to red-violet, it was confirmed that the Soret band appeared at 410 nm. NEt_3 (0.2 mL; 1.43 mmol) and chloranil (2.2 g; 9.0 mmol) were added to the solution and then heated at 60°C for 1 h under dark conditions. After evaporation, the condensed solution was filtrated and washed with $CHCl_3$. The crude H_2tcp was purified by a column chromatography on silica gel (Fuji Silysia BW 300) using $CHCl_3$-MeOH (50:1) as eluent.

meso-Tetra(p-cyanophenyl)porphyrin. Yield 1.2%. ^1H NMR δ = 8.10 (d, J = 8.6 Hz, 8H), 8.33 (d, J = 8.3 Hz, 8H), 8.60 (s, 8H). HRMS: m/z calcd for $C_{48}H_{27}N_8$: ($[M + H]^+$) 715.2359, found 715.2350.

2.4. Measurement of Acidity of the $Mg(NO_3)_2$-SiO_2 Composite

According to the reported method [18] , the acidity of the $Mg(NO_3)_2$-SiO_2 composite was measured as follows. At first, absorption spectra of H_2tcp were measured in $CHCl_3$-MeOH (1:2, v/v) under different pHs which were adjusted by $HClO_4$. The absorptions were observed at 514 and 645 nm due to purple free base porphyrin (H_2tcp) and the greenish protonated porphyrin (H_4tcp^{2+}), respectively. The absorbances (A_P and A_G) were measured at 514 nm and 645 nm, respectively. Fraction ($F = A_P/(A_G + A_P)$) was calculated at every pH and plotted against the pH to make the pH-profile of F values (Figure 1). The pH-profile was fitted by sigmoid curves (Equation (1)) which was presented by three parameters, F_{max}, pKa, and S, which denotes maximum F values, acid dissociation constant of H_4tcp^{2+}, and slope of the fitting curve at pKa, respectively. The relative standard deviation (RSD) was 0.9996. Each value of F_{max}, pKa, and S for H_2tcp were determined to be 0.875, 1.376, and 6.220, respectively.

$$F = \frac{F_{\max}}{1 + \exp\{(pK_a - pH)\}} \quad (1)$$

When the aromatic hydrocarbon was not added the $Mg(NO_3)_2$-SiO_2 composite, acidity of the $Mg(NO_3)_2$-SiO_2 composite was measured in a solid state using the pH-profile as follows. A $CHCl_3$solution (10 mL) of H_2tcp (0.7 μmol) and an aqueous solution (2 mL) of $Mg(NO_3)_2$ (173 mg) were mixed with SiO_2 (1.0 g, Fuji silysia A type, 1.8 - 5.0 mm). Here, large sizes of silica gel were used for CLSM analysis. After standing for 2 h until almost all of the H_2tcp had been adsorbed on $Mg(NO_3)_2$-SiO_2 composite, the solvent was evaporated and the H_2tcp-$Mg(NO_3)_2$-SiO_2 composite was dried under reduced pressure. Figure 2 shows the absorption spectra of H_2tcp adsorbed on a $Mg(NO_3)_2$-SiO_2 composite which was measured for five beads on CLSM. The F-values were determined to be 0.06 ± 0.04 by averaging five spectra. Using fitting curves, F-values were converted to pH, which was determined to be 0.96 for the $Mg(NO_3)_2$-SiO_2 composite under dry conditions.

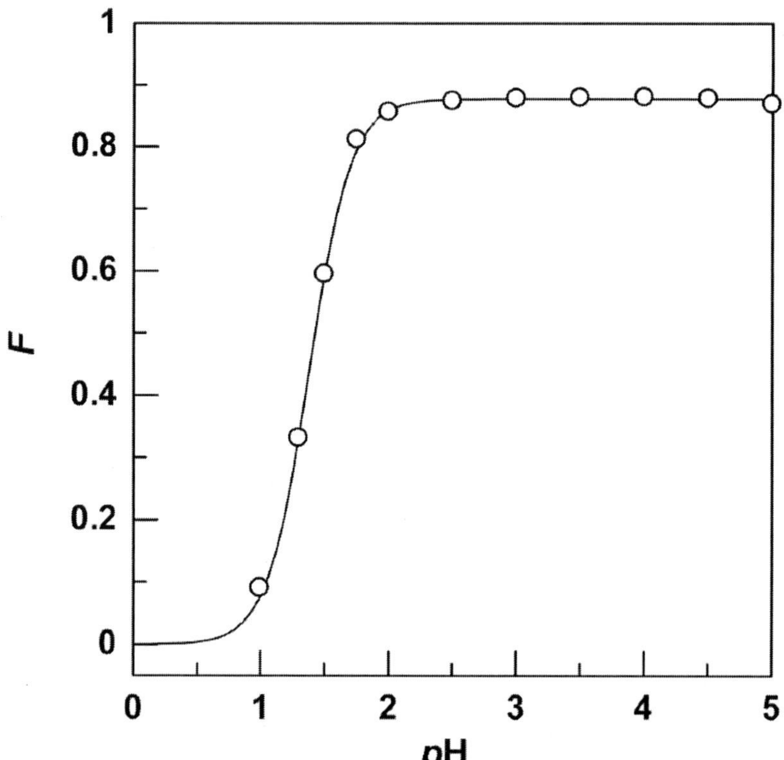

Figure 1. The pH-dependence of F values of H_2tcp in $CHCl_3$-MeOH. The pH was adjusted by $HClO_4$.

3. RESULTS AND DISCUSSION

3.1. Solid-Phase Nitration of Aromatic Hydrocarbons (1) with $Mg(NO_3)_2$ on Silica Gel

Solid-phase nitration of aromatic hydrocarbons (1) was performed by heating a mixture of 1 (3.62 mmol), $Mg(NO_3)_2 \cdot 6H_2O$ (504.9 mg, 1.97 mmol), and silica gel (2.92 g, 48.7 mmol) at a given temperature under N_2 atmosphere under magnetic steering. The solid-phase nitration of 1,4-dimethoxybenzene (1a) with $Mg(NO_3)_2$ on silica gel produced 1,4-dimethoxy-2-nitrobenzene (2a). Figure 3 showed the time- conversion plots of 2a at various temperatures. From the plots, the optimized temperature was determined to be 150°C. Therefore, the reaction temperature was fixed at 150°C in the solid-phase nitration of other aromatics (1b-1j). The results are summarized in Table 1.

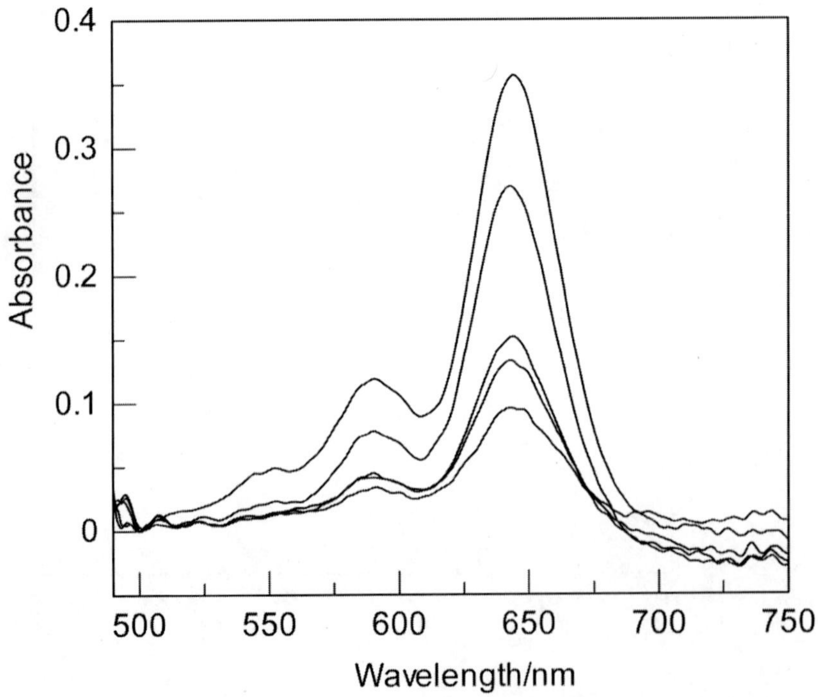

Figure 2. Measurement of absorption spectra of five beads of the H_2tcp-adsorbed $Mg(NO_3)_2$-SiO_2 under dry conditions using CLSM.

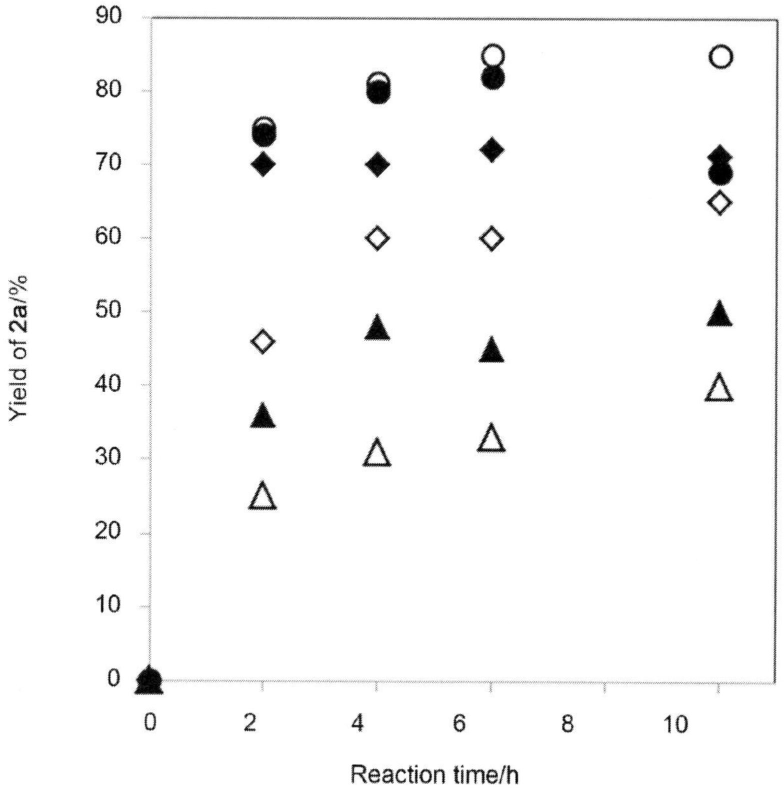

Figure 3. Time-conversion of 2a in solid-phase nitration of 1a with $Mg(NO_3)_2$ on silica gel: Reaction temperature = 70 (△), 90 (▲), 110 (◇), 130 (◆), 150 (○), and 170°C (●).

The solid-phase nitration of 1,2- and 1,3-dimethoxybenzenes (1b and 1c) and 4-methylanisole (1e) gave the nitroaromatics (2b and 2c, 2e and 2e') but yields were low. In the cases of naphthalene derivatives (1g-1i), the α-nitrated compounds (2g-2j) were obtained. Moreover, it was noteworthy that the nitration of 1,3,5-trimethoxybenzene (1d) produced 2,4,6,2',4',6'-hexamethoxy-3-nitrobiphenyl (3d) which was the nitrated dimer of 1d. However, 1,3,5-trimethoxy-2- nitrobenzene was not formed, though it was reported that the reaction of 1d with HNO_3 in the presence of $HClO_4$gave 1,3,5-trimethoxy-2-nitrobenzene [24]. The structure of 3d was undoubtedly confirmed by MS and NMR spectra. In the cases of p-cresol (1f) and 2-naphthol (1j), the nitration occurred at the hydroxyl group to give 4-tolyl nitrate (4f) and 2-naphthyl nitrate (4j), respectively. Usual nitration of 4f and 4j in solution occurred at aromatic ring [25]. Recently it was reported that the efficient nitration of phenol and phenol derivatives with $Al(NO_3)_3 \cdot 9H_2O$ [12] [13] and $Bi(NO_3)_3 \cdot 5H_2O$ [26] on silica gel occurred at aromatic ring. Therefore, it was suggested that the nitration ability of SiO_2-$Mg(NO_3)_2$ was not so strong, because the nitration was restricted to the electron rich substrates. In the case of phenol derivatives, silica gel supported the dehydration between OH group and HNO_3 to give Ar-ONO_2. Since aromatic nitrate had been recognized to be unstable [27] , the present synthesis had synthetic worthy.

3.2. Reaction Pathways

The silica gel is constructed by Si-O bonds such as Si-O-Si, Si=O, and Si-OH. It is well-known that the Si-OH group remained on the surface under heating below 300°C [7]. We previously elucidated that the Si-OH on silica gel could react with $MgCl_2$ and $MgSO_4$ under dry conditions to release proton [18]. Their acidities were determined by the colorimeter analysis using meso-tetraarylporphyrin as a pH-indicator. In the present study, the reaction of $Mg(NO_3)_2$ with silica gel generated HNO_3 along with the adsorption of Mg^{2+} ion on silica gel (Equation (2)). The acidity (pH) of $Mg(NO_3)_2$-SiO_2 composite was determined to be 0.96, which was more acidic compared with $MgCl_2$-SiO_2 and $MgSO_4$-SiO_2 composites whose pH were 1.73 and 1.61, respectively. The presence of excess water made the pH of $Mg(NO_3)_2$-SiO_2 neutral. Aromatic nitration usually occurs under acidic conditions. It was suggested that HNO_3 reacted with the proton to form the NO_2^+ (Equation (3)). More electron-rich aromatic hydrocarbons (1a and 1d) were allowed the efficient nitration. Therefore, the nitration proceeded through electrophilic attack of NO_2^+ to the aromatic ring (Equation (4)). $Mg(NO_3)_2$ and silica gel played as nitration reagent and desiccant, respectively.

Generation of H^+

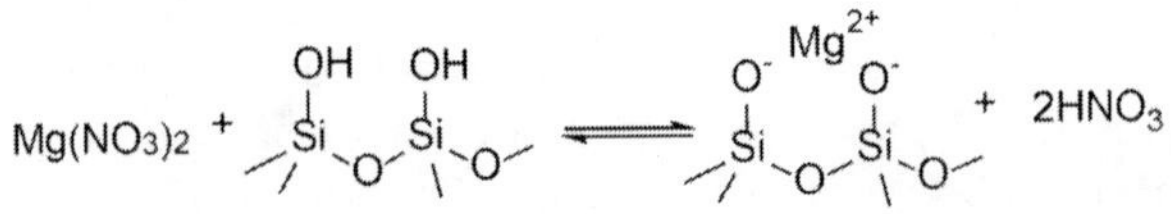

(2)

Generation of NO_2^+

$$HNO_3 + H^+ \longrightarrow NO_2^+ + H_2O$$

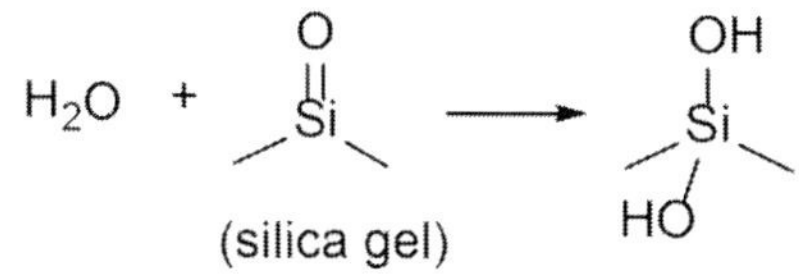

(3)

Nitration

$$NO_2^+ + Ar\text{-}H \longrightarrow Ar\text{-}NO_2 + H^+$$

(4)

In order to elucidate the mechanism of the formation of 3d, acid-catalyzed dimerization of 1d was attempted by the heating of 1d on silica gel at 150°C for 7 h with $MgCl_2$ which can release proton but has no electrophilicity. The dimerization of 1d did not occur. It is well known that the dimerization of aromatic hydrocarbons in the presence of electrophiles has been reported [28]. The electrophilic attack of NO_2^+ to 1d gave intermediates 5A and/or 5B. The 5A is less stable and more reactive than 5B, leading to the reaction of 5Awith another 1d at ortho and/or para positions (Friedel-Crafts acylation) to give the dimer which allowed the nitration to form 3d (Scheme 2).

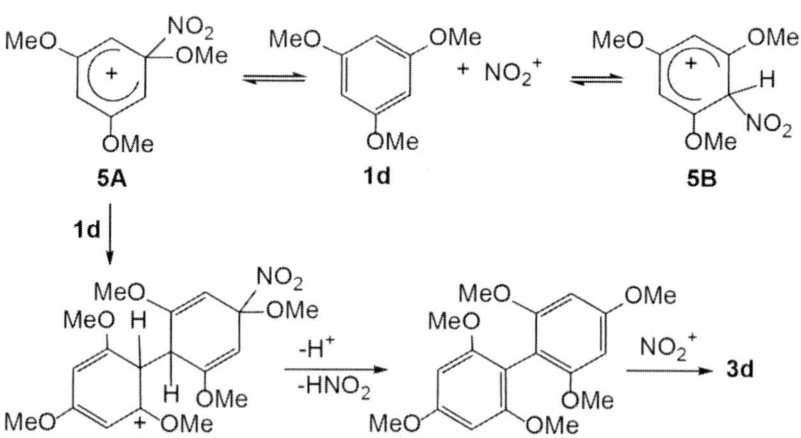

Scheme 2. Possible mechanism of the formation of 3d.

Table 1. The solid-phase nitration of benzene (1a-1f) and naphthalene derivatives (1g-1j) with $Mg(NO_3)_2$ on solid state of silica gel[a].

Run	Arenes (1)	Conditions	Conv./%[b]	Product (Yield/%)[c]
1	1,4-Dimethoxybenzene (**1a**)	150°C, 6 h	99	**2a** (82)
2	1,2-Dimethoxybenzene (**1b**)	150°C, 4 h	100	**2b** (24)
2	1,3-Dimethoxybenzene (**1c**)	150°C, 4 h	100	**2c** (9)
3	1,3,5-Trimethoxybenzene (**1d**)	150°C, 4 h	100	**3d** (56)
4	4-Methylanisole (**1e**)	150°C, 4 h	100	**2e** (8), **2e'** (3)
5	*p*-Cresol (**1f**)	150°C, 4 h	100	**2f** (4), **4f** (41)
6	1-Methoxynaphthalene (**1g**)	150°C, 4 h	100	**2g** (17)
7	2-Methoxynaphthalene (**1h**)	150°C, 4 h	70	**2h** (71)
8	2,3-Dimethylnaphthalene (**1i**)	150°C, 4 h	82	**2i** (61)
9	2-Naphthol (**1j**)	150°C, 4 h	100	**4j** (79)

a) The solid-phase nitration was performed by heating **1a-1j** (3.62 mmol) with $Mg(NO_3)_2$ (1.97 mmol) on silica gel (2.92 g) at 70°C - 170°C; b) Conversion of **1**; c) Isolated yields based on the **1** used.

4. CONCLUSION

In conclusion, the solid phase nitration of electron-rich aromatic hydrocarbons such as 1,4-dimethoxybenzene (1a) proceeded using $Mg(NO_3)_2$ on silica gel. Unique nitration occurred in 1,3,5-trimethoxybenzene (1d) which afforded nitrated dimer. In the cases of p-cresol (1f) and 2-naphthol (1j), esterification occurred to give aromatic nitrates. Thus the combination of $Mg(NO_3)_2$ with silica gel can eliminate the use of sulfuric acid from the aromatic nitration. Moreover, the acidic $Mg(NO_3)_2$-SiO_2 composite could be turned into neutrality by only exposing wet conditions and disposed safely since the composite did not involve harmful elements. Thus the solid-phase nitration using $Mg(NO_3)_2$-SiO_2 composite will provide safety and environmentally conscious chemical process.

REFERENCES

1. Yan, G. and Yang, M. (2013) Recent Advances in the Synthesis of Aromatic Nitro Compounds. Organic & Biomolecular Chemistry, 11, 2554-2566. https://doi.org/10.1039/c3ob27354g

2. Riego, J.M., Sedin, Z., Zaldívar, J.M., Marziano, N.C. and Tortato, C. (1996) Sulfuric Acid on Silica-Gel: An Inexpensive Catalyst for Aromatic Nitration. Tetrahedron Letters, 37, 513-516. https://doi.org/10.1016/0040-4039(95)02174-4
3. Iler, R.K. (1979) The Chemistry of Silica. John Wiley & Sons, New York.
4. Badgujar, D.M., Taiwar, M.B., Asthana, S.N. and Mahulikar, P.P. (2007) Environmentally Benign Synthesis of Aromatic Nitro Compounds Using Silica Supported Inorganic Nitrates. Journal of Scientific and Industrial Research, 66, 250-251.
5. Hajipour, A.R. and Ruoho, A.E. (2005) Nitric Acid in the Presence of P2O5 Supported on Silica Gel—A Useful Reagent for Nitration of Aromatic Compounds under Solvent-Free Conditions. Tetrahedron Letters, 46, 8307-8310. https://doi.org/10.1016/j.tetlet.2005.09.178
6. Grenier, J.-L., Catteau, J.-P. and Cotelle, P. (1999) Nitration of Electron-Rich Aromatic Compounds by Cerium Ammonium Nitrate Coated on Silica. Synthetic Communications, 29, 1201-1208. https://doi.org/10.1080/00397919908086091
7. Augusto, J., Rodrigues, R., De Oliveira Filho, A.P., Moran, P.J.S. and Custódio, R. (1999) Regioselectivity of the Nitration of Phenol by Acetyl Nitrate Adsorbed on Silica Gel. Tetrahedron Letters, 55, 6733-6738.
8. Samajdar, S., Becker, F.F. and Banik, B.K. (2000) Surface-Mediated Highly Efficient Regioselective Nitration of Aromatic Compounds by Bismuth Nitrate. Tetrahedron Letters, 41, 8017-8020. https://doi.org/10.1016/S0040-4039(00)01397-6
9. Iranpoor, N., Firouzabadi, H., Heydari, R. and Shiri, M. (2005) Nitration of Aromatic Compounds by Zn(NO3)2· 2N2O4 and Its Charcoal-Supported System. Synthetic Communications, 35, 263-270. https://doi.org/10.1081/SCC-200048450
10. Al-Masum, M. and Welch, R.L. (2014) Catalyst Free, Base Free Microwave Irradiated Synthesis of Aryl Nitrites from Potassium Aryltrifluoroborates and Bismuth Nitrate. Tetrahedron Letters, 55, 1726-1728. https://doi.org/10.1016/j.tetlet.2014.01.102
11. Al-Masum, M., Saleh, N. and Islam, T. (2013) A Novel Route to Organonitrites by Pd-Catalyzed Cross-Coupling of Sodium Nitrite and Potassium Organotrifluoroborates. Tetrahedron Letters, 54, 1141-1144. https://doi.org/10.1016/j.tetlet.2012.12.047
12. Patil, M.R., Mohite, P.H., Shisodia, S. and Keri, R.S. (2015) Regioselective Nitration of Phenols and Phenyl Ethers Using Aluminium Nitrate on Silica as a Nitrating System. Letters in Organic Chemistry, 12, 129-135. https://doi.org/ 10.2174/1570178612666150108000402
13. Ghorbani-Choghamarani, A., Goudarziafshar, H., Nikoorazm, M. and Yousefi, S. (2009) Aluminum Nitrate and Silica Sulfuric Acid as Efficient Nitrating Media for the Mononitration of Phenols under Mild and Heterogeneous Conditions. Canadian Journal of Chemistry, 87, 1144-1147. https://doi.org/10.1139/V09-081
14. Fueda, Y., Matsumoto, J., Shiragami, T., Nobuhara, K. and Yasuda, M. (2007) Porphyrin/MgCl2/Silica Gel Composite as a Cobalt-Free Humidity Indicator. Chemistry Letters, 36, 1246-1247. https://doi.org/10.1246/cl.2007.1246
15. Consolidated Version of Directive EU/67/548/EEC.
16. Gordeeva, L.G., Glaznev, I.S., Savchenko, E.V., Malakhov, V.V. and Aristov, Y.I. (2006) Impact of Phase Composition on Water Adsorption on Inorganic Hybrids "Salt/Silica". Journal of Colloid and Interface Science, 301, 685-691. https://doi.org/10.1016/j.jcis.2006.05.009
17. Matsumoto, T., Mitsumura, Y., Miyamoto, M., Matsumoto, J., Shiragami, T., Fueda, Y., Nobuhara, K. and Yasuda, M. (2011) Quantitative Analysis for a Color-Change of Humidity Indicator by Microscopic Absorption Spectrometry. Analytical Sciences, 27, 623-628.
18. Matsumoto, T., Hirose, D., Yamauchi, A., Matsumoto, J., Shiragami, T., Fueda, Y. and Yasuda, M. (2012) Measurement of Acidity of Magnesium Salts-Silica Gel Composites by Microscopic Absorption Spectrometry Using Porphyrin as pH-In- dicator. Bunseki Kagaku, 61, 851-856.
19. Tanemura, K., Suzuki, T., Nishida, Y., Satsumabayashi, K. and Horaguchi, T. (2003) A Mild and Efficient Method for the Mononitration of Aromatic Compounds by Cerium

(III) Ammonium Nitrate in Acetic Anhydride. Journal of Chemical Research, 2003, 497-499. https://doi.org/10.3184/030823403103174696

20. Clinton, R.O. and Page, D.F. (1963) US Patent 3,076,845.
21. Chemical Synthesis Database. http://www.chemsynthesis.com/
22. Mellor, J.M., Mittoo, S., Parkes, R. and Millar, R.W. (2000) Improved Nitrations Using Metal Nitrate-Sulfuric Acid Systems. Tetrahedron, 56, 8019-8024. https://doi.org/10.1016/S0040-4020(00)00720-1
23. Matsumoto, J., Matsumoto, T., Senda, Y., Shiragami, T. and Yasuda, M. (2008) Preparation and Characterization of Porphyrin Chromophores Immobilized on Micro-Silica Gel Beads. Journal of Photochemistry and Photobiology A: Chemistry, 197, 101-109. https://doi.org/10.1016/j.jphotochem.2007.12.010
24. Moodie, R.B., Schofield, K. and Thomas, P.N. (1978) Electrophilic Aromatic Substitution. Part 19. The Nitration of Some Reactive Aromatic Compounds in Perchloric Acid. Journal of the Chemical Society, Perkin Transactions, 2, 318-323. https://doi.org/10.1039/p29780000318
25. Rajanna, K.C., Chary, V.S., Kumar, M.S., Krishnaiah, G., Srinivas, P., Venkanna, P., Venkateswarlu, M., Ramesh, K., Reddy, K.R. and Suresh, B. (2015) Ultrasonic and Microwave Effects in Polyethylene Glycol-Bound Metal Nitrate Initiated Nitration of Aromatic Compounds under Acid Free Conditions. Green Chemistry Letters and Reviews, 8, 50-55. https://doi.org/10.1080/17518253.2015.1105309
26. Sun, H.-B., Hua, R. and Yin, Y. (2005) Highly Efficient Nitration of Phenolic Compounds in Solid Phase or Solution Using Bi(NO3)3·5H2O as Nitrating Reagent. The Journal of Organic Chemistry, 70, 9071-9073. https://doi. org/10. 1021/jo051466927.
27. Crivello, J.V. (1981) Nitrations and Oxidations with Inorganic Nitrate Salts in Trifluoroacetic Anhydride. The Journal of Organic Chemistry, 46, 3056-3060. https://doi.org/10.1021/jo00328a013
28. Puskas, I. and Fields, E.K. (1966) Synthesis of Nitropolyalkylbiphenyls by Nitrative Coupling of Di- and Trialkylbenzenes. The Journal of Organic Chemistry, 31, 4204-4210. https://doi.org/10.1021/jo01350a077

CHAPTER 2

Study of the $(Ca_{1-x}Sr_x)$ RuO_3 System with Nano-Crystals Prepared by the Solid-State Reaction Method

Adolfo Quiroz[1,2*], Elizabeth Chavira[2], José Eduardo Espinosa[1], Rodolfo Palomino-Merino[1], Ernesto Esteban Marinero[3], Masaya Nishioka[4], Valentín García-Vázquez[5]

[1]Facultad de Ciencias Físico-Matemáticas, Postgrado en Física Aplicada, Benemérita Universidad Autónoma Puebla, Ciudad Universitaria, Puebla, México

[2]Instituto de Investigaciones en Materiales, Universidad Nacional Autónoma de México, México, D.F., México

[3]Purdue University, West Lafayette, IN, USA

[4]Hitachi San José Research Center, San José, California, USA

[5]Instituto de Física LRT, Benemérita Universidad Autónoma de Puebla, Puebla, México

ABSTRACT

We present a study of their structure, morphology, electrical and magnetic properties on the $(Ca_{1-x}Sr_x)$ RuO_3 system for x = 0.0, 0.07, 0.10, 0.15 and 1.0. The samples were prepared by the solid- state reaction method in air at ambient pressure and heat in the 700°C - 800°C range for 48 h. By X-ray powder diffraction (XRD), we determine a solid solution until x = 0.15. Scanning electron mi- croscopy (SEM) indicates that the particle size is 77 - 266 nm. The resistance measurements, as a function of temperature measurements from 7 to 300 K the $(Ca_{1-x}Sr_x)$ RuO_3 system for x = 0.0, 0.07, 0.10, 0.15 and 1.0 show a metallic behaviour. We can even observe that the resistance of the sam- ples is due to the partial substitution of Sr^{2+} ions and Ru ion valence. Finally, the sample x = 0.07 has a magnetization applied high field to 10 K, whereas that to 300 K does not have a magnetiza- tion.

Keywords: Solid-State Reaction, XRD, SEM, Electric Resistivity, Magnetic Properties

1. INTRODUCTION

The coexistence of superconductivity and magnetic order in the rutheno-cuprates compounds like $RuSr_2GdCu_2O_8$ (Ru-1212) and their properties has been extensively studied [1] -[4]. Motivated by the discovery of high-T_c superconductivity in cuprates and the colossal magneto resistance effects in manganites, the research was initially focused on 3d transition-metal compounds [5]. However, as it has become increasingly clear that interesting physical phenomena of similar origin also happen in 4d and 5d electron systems, they have been getting a fair amount of attention recently. Among the 4d or 5d transition-metal compounds, ruthenium oxides probably have attracted most attention because of the discovery of superconductivity in Sr_2RuO_4 compound [6] and the poten- tial of unique perovskite ferromagnetic metal $SrRuO_3$ compound [7] for thin-film applications such as tunneling magneto-resistance or ferroelectric random access memory. These ruthenites also show diverse physical proper- ties depending on the composition or crystal structures. For example, when Sr ion is replaced by Ca ion in the above-mentioned ruthenates, the metallic and magnetic properties are significantly suppressed [8] [9]. From all these studies, it becomes evident that the preparation method plays a very important role, particularly for ob- taining the different physical properties [10]. The research into electrical properties of these compounds requires very pure materials in order to optimize the particular properties of the prepared materials. In this paper, we pro- vide information about the structural, electrical and magnetic properties of the prepared samples using the stoichiometric composition of $(Ca_{1-x}Sr_x)$ RuO_3 system (CSRO) for $x = 0.0$, 0.07, 0.10, 0.15 and 1.0.

2. EXPERIMENTAL PROCESS

2.1. Synthesis

The nano-crystalline samples of the CSRO system were synthesized by solid-state reaction technique at ambient pressure. The starting materials were: RuO_2 anhydrous (99.9% STREM), $SrCO_3$ (99.5% CERAC) and $CaCO_3$ (99.99% BAKER). The structure of each reagent was corroborated by XRD. Prior to weighing, $SrCO_3$ and $CaCO_3$ were pre-heated during 10 - 20 min at 120°C, in order to be dehydrated. The stoichiometric mixture of these compounds was done in an agate mortar in air, during 15 min, resulting inhomogenous slurry. The milled polycrystals were annealed between 700°C and 800°C in a thermolyne 46,100 furnace (±4°C) during two days in air, to decompose the carbonates. The resultant nano-crystals of the samples with $0 \leq x \leq 1.0$ were compressed into pellets (diameter 13 mm thickness 1.0 - 1.5 ± 0.05 mm), by the application of a pressure of 1/4 ton/cm^2 for 15 min in vacuum. Specimens compacted were sintered at 800°C during four days in air.

2.2. Characterization

All reagents and samples were characterized by (XRD), using a Bruker-AXS D8-Advance diffractometer with λ (CuK_α) = 1.54 Å radiation and graphite monocromator. Diffraction patterns were collected at room temperature on the 5° - 70° in a 2θ-range with a step size of 0.017 and time per step of 397 s. The change in morphology grain size in CSRO system obtained by different heat treatments, was observed by scanning electron microscopy (SEM) on a JOEL JSM-6610LV. The micrographs 50.00 KX, were taken with a voltage of 20 KV, current intensity of 1000 pA and WD = 10 mm. The Energy Dispersive X-Ray (EDX) was performed on the same equipment equipped with an Oxford/Link System electron probe microanalyser (EPMA). The standard four-probe method with DC resistance measurement was used as a function of temperature. The system is made up in a close-cycle refrigerator tool with conventional equipment for low-level electrical measurements. Continuous monitoring of all electrical parameters during a measurements cycle allows systematic errors in the resistance values to be detected in real-time, permitting clean R vs. T profiles to be obtained with no need of additional mathematical treatment to the experimental data [11]. The magnetization was obtained on a VSM-P525 vibrating sample magnetometer. To measure the zero field-cooled (ZFC) and the field-cooled (FC) magnetization, the samples were cooled down to 2 K at zero field and 100 Oe, respectively [12]. Then the samples were measured upon heating at 100 Oe.

3. RESULTS AND DISCUSSION

The XRD patterns of $(Ca_{1-x}Sr_x)$ RuO_3 system are shown in the Figure 1. From those spectra, samples with x = 0.07, 0.10 and 0.15 show a solubility up to x

= 0.15. Meanwhile, for $x = 0.0$ and $x = 1.0$, we observed a single orthorhombic phase identified as $CaRuO_3$ PDF (70-2790) when $x = 0$, and $SrRuO_3$ PDF (70-2791), when $x = 1.0$ [13] [14]. The samples with $x = 0.07$ - 0.15 (that have not been reported before in the literature) show very weak reflections of a tetragonal secondary phase (identified as RuO_2 PDF (43-1027)) [15]. In the same referred samples, the Ca ion content variation shows shifts in the peaks. Then, with the calculated lattice parameters a, b, c, and the unit cell volume (V) of each of the compounds considering the following Miller index (h k l): (4 0 0), (0 4 0) and (0 0 4), which are shown in the following Table 1, can be observed that little peaks shift.

Finally, the net lattice parameters of $(Ca_{1-x}Sr_x)$ RuO_3, x = 0.0, 0.07, 0.10, 0.15 and 1.0 system vary with the inclusion of the Sr-ion content and Ru ion coordination. Since the ionic radius of Ca^{2+}ion (Ca^{2+} = 1.34 Å) is lower than the ionic radius of Sr^{2+} ion (Sr^{2+} = 1.44 Å) [16] , we conclude that the Ca ions are substituted by Sr ions with the observed unit cell variation in the volume (V) of each of the compounds, see Table 1. For the behavior of the lattice parameters, the solid solution has a substitution mechanism, where the Sr^{2+} ions substitute Ca^{2+} ions till x = 0.15. With respect to the examined diffractograms, it is worth to mention that the samples with $x = 0.0$ and $x = 1.0$ present a single orthorhombic phase identified as $CaRuO_3$ and $SrRuO_3$, respectively. Is other synthesis way to obtained this compounds reported in the literature, PDF (70-2790) and PDF (70-2791), with low temperature reaction. In contrast, the samples in the $x = 0.07$ - 0.15 range present reflections of a secondary phase identified as RuO_3 [16]. That's way the idea that the anions contributed to the formation of the mechanism of the solid solution.

The next step was the characterization of the samples achieved by SEM. The observed morphology is presented in Figure 2, shows considerable variations in sizes, very few secondary phases and shapes of particles. The grain size varies between 77 to 266 nm. The presented micrographs in the mentioned Figure 2 were taken on the surface of the representative pellets of the CSRO samples with a magnification of 50 KX. Also, in some regions we observe semi-fusion that can be attributed to the ruthenium content. We can observe the secondary phase in the other gray color.

Figure 3 shows the observed resistance of the sintered materials in relation to the temperature. All the samples of $Ca_{1-x}Sr_xRuO_3$, $x = 0.0$, 0.07, 0.10, 0.15 and 1.0 system show a metallic behavior. The observed resistance in $CaRuO_3$ compound (2.009×10^{-2} Ω) is much higher than that of $SrRuO_3$ compound (9.367×10^{-3} Ω). The

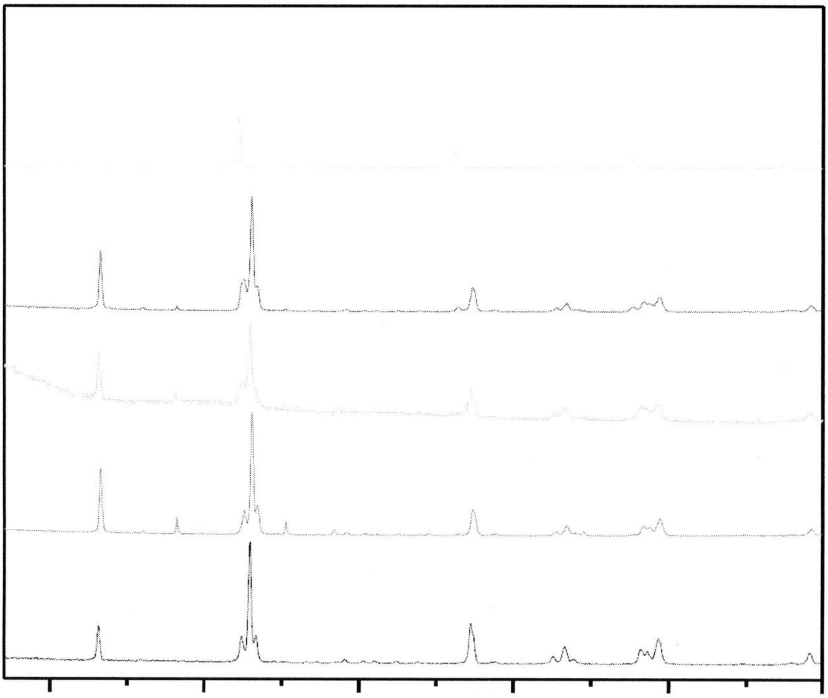

Figure 1. XRD Patterns evolution of CSRO system.

Table 1. Lattice parameters of CSRO system.

X	*a* (Å)	*b* (Å)	*c* (Å)	*V* (Å)
0.0	5.519(0)	7.664(9)	5.364(0)	226.9(1)
0.07	5.556(6)	7.839(8)	5.530(5)	240.9(3)
0.10	5.524(4)	7.843(8)	5.432(8)	235.4(2)
0.15	5.556(7)	7.835(1)	5.530(6)	241.0(2)
1.0	5.574(0)	7.859(4)	5.541(0)	242.7(4)

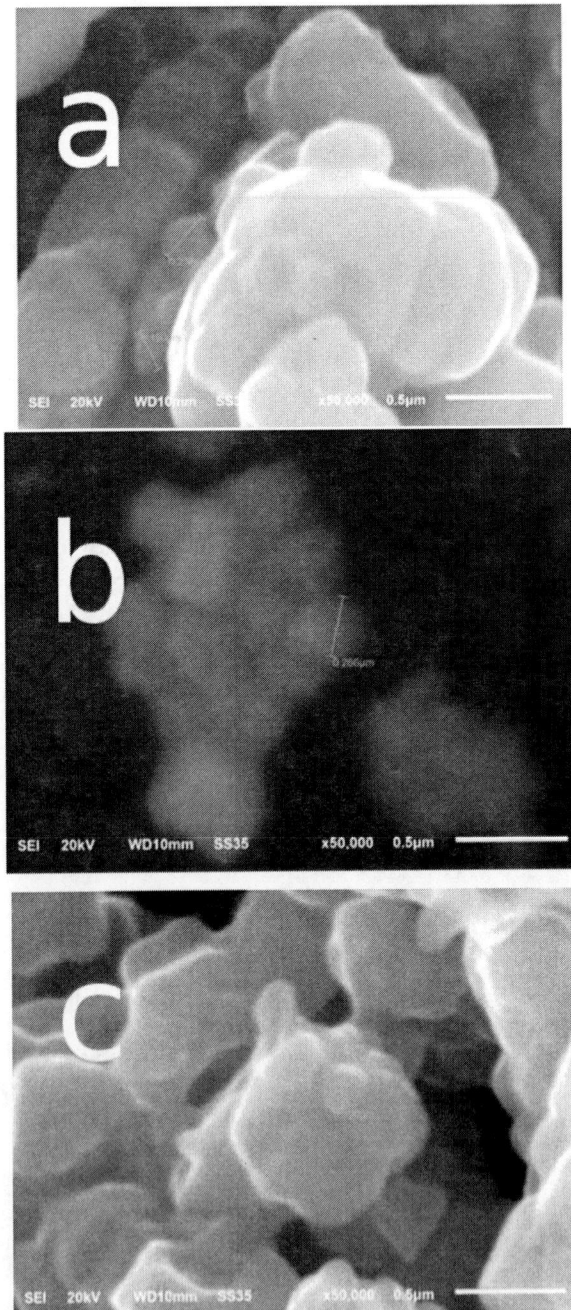
a
SEI 20kV WD10mm 0.5μm
b
SEI 20kV WD10mm SS35 x50,000 0.5μm
c
SEI 20kV WD10mm SS35 x50,000 0.5μm

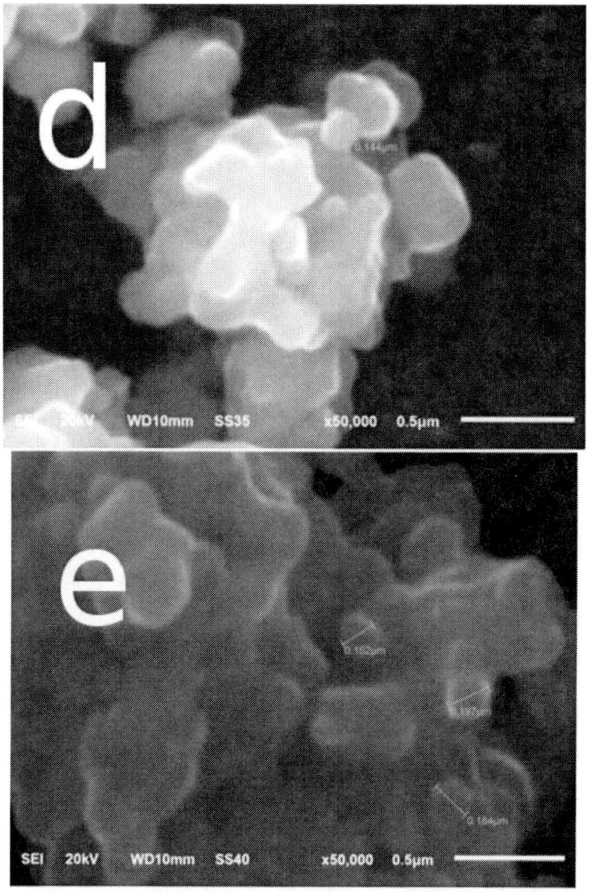

Figure 2. SEM of the CSRO system. (a) Pellet surface morphology of the $CaRuO_3$ sample with particle size of 190 - 238 nm; (b) Surface morphology of the $Ca_{0.93}Sr_{0.07}RuO_3$ sample with particle size of 77 - 266 nm; (c) Surface morphology of the $Ca_{0.90}Sr_{0.10}RuO_3$ sample with particle size of 199 nm; (d) Surface morphology of the $Ca_{0.85}Sr_{0.15}RuO_3$ sample with particle size of 144 nm; (e) Surface morphology of the $SrRuO_3$ sample with particle size of 162 - 197 nm.

$CaRuO_3$ compound does not exhibit ferromagnetism and its magnetic properties are still under discussion [17] together with long-range magnetic order. Moreover, as seen with the resistance curve, it does not present any anomaly in the measured range of temperature. The $SrRuO_3$compound shows long-range magnetic ordering. The slope changes of 7 to 155 K, which is called Kondo effect. The temperature at which the pronounced break occurs agrees with the reported one, where T_{curie} (150 K) is related to spin scattering [18].

In the $Ca_{1-x}Sr_xRuO_3$ samples within the range of $0.07 \leq x \leq 0.15$, the short-range ferromagnetic interactions appear. This indicates that the ferromagnetism has been suppressed through the process of substitution of Sr^{2+} ions by Ca^{2+} ions. For the compounds with large Ca^{2+} ions doping ($x \geq 0.7$), no clear phase transition is discerned, and only some irreversibility is observed in the magnetization curves of these materials. The disappearance of the long-range magnetic order is commonly related to the distortion of the RuO_6octahedra associated with the partial or total replacement of Sr^{2+} ions by Ca^{2+} ions, and the corresponding narrowing of the 4d bandwidth [19]. The samples mentioned before show variations in each one of the profiles. This is because each one has a different chemical composition. Therefore, we can even observe that the resistance of the samples is due to the partial or total composition of Ca^{2+} ions. Since the substitution of Ca^{2+} ions for Sr^{2+} ions introduces and induces the local distortion in the vicinity of calcium ions [20] changing the Ru-O-Ru bond angle and the bandwidth, which weakens the ferromagnetism.

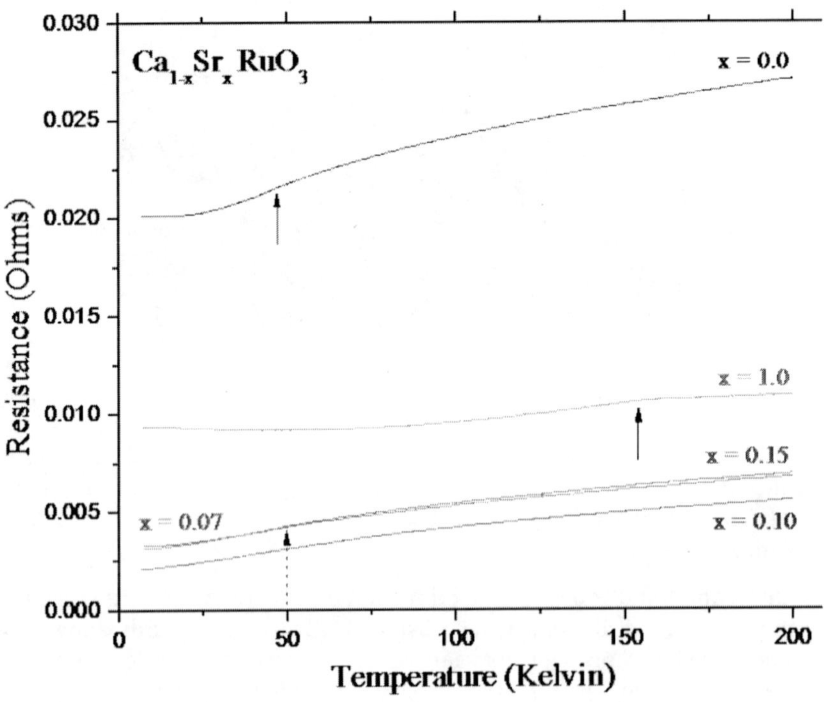

Figure 3. Temperature dependence of resistance of $Ca_{1-x}Sr_xRuO_3$. The arrows indicate the inflexions.

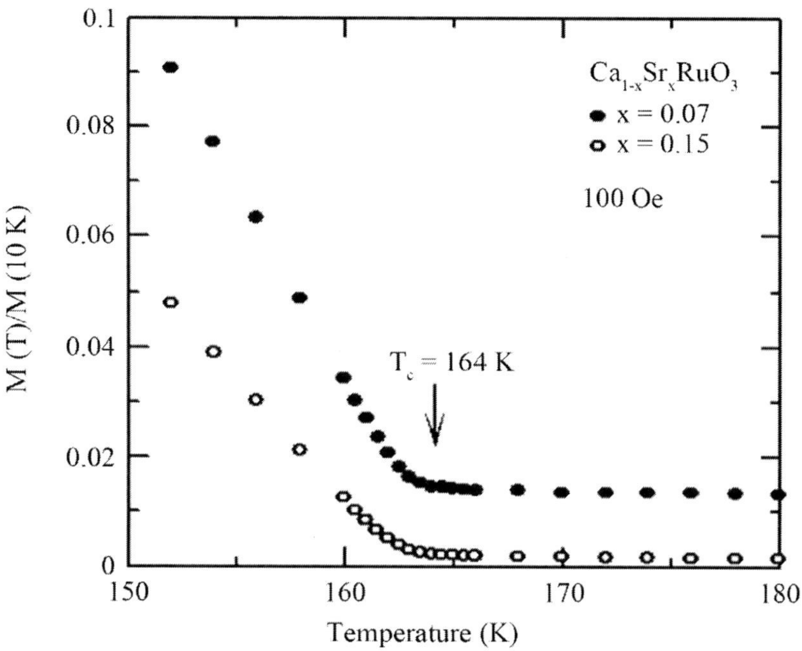

Figure 4. Magnetization measurements to 10 K ((M(T)/M(10 K)) vs curves temperature. $Ca_{1-x}Sr_xRuO_3$ samples with x = 0.07 and 0.15.

However, for $Ca_{1-x}Sr_xRuO_3$ ($0.07 \leq x \leq 0.15$) samples, the electrical resistance decreases with the incorporation of Sr^{2+} ions ($2.1 \times 10^{-3}\ \Omega$), giving less resistance than that of $SrRuO_3$ ($9.3 \times 10^{-3}\Omega$). With the sample preparation described above, we did not find any superconducting phase. We observe the same behaviors that were reported for other similar compounds [7] [13] [20] - [22].

The magnetization at 10 K for $x = 0.07$ and 0.15 samples are based on the application of a magnetic field of 100 Oe, as shown in Figure 4. The values T_c obtained for different samples are at a temperature of ~164 K, which is very close to those reported in the literature [6]. These samples exhibit such behavior because they have two phases $CaRuO_3$ $(1 - x)$ + ($SrRuO_3$) (x), not one, which explains why both samples have the same transition FM to temperature at T_c—164 K.

Finally, as shown in Figure 5, we measured the magnetic moment (emu) vs. the magnetic field (Oe) for the $x = 0.07$ sample. This sample has a magnetization applied by a high field at 10 K, whereas at 300 K, it does not present magnetization. A narrow hysteresis involves a small amount of energy dissipated repeatedly, reversing the magnetization. This material could be useful in transformers and other devices for alternating current, where a zero hysteresis would be optimal.

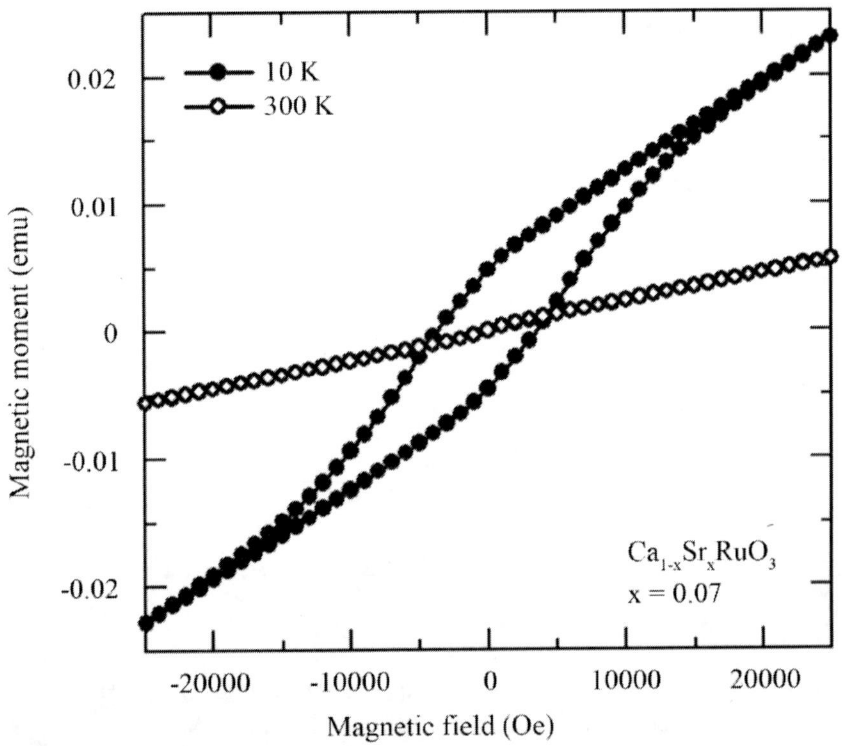

Figure 5. The curves of magnetic moment (emu) vs magnetic field (Oe) of sample $Ca_{0.93}Sr_{0.07}RuO_3$.

4. CONCLUSION

In this work, we obtained nano-crystalline samples of CSRO system by solid-state reaction in air at room temperature, in which a solubility up to $x = 0.15$ was observed. The SEM micrographs exhibit an almost-spherical grain size distribution from 77 to 266 nm. We also observed that the compounds of the CSRO system exhibit metallic behavior. However, for $x = 0.07$ and 0.15, the samples exhibit a FM transition to temperature T_c at ~164 K, indicating that the transition temperature decreases with increasing Sr ions concentration. Finally, we found that the $x = 0.07$ sample has a magnetization at temperature of 10 K, whereas at 300 K the sample does not present a hysteresis behavior.

ACKNOWLEDGEMENTS

Adolfo Quiroz, Elizabeth Chavira, José Eduardo Espinosa, Rodolfo Palomino-Merino, Ernesto Esteban Marinero, Masaya Nishioka, Valentín García-Vázquez This work was partially supported by CONACYT-80380, UNAM-IN109308.

REFERENCES

1. Bauernfeind, L., Widder, W. and Braun, H.F. (1995) Ruthenium-Based Layered Cuprates $RuSr_2LnCu_2O_8$ and $RuSr_2(Ln_{1+x}Ce_{1-x})Cu_2O_{10}$ (LnSm, Eu and Gd). Physica C: Superconductivity, 254, 151-158. http://dx.doi.org/10.1016/0921-4534(95)00574-9
2. Bauernfeind, L., Widder, W. and Braun, H.F. (1996) Superconductors Consisting of $CuO_2$and RuO_2 Layers. Journal of Low Temperature Physics, 105, 1605-1610. http://dx.doi.org/10.1007/BF00753929
3. Bernard, C., Tallon, J.L., Brucher, E. and Kremer, R.K. (2000) Evidence for a Bulk Meissner State in the Ferromagnetic Superconductor $RuSr_2GdCu_2O_8$ from dc Magnetization. Physical Review B, 61, Article ID: R14960.
4. Tallon, J.L., Bernhard, C., Bowden, M., Gilberd, P., Stoto, T., Pringle, D., et al. (1999) Coexisting Ferromagnetism and Superconductivity in Hybrid Rutheno-Cuprate Superconductors. IEEE Transactions on Applied Superconductivity, 9, 1696-1699. http://dx.doi.org/10.1109/77.784779
5. Park J., Oh, J.-S., Park, J.-H., Kim, D.M. and Eom, C.-B. (2004) Electronic Structure of Epitaxial $(Sr,Ca)RuO_3$ Films Studied by Photoemission and X-Ray Absorption Spectroscopy. Physical Review B, 69, Article ID: 085108. http://dx.doi. org/ 10.1103/PhysRevB.69.085108
6. Maeno, Y., Hashimoto, H., Yoshida, K., Nishizaki, S., Fujita, T., Bednorz, J.G. and Lichtenberg, F. (1994) Supercon- ductivity in a Layered Perovskite without Copper. Nature (London), 372, 532-534. http://dx.doi.org/10.1038/372532a0
7. Eom, C.B., Cava, R.J., Fleming, R.M., Phillips, J.M., van Dover, R.B., Marshall, J.H., Hus, J.W.P., Krajewski, J.J. and Peck Jr., W.F. (1992) Single-Crystal Epitaxial Thin Films of the Isotropic Metallic Oxides $Sr_{1-x}Ca_x RuO_3$ ($0 \leq x \leq 1$). Science, 258, 1766. http://dx.doi.org/10.1126/science.258.5089.1766
8. Cox, P.A., Egdell, R.G., Goodenough, J.B., Hamnett, A. and Naish, C.C. (1983) The Metal-to-Semiconductor Transition in Ternary Ruthenium (IV) Oxides: A Study by Electron Spectroscopy. Journal of Physics C: Solid State Physics, 16, 6221. http://dx.doi.org/10.1088/0022-3719/16/32/014
9. Shepard, M., McCall, S., Cao, G. and Crow, J.E. (1997) Thermodynamic Properties of Perovskite $ARuO_3$ (A = Ca, Sr, and Ba) Single Crystals. Journal of Applied Physics, 81, 4978. http://dx.doi.org/10.1063/1.365018
10. Felner, I., Asaf, U., Reich, S. and Tsabba, Y. (1999) Magnetic Properties of $RSr_2RuCu_2O_{8+\delta}$ (R = Eu and Gd). Physica C: Superconductivity, 311, 163-171. http://dx.doi.org/10.1016/S0921-4534(98)00642-X
11. García-Vázquez, V., Pérez-Amaro, N., Canizo-Cabrera, A., Cumplido-Espíndola, B., Martínez-Hernández, R. and Abarca-Ramírez, M.A. (2001) Selected Error Sources in Resistance Measurements on Superconductors. Review of Scientific Instruments, 72, 3332. http://dx.doi.org/10.1063/1.1386899, http://www.qdusa.com

12. Fumihiko, F. and Tsuda, N. (1994) On the Magnetism and Electronic Conduction of Itinerant Magnetic System $Ca_{1-x}Sr_xRuO_3$. Journal of the Physical Society of Japan, 63, 3798-3807. http://dx.doi.org/10.1143/JPSJ.63.3798
13. Rama Rao, M.V., Sathe, V.G., Sornadurai, D., Panigrahi, B. and Shripathi, T. (2001) Electronic Structure of $ARuO_3$ (A = Ca, Sr and Ba) Compounds. Journal of Physics and Chemistry of Solids, 62, 797-806. http://dx.doi.org/10.1016/S0022-3697(00)00262-6
14. Takeda, T., Nagata, M., Kobayashi, H., Kanno, R., Kawamoto, Y., Takano, M., Kamiyama, T., Izumi, F. and Sleight, A.W. (1998) High-Pressure Synthesis, Crystal Structure, and Metal-Semiconductor Transitions in the $Tl_2Ru_2O_{7-\delta}$ Pyrochlore. Journal of Solid State Chemistry, 140, 182-193.
15. Shannon, R.D. (1976) Revised Effective Ionic Radii and Systematic Studies of Interatomic Distances in Halides and Chalcogenides. Acta Crystallographica A, 32, 751-767.
16. He, T. and Cava, R.J. (2001) Disorder-Induced Ferromagnetism in $CaRuO_3$. Physical Review B, 63, Article ID: 172403. http://dx.doi.org/10.1103/PhysRevB.63.172403
17. Gan, Q., Rao, R.A., Eom, C.B., Garrett, J.L. and Lee, M. (1998) Direct Measurement of Strain Effects on Magnetic and Electrical Properties of Epitaxial $SrRuO_3$ Thin Films. Applied Physics Letters, 72, 978-980. http://dx.doi.org/10.1063/1.120603
18. Cao, G., McCall, S., Shepard, M., Crow, J.E. and Guertin, R.P. (1997) Thermal, Magnetic, and Transport Properties of Single-Crystal $Sr_{1-x}Ca_xRuO_3$ ($0 \leq x \leq 1.0$). Physical Review B, 56, 321-329. http://dx.doi.org/10.1103/PhysRevB.56.321
19. Shikano, M., Huang, T.K., Inaguma, Y., Itoh, M. and Nakamura, T. (1994) Pressure Dependence of the Magnetic Transition Temperature for Ferromagnetic $SrRuO_3$. Solid State Communications, 90, 115-119. http://dx.doi.org/10.1016/0038-1098(94)90942-3
20. Mathieu, R., Asamitsu, A., Yamada, H., Takahashi, K.S., Kawasaki, M., Fang, Z., Nagaosa, N. and Tokura, Y. (2004) Scaling of the Anomalous Hall Effect in $Sr_{1-x}Ca_xRuO_3$. Physical Review Letters, 93, Article ID: 016602. http://dx.doi.org/10.1103/PhysRevLett.93.016602
21. Pi, L., Fan, E.-H., Xiao, Y. and Zhang, Y.-H. (2006) Magnetic and Electrical Transport Properties of $Sr_{1-x}La_xRuO_3$ ($0 \leq x \leq 0.10$). Chinese Physics Letters, 23, 2225-2228. http://dx.doi.org/10.1088/0256-307X/23/8/072

CHAPTER 3

Capacitive Properties of Mesoporous Mn-Co Oxide Derived from a Mixed Oxalate

Yong Yan, Bingcai Wu, Cuihong Zheng, Daolai Fang*

School of Materials Science and Engineering and Anhui Key Laboratory of Metal Materials and Processing, Anhui University of Technology, Ma'anshan, China.

ABSTRACT

A mesoporous Mn-Co oxide for supercapacitors was derived from a mixed oxalate $Mn_{0.8}Co_{0.2}C_2O_4 \times nH_2O$, which was synthesized by a solid-state coordination reaction at room temperature. The synthesized mixed Mn-Co oxalate was decomposed in air at 250°C, resulting in a tetragonal spinel Mn-Co oxide with a primary particle size less than 100 nm. The obtained Mn-Co oxide presents a mesoporous texture with a specific surface area of 120 $m^2 \cdot g^{-1}$. Electrochemical properties of the Mn-Co oxide electrode were investigated by cyclic voltammetry and galvanostatic charge/discharge in 6 $mol \cdot L^{-1}$ KOH electrolyte. The Mn-Co oxide electrode delivered specific capacitances of 383 and 225 $F \cdot g^{-1}$ at scan rates of 2 and 50 $mV \cdot s^{-1}$, respectively. Subjected to 500 cycles at a current density of 1.34 $A \cdot g^{-1}$, the symmetrical Mn-Co oxide capacitor showed specific capacitance of 179 $F \cdot g^{-1}$, still retaining ~85% of its initial capacitance. The obtained Mn-Co oxide material showed good capacitive performance, which was promising for supercapacitor applications.

Keywords: Supercapacitors; Manganese Oxide; Solid-State Reaction; Cyclic Voltammetry; Capacitive Performance

1. INTRODUCTION

Electrochemical supercapacitors possess the unique energy-storage performance, such as greater power density and longer cycle life than secondary batteries, as well as higher energy density than conventional capacitors [1], showing great potential to be used in the areas of hybrid power sources, peak power sources, backup power storage, lightweight electronic fuses, starting power of fuel cells [2,3]. Generally the supercapacitors can be classified into two categories: electrical-double-layer capacitors (EDLCs), which build up electrical charge at the electrode/electrolyte interface [4,5], and pseudocapacitors, which are based on reversible faradic redox reactions at the interfaces at certain potentials [6-10].

Recently, with the purpose of achieving high energy density supercapacitors, great efforts have been devoted to search for excellent pseudocapacitive materials [11- 20]. Amorphous hydrated ruthenium oxide has been demonstrated to be an excellent pseudocapacitor material, which exhibits a high conductivity, good electrochemical stability, and a large specific capacitance (SC) of 720 - 760 $F \cdot g^{-1}$ [21]. However, high cost and toxicity of the ruthenium oxide greatly limit its commercial applications. Thus, other less expensive and less toxic transition metal oxides, such as amorphous and crystallized MnO_2 [11- 14], nanostructured NiO [15,16], and mesoporous and nanowire-array Co_3O_4 [17,18], have been developed as electrode materials for supercapacitors. The electrode materials of the cheaper transition metal oxides exhibit remarkable capacitive nature, delivering a considerable SC. However, the resistivity and the equivalent series resistance (ESR) of these cheaper transition metal oxides are very large, which greatly limits their capacity and power density available.

On the other hand, it is reported that mixed oxide composites exhibit superior capacitive performance to single transition metal oxides. It has been demonstrated that the incorporation of Ni [19], Pb [20], Mo [22], and V [23] oxides into pristine manganese oxide can enhance specific capacitance and electrical conductivity of pristine manganese oxide. Nevertheless, the effect of Co oxide addition into pristine manganese oxide has not been extensively studied yet, and the results from different research groups vary significantly [24-26].

In this work, a mesoporous Mn-Co oxide was derived from a mixed $Mn_{0.8}Co_{0.2}C_2O_4 \times nH_2O$, which was synthesized by a solid-state coordination reaction. Its phase compositions and microstructures were characterized by X-ray diffraction (XRD), scanning electron microscope (SEM) and N_2adsorption-desorption measurements, and its electrochemical properties were analyzed by cyclic voltammetry and galvanostatic charge/discharge cycling. Considering the considerable content of cobalt added in the prepared Mn-Co oxides, and a remarkable pseudocapacitive behavior of cobalt oxide in alkaline aqueous electrolyte [17,18], KOH aqueous electrolyte was used to investigate the electrochemical properties of the Mn-Co oxide. The obtained Mn-Co oxide shows excellent capacitive performance in terms of specific capacitance (SC), power capability and cycling stability, which is promising for supercapacitor applications.

2. EXPERIMENTAL

2.1. Synthesis of the Mn-Co Oxide Powder

Analytical-grade manganese acetate $Mn(CH_3COO)_2 \times 4H_2O$, cobalt acetate $Co(CH_3COO)_2 \times 4H_2O$, and oxalic acid H_2- $C_2O_4 \times 2H_2O$ were used as starting materials. All the reagents were purchased from Sinopharm Group Chemical Reagent Company in China, and used as received. A proper amount of the starting materials was weighed accurately with a molar ratio of $Mn(CH_3COO)_2 \times 4H_2O$:Co- $(CH_3COO)_2 \times 4H_2O$:$H_2C_2O_4 \times 2H_2O$ of 0.8:0.2:1.1, then mixed and milled in a mortar to form a powder mixture. Subsequently, the obtained powder mixture was ball milled at room temperature for 5 h in a polyethylene container using zirconia balls as milling medium. The milled mixture was dried at 80°C, resulting in a Mn-Co oxalate precursor. Finally, the Mn-Co oxalate precursor was calcined in air at 250°C for 10 h to obtined Mn-Co oxide.

2.2. Structural Characterization

A Philips X'pert Pro X-ray diffractometer with CuK_a radiation (λ = 1.5406 Å) was used to analyze the phase compositions of the dried milled mixture and the calcined Mn-Co oxide. Diffraction data were collected in the 2θ range from 15° to 80°, using the step-scan mode with a scanning speed of 0.02° step-size and 1 s per step. The morphology of the obtained Mn-Co oxide was observed by using a JSM 6490 scanning electron microscope (SEM). N_2 adsorption-desorption measurements were carried out at a Coulter SA 3100 system at 77.3 K. The surface area was calculated using the Brunauer-Emmett-Teller (BET) equation. Pore size distributions were calculated by the Barrett-Joyner-Halenda (BJH) method. Samples were dried and degassed for 2 h at 150°C to remove any surface organic or moisture contaminants.

2.3. Electrochemical Characterization

The electrodes for electrochemical measurements consisted of the prepared Mn-Co oxide, acetylene black (AB) and polytetrafluoroethylene (PTFE), whose weight ratio was 75:20:5. First, the Mn-Co oxide and the acetylene black were fully mixed and ground, then a proper amount of PTFE binder was added into the powder mixture to achieve a homogenous slurry, followed by painting the prepared slurry onto a nickel foam current collector with an area of 1 cm ´ 1 cm. Finally, the painted current collector was dried for 10 h at 80°C, and pressed at a pressure of 10 MPa to form a reliable electrode. Each of the fabricated electrodes contained ~4 mg of the Mn-Co oxide electroactive material.

All the electrochemical measurements concerned were carried out in 6 mol·L^{-1} KOH aqueous electrolyte. Cyclic voltammetric measurements were performed on a CHI6- 04C electrochemical workstation in a three-electrode cell set-up, using a Mn-Co oxide electrode as the working electrode, a platinum foil as the counter electrode, a saturated calomel electrode (SCE) as the reference electrode. Cyclic voltammograms at various scan rates were recorded between -0.4 and 0.4 V vs. SCE.

For evaluation of galvanostatic charge/discharge performance of the obtained Mn-Co oxideelectrodes, a symmetrical capacitor was assembled using a couple of the identical electrodes separated by a porous polypropylene separator. The galvanostatic charge/discharge performance at different current densities was determined by a battery test system of Neware BTS-3008W in a potential range from 0 to 0.8 V.

3. RESULTS AND DISCUSSION

3.1. Phase Compositions and Microstructures

Figure 1(a) gives the XRD pattern of the Mn-Co oxalate synthesized by a solid-state coordination reaction. The XRD pattern of the synthesized Mn-Co oxalate matches that of α-form manganese oxalate (JCPDS, No. 25-0544), suggesting that the synthesized oxalate is a binary Mn-Co mixed oxalate [27]. Considering a molar ratio of Mn^{2+}: Co^{2+} of 4:1 in the mixture of the starting materials used, and almost no loss of metal ions during ball-milling, the obtained Mn-Co oxalate can be expressed as a chemical formula of $Mn_{0.8}Co_{0.2}C_2O_4 \times nH_2O$. **Figure 1(b)** shows the XRD pattern of the Mn-Co oxide powder.

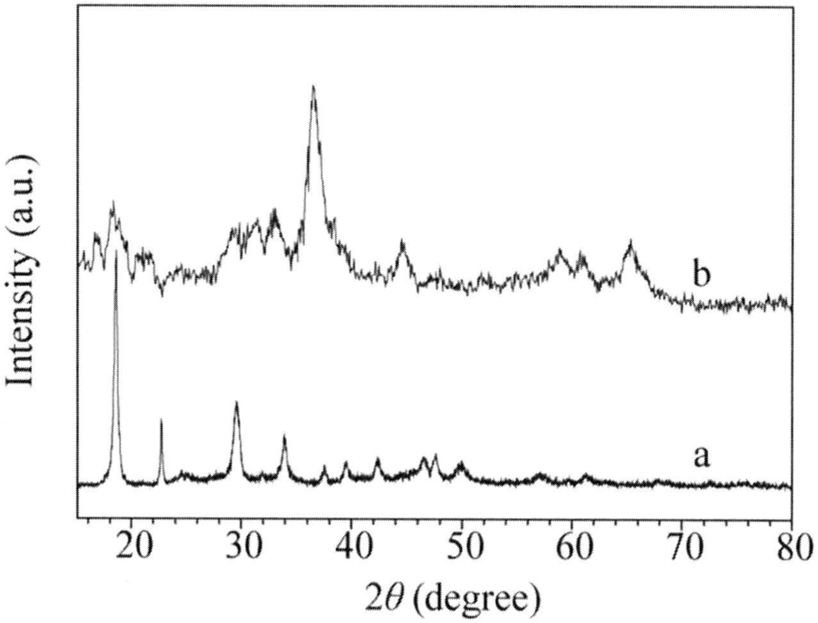

Figure 1. XRD patterns of (a) the Mn-Co oxalate and (b) the obtained Mn-Co oxide.

No characteristic diffraction peaks of the Mn-Co oxalate are observed in the XRD patttern of the calcined Mn-Co oxide in **Figure 1**(**b**), indicating that thermal decomposition of the Mn-Co oxalate is nearly completed in air at 250°C in the duration of 10 h. The low intensity and remarkable broadening of the diffraction peaks disclose that the obtained Mn-Co oxide is poorly crystallized. The discernible peaks around 2θ = 18.2°, 29.3°, 31.4°, 32.9°, 36.5°, 44.6°, 58.9°, 60.3° and 65.3° are well in accordance with those of tetragonal spinel Mn_3O_4 (JCPDS, No. 89-4837). Nevertheless, compared with the tetragonal spinel Mn_3O_4, the positions of the diffraction peaks for Mn-Co oxide shift positively, indicating the smaller lattice parameters of the obtained Mn-Co oxide. It has been demonstrated that the length of the Co^{n+}-O bond is smaller than that of the Mn^{n+}-O bond in the identical sublattice of a spinelstructured mixed oxide [28]. The smaller lattice parameters of the obtained Mn-Co oxide imply that the Co cations are incorporated into the crystal lattice of the tetragonal spinel Mn_3O_4, forming a single-phase Mn-Co mixed oxide. Therefore, the chemical formula of the obtained spinel Mn-Co oxide can be denoted as $Mn_{2.4}Co_{0.6}O_4$. Formation of the spinel Mn-Co oxide at a low calcination temperature of 250°C is probably related to the compositional homogeneity of the synthesized Mn-Co mixed oxalate [27].

Figure 2 shows the SEM images of the obtained Mn-Co oxide powder under different magnification. As seen from the lower-magnification SEM image in **Figure 2(a)**,

Figure 2. SEM images of the Mn-Co oxide powder under (a) lower and (b) higher magnification.

the obtained Mn-Co oxide powder is composed of irregular-shaped agglomerates with a size less than 5 mm. The higher-magnification SEM image in **Figure 2(b)** shows that each of the irregular agglomerates observed under lower magnification actually contains a great number of ultrafine primary particles with a particle size less than 100 nm. The ultra-fine particles of the prepared MnCo oxide certainly lead to a larger surface area available for charge storage, which is beneficial to en hancing its supercapacitive performance.

The mesoporosity and textural properties of the asprepared Mn-Co oxide was investigated by N_2adsorption-desorption measurement at liquid N_2 temperature. **Figure 3** gives the N_2 adsorption-desorption isotherm and BJH pore-size distribution of the Mn-Co oxide. It can be seen that the isotherm presents a typical type IV sorption behavior, indicating a mesoporous structure of the obtained Mn-Co oxide according to the classification of IUPAC [29]. Furthermore, there is an abrupt increase in adsorption volume of adsorbed N_2 at a certain P/P_0 value of 0.52. The sharp increase is generally attributed to capillary condensation, and the P/P_0 position of the inflection is associated with the pore size [30]. The BJH pore-size distribution, as given in the inset of **Figure 3**, reveals that the pore size distribution for the Mn-Co oxide is narrow and situated in the mesoporous region (2 - 50 nm). Also the textural properties of the Mn-Co oxide are given in **Table 1**. This demonstrates that the synthesis route of the solid-state coordination reaction we adopted allows the obtained Mn-Co oxide with a mesoporous structure and a considerable BET surface area, which are the desired textural properties for high-performance supercapacitor materials.

Table 1. Textural properties of the as-prepared Mn-Co oxide from N_2adsorption-desorption measurements.

BET surface ($m^2 \cdot g^{-1}$)	Pore volume ($cm^3 \cdot g^{-1}$)	Average pore size (nm)
120	0.26	19

3.2. Cyclic Voltammograms

Cyclic voltammetry is considered to be a powerful tool to indicate the capacitive behavior of electrode materials. **Figure 4** shows the cyclic voltammograms of the Mn-Co oxide electrode measured in a scan rate

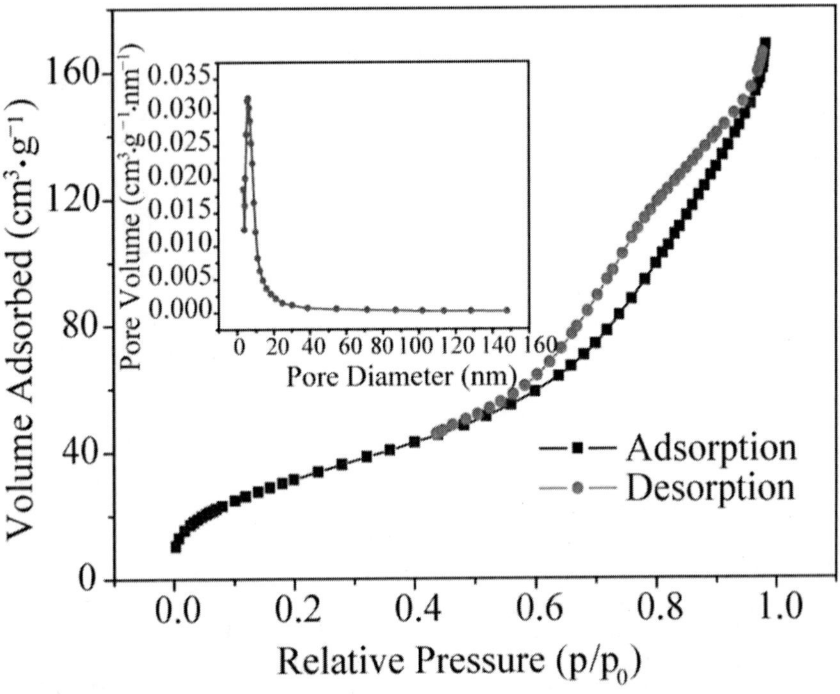

Figure 3. N_2 adsorption-desorption isotherm and BJH poresize distribution of the as-prepared Mn-Co oxide.

range from 2 to 50 mV·s^{-1} between -0.4 and 0.4 V vs. SCE in 6 mol·L^{-1}KOH aqueous electrolyte. Though slightly distorted at higher scan rates, all the cyclic voltammograms obtained are roughly rectangular in shape, and approximately symmetrical in anodic and cathodic directions, suggesting good capacitive behavior of the Mn-Co oxide electrode. The cyclic voltammograms exhibit a couple of approximately symmetric current peaks, i.e. an oxidation one around 0.16 V and a reduction one around 0.09 V, which are due to intercalation/detercalation of protons or alkaline metal cations in the Mn-Co oxide electrode in KOH aqueous electrolyte, indicating a remarkable faradic pseudocapacitance contribution [31]. The small potential difference between the anodic peak and the cathodic peak suggests the high reversibility of the Mn-Co oxide electrode.

Specific capacitance of the Mn-Co oxide electrode can be calculated from the cyclic voltammograms according to the formula

$$C_S = \frac{I}{mv} \tag{1}$$

where C_S is the specific capacitance ($F \cdot g^{-1}$), m the mass (g) of the electroactive material, v the potential scan rate ($V \cdot s^{-1}$), and I the even current response (A) defined by

$$I = \frac{1}{2(V_c - V_a)} \int i \mathrm{d}V \tag{2}$$

(V_a and V_c represent the lowest and highest potentials (V), respectively). Specific capacitance and specific current of the Mn-Co oxide electrode at ovarious scan rates were calculated from the cyclic voltammograms, as given in Figure 5. Figure 5(a) shows that specific capacitance decreases with increasing scan rate. The maximum spe cific capacitance of 383 $F \cdot g^{-1}$ is obtained for the Mn-Co oxide electrode at a scan rate of 2 $mV \cdot s^{-1}$. Importantly, a considerable specific capacitance of 225 $F \cdot g^{-1}$ is delivered

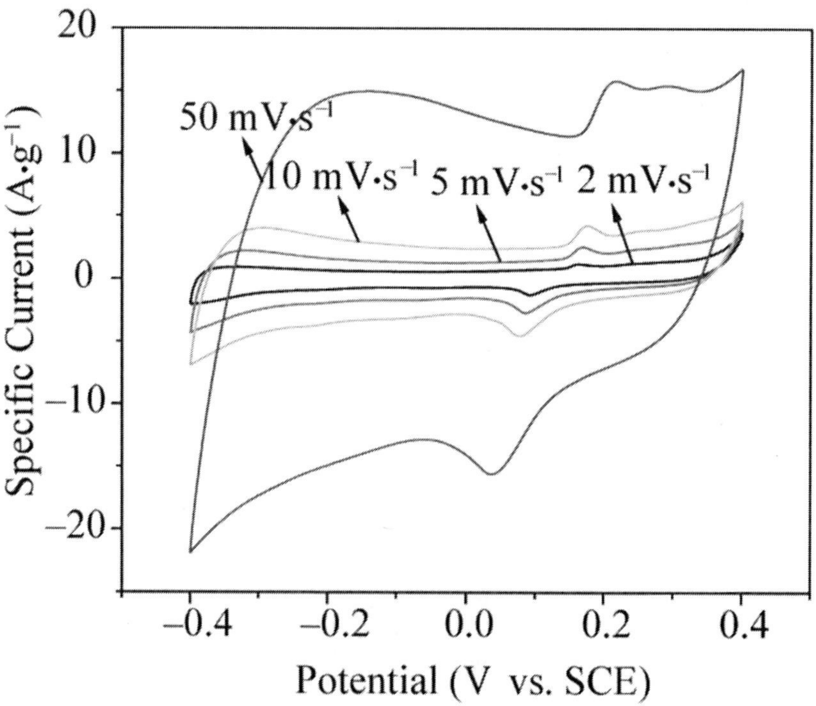

Figure 4. Cyclic voltammograms with various scan rates of the Mn-Co oxide electrode.

even at a high scan rate of 50 mV·s^{-1}, indicating high power density of the Mn-Co oxide electrode. **Figure 5(b)** demonstrates that the specific current increases approximately linearly with increasing scan rates, confirming a high power characteristic of the Mn-Co oxide electrode material. Compared with the pristine manganese oxide consisting of *a* and *γ*-MnO_2 phases derived from manganese oxalate in our previous work [32], the capacitive performance of the prepared Mn-Co oxide is much improved in the SC and power property. Notably, the specific capacitance of the obtained Mn-Co oxide electrode material is larger than that of the pristine manganese oxide electrode materials derived from sol-gel route [12], coprecipitation [33] and improved reduction reaction [34]. Moreover, the facility of the solid-state coordination reaction we adopted has the advantages of a relatively low number of processing steps, low process costs and a great convenience of obtaining mixed oxides with desired chemical compositions [35]. From a commercial and economical point of view, it is a promising method for preparation of oxide electrode materials, especially, of multi-component mixed oxide electrode materials.

3.3. Galvanostatic Charge-Discharge Analysis

A symmetrical capacitor was assembled using a couple of the identical Mn-Co oxide electrodes separated by a porous polypropylene separator. **Figure 6** shows the galvanostatic charge/discharge profiles of the Mn-Co oxide capacitor at different current densities. It can be seen that the charge/discharge profiles are somewhat curved, suggesting a pseudocapacitive characteristic of the Mn-Co oxide material. This is in accordance with the cyclic voltammetry result in **Figure 4**, where a couple of redox current peaks can be observed. The efficiency of the charge/discharge cycling is in the range of 0.88 - 0.92.

The specific capacitance of the symmetrical Mn-Co oxide capacitor, expressed as the one of a single electrode, can be calculated by the formula, where C_s is

$$C_S = \frac{4I\Delta t}{m\Delta V} \quad (3)$$

the discharge specific capacitance (in F·g^{-1}), I the discharge current (in A), Dt the discharge time (in s), m the total mass (in g) of the electroactive material contained in the symmetrical capacitor, DV the discharge potential

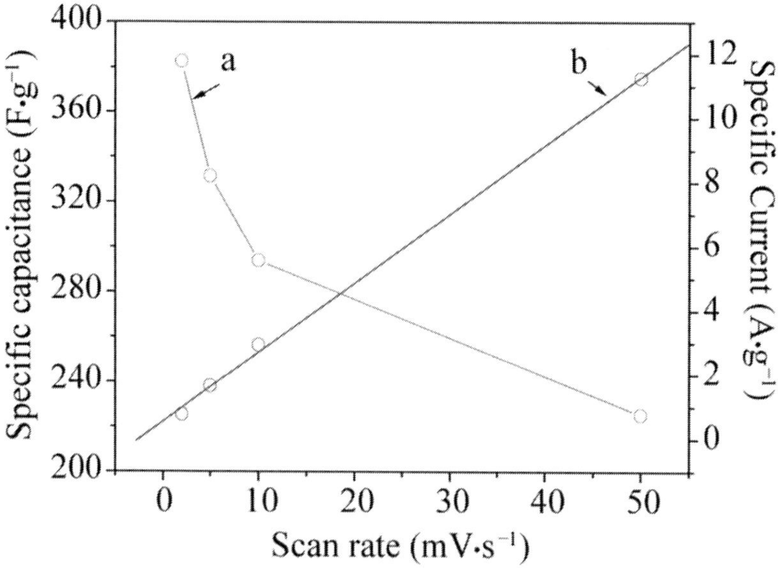

Figure 5. Scan rate dependence of (a) specific capacitance and (b) specific current of the obtained Mn-Co oxide electrode.

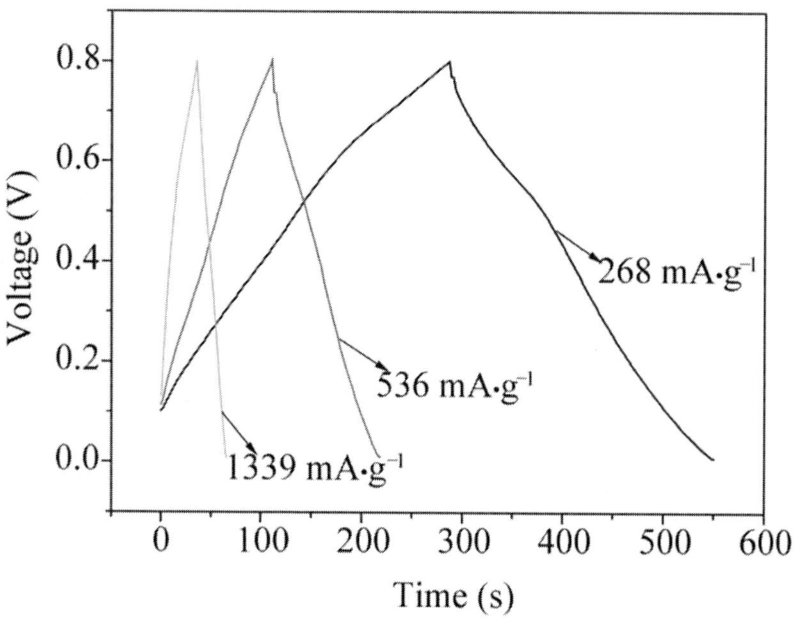

Figure 6. Galvanostatic charge/discharge profiles of the symmetrical Mn-Co oxide capacitor at various current densities.

range (in V) [36]. The specific capacitance of the symmetrical Mn-Co oxide capacitor at current densities of 0.27, 0.54 and 1.34 A·g^{-1} was calculated to be 354, 289 and 208 F·g^{-1}, respectively. Apparently, the specific capacitance reduction is relatively small when increasing current density from 0.27 to 1.34 A·g^{-1}, and a considerable specific capacitance of 208 F·g^{-1} is still retained at a high current density of 1.34 A·g^{-1}, demonstrating a high power characteristic of the obtained Mn-Co oxide material. This is in agreement with the cyclic voltammetry result in **Figure 5(b)**, where the specific current rises almost linearly with increasing scan rates.

For examination of the long-term cycling stability of the obtained Mn-Co oxide electrode material, the symmetrical Mn-Co oxide capacitor was subjected to charge/ discharge cycling at a current density of 1.34 A·g^{-1} for consecutive 500 cycles. The variation of the specific capacitance of the capacitor with cycle number is shown in **Figure 7**. During the first 100 cycles, the specific capacitance of the capacitor fades fast, decreasing by ~12% of its initial specific capacitance. During the next 400 cycles, the specific capacitance fades slightly and becomes stable, decreasing by ~3% of its initial specific capacitance. A considerable specific capacitance of 179 F·g^{-1}, about ~85% of its initial specific capacitance, is still remained for the capacitor after 500 cycles, suggestive of good cycling stability, which is of practical importance.

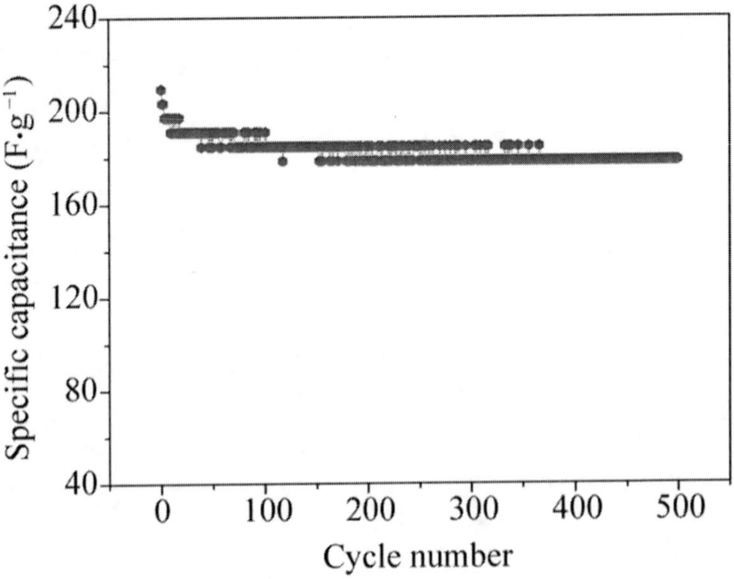

Figure 7. Cycling life of the symmetrical Mn-Co oxide capacitor at a current density of 1.34 A·g^{-1} between 0 and 0.8 V.

4. CONCLUSION

After the text edit has been completed, the paper is ready for the template. Duplicate the template file by using the Save As command, and use the naming convention prescribed by your journal for the name of your paper. In this newly created file, highlight all of the contents and import your prepared text file. You are now ready to style your paper. A mesoporous Mn-Co oxide powder with a particle size less than 100 nm was obtained by calcining the solid-state coordination reaction derived mixed oxalate $Mn_{0.8}Co_{0.2}C_2O_4 \times nH_2O$ in air at 250°C. The obtained Mn-Co oxide was poorly crystallized with a tetragonal spinel structure. The Mn-Co oxide electrode possesses the maximum capacitance of 383 $F\cdot g^{-1}$ at a scan rate of 2 $mV\cdot s^{-1}$, still retaining a considerable specific capacitance of 225 $F\cdot g^{-1}$ at a high scan rate of 50 $mV\cdot s^{-1}$. The specific current of the cyclic voltammograms varies roughly linearly with scan rates in the range of 2 $mV\cdot s^{-1}$ to 50 $mV\cdot s^{-1}$, demonstrating a high power characteristic of the obtained Mn-Co oxide electrode. At a charge/discharge current density of 1.34 $A\cdot g^{-1}$, the symmetrical Mn-Co oxide capacitor exhibits a specific capacitance of 210 $F\cdot g^{-1}$ during the first cycle, remaining ~85% of its initial capacitance after 500 cycles. In comparison with powder-based pristine manganese oxide material, the Mn-Co oxide material we prepared shows better supercapacitive performance, which is a promising candidate for superacpacitor materials.

5. ACKNOWLEDGEMENTS

Yong Yan, Bingcai Wu, Cuihong Zheng, Daolai Fang This work is financially supported by the Natural Science Foundation of Education Commission of Anhui Province under Grant no. KJ2012A045.

REFERENCES

1. B. E. Conway, "Electrochemical Supercapacitors," Kluwer Academic Publishers, New York, 1999.
2. Burke, "Ultracapacitors: why, how, and where is the technology," Journal of Power Sources, Vol. 91, No. 1, 2000, pp. 37-50. doi:10.1016/S0378-7753(00)00485-7
3. J. R. Miller and A. Burke, "Electrochemical Capacitors: Challenges and opportunities for Real-World Applications," Electrochemical Society Interface, Vol. 17, No. 1, 2008, pp. 53-57.
4. Y. Liu, K. X. Li, J. L. Wang, G. H. Sun and C. G. Sun, "Preparation of Spherical Activated Carbon with Hierarchical Porous Texture," Journal of Materials Science, Vol. 44, No. 17, 2009, pp. 4750-4753. doi:10.1007/s10853-009-3710-6

5. G. Pandolfo and A. F. Hollenkamp, “Carbon Propertites and Their Role in Supercapacitors,” Journal of Power Sources, Vol. 157, No. 1, 2006, pp. 11-27.doi:10.1016/j.jpowsour.2006.02.065
6. T. Nathan, A. Aziz, A. F. Noor and S. R. S. Prabaharan, “Nanostructured NiO for Electrochemical Capacitors: Synthiesis and Electrochemical Properties,” Journal of Solid State Electrochemistry, Vol. 12, No. 7-8, 2007, pp. 1003-1009. doi:10. 1007/ s10008-007-0465-3
7. J. Cheng, G. P. Cao and Y. S. Yang, “Characterization of Sol-Gel-Derived NiO_x Xerogels Supercapacitors,” Journal of Power Sources, Vol. 159, No. 1, 2006, pp. 734-731. doi:10.1016/j.jpowsour.2005.07.095
8. D. Bélanger, T. Brousse and J. W. Long, “Manganese Oxides: Battery Materials Make the Leap to Electrochemical Capacitors,” Electrochemical Society Interface, Vol. 17, No. 1, 2008, pp. 47-52.
9. D. K. Bhat and M. S. Kumar, “N and P Doped Poly(3,4- Ethylenedioxythiophene) Electrode Materials for Symmetric Redox Supercapacitors,” Journal of Materials Science, Vol. 42, No. 19, 2008, pp. 8158-8162 doi:10.1007/s10853-007-1704-9
10. H. Guan, L. Z. Fan, H. C. Zhang and X. H. Qu, “Polyaniline Nanofiber Obtained by Interfacial Polymerization for High-Rate Supercapacitors,” Electrochimica Acta, Vol. 56, No. 2, 2010, pp. 964-968. doi:10.1016/j.electacta.2010.09.078
11. E. H. Liu, R. Ding, X. Y. Meng, S. T. Tan and J. C. Zhou, “Potentiodynamical deposition of Nanosized Manganese Oxides as High Capacitance Electrochemical Capacitors,” Journal of Materials Science Materials in Electronics, Vol. 18, No. 12, 2007, pp. 1179-1182doi:10.1007/s10854-007-9369-3
12. R. N. Reddy and R. G. Reddy, “Synthesis and Electrochemical Characterization of Amorphous MnO_2 Electrochemical Capacitor Electrode Material,” Journal of Power Sources, Vol. 132, No. 1-2, 2004, pp. 315-320 doi:10.1016/j.jpowsour.2003.12.054
13. T. Brousse, M. Toupin, R. Dugas, L. Athouël, O. Crosnier and D. Bélange, “Crystalline MnO_2 as Possible Alternatives to Amorphous Compounds in Electrochemical Supercapacitors,” Journal of the Electrochemical Society, Vol. 153, No. 12, 2006, pp. A2171-A2180. doi:10.1149/1.2352197
14. L. Athouel, F. Moser, R. Dugas, O. Crosnier, D. Bélanger and T. Brousse, “Variation of the MnO_2 Birnessite Structure upon Charge/Discharge in an Electrochemical Supercapacitor Electrode in Aqueous Na_2SO_4 Electrolyte,” Journal of Physical Chemistry, Vol. 112, No. 18, 2008, pp. 7270-7277. doi:10.1021/jp0773029
15. F. B. Zhang, Y. K. Zhou and H. L. Li, “Nanocrystalline NiO as an Electrode Material for Electrochemical Capacitor,” Materials Chemistry and Physics, Vol. 83, No. 2-3, 2004, pp. 260-264. doi:10.1016/j.matchemphys.2003.09.046
16. X. M. Liu, X. G. Zhang and S. Y. Fu, “Preparation of Urchinlike NiO Nanostructures and Their Electrochemical Capacitive Behaviors,” Materials Research Bulletin, Vol. 41, No. 3, 2006, pp. 620-627. doi:10.1016/j.materresbull.2005.09.006
17. T. Y. Wei, C. H. Chen, K. H. Chang, S. Y. Lu and C. C. Hu, “Cobalt Oxide Aerogels of Ideal Supercapacitive Properties Prepared with an Epoxide Synthetic Route,” Chemistry of Materials, Vol. 21, No. 14, 2009, pp. 3228-3233. doi:10.1021/cm9007365
18. Y. Y. Gao, S. L. Chen, D. X. Cao, G. L. Wang and J. L. Yin, “Electrochemical Capacitance of Co_3O_4 Nanowire Arrays Supported on Nickel Foam,” Journal of Power Sources, Vol. 195, No. 6, 2010, pp. 1757-1760. doi:10.1016/j.jpowsour.2009.09.048
19. E. H. Liu, W. Li, J. Li, X. Y. Meng and S. T. Tan, “Preparation and Characterization of Nanostructured NiO/MnO_2 Composite Electrode for Electrochemical Supercapacitors,” Materials Research Bulletin, Vol. 44, No. 5, 2009, pp. 1122-1126.doi:10.1016/j.materresbull.2008.10.003
20. H. Kim and B. N. Popov, “Synthesis and Characterization of MnO_2 Based Mixed Oxides as Supercapacitors,” Journal of The Electrochemical Society, Vol. 150, No. 3, 2003, pp. D56-D62. doi:10.1149/1.1541675
21. J. P. Zheng, “Ruthenium Oxide-Carbon Composite Electrodes for Electrochemical Capacitors,” Electrochemical and Solid-State Letters, Vol. 2, No. 8, 1999, pp. 359-361doi:10.1149/1.1390837

22. M. Nakayama, A. Tanaka, Y. Sato, T. Tonosaki and K. Ogura, "Electrodeposition of Manganese and Molybdenum Mixed Oxide Thin Films and Their Charge Storage Properties," Langmuir, Vol. 21, No. 13, 2005, pp. 5907- 5913. doi:10 .1021/la050114u
23. X. Y. Xie, W. W. Liu, L. Y. Zhao and C. D. Huang, "Structural and Electrochemical Behavior of Mn-V Oxide Synthesized by a Novel Precipitation Method," Journal of Solid State Electrochemistry, Vol. 14, No. 1, 2010, pp. 1585-1594. doi:10.1007/s10008-009-0987-y
24. P. Y. Chuang and C. C. Hu, "The electrochemical Characteristics of Binary Manganese-Cobalt Oxides Prepared by Anodic Deposition," Materials Chemistry and Physics, Vol. 92, No. 1, 2005, pp. 138-145. doi:10.1016/j.matchemphys.2005.01.004
25. G. Y. Zhao, C. L. Xu and H. L. Li, "Highly Ordered Cobalt-Manganese Oxide (CMO) Nanowire Array Thin Film on Ti/Si Substrate as an Electrode for Electrochemical Capacitor," Journal of Power Sources, Vol. 163, No. 2, 2007, pp. 1132-1136.doi:10.1016/j.jpowsour.2006.09.085
26. E. Machefauxa, T. Brousseb, D. Bélangerc and D. Guyomard, "Supercapacitor Behavior of New Substituted Manganese Dioxides," Journal of Power Sources, Vol. 165, No. 2, 2007, pp. 651-655 doi:10.1016/j.jpowsour.2006.10.060
27. D. L. Fang, Z. B. Wang, P. H. Yang, W. Liu and C. S. Chen, "Preparation of Ultra-Fine Nickel Manganite Powders and Ceramics by a Solid-State Coordination Reaction," Journal of the American Ceramic Society, Vol. 89, No. 1, 2006, pp. 230-235. doi:10.1111/j.1551-2916.2005.00666.x
28. D. L Fang, C. S Chena and A. J. A. Winnubstc, "Preparation and Electrical Properties of $Fe_xCu_{0.10}Ni_{0.66}Mn_{2.24-x}O_4$ ($0 \leq x \leq 0.90$) NTC ceramics," Journal of Alloys and Compounds, Vol. 454, No. 1-2, 2008, pp. 286-291. doi:10.1016/ j.jallcom. 2006.12.059
29. F. Rouquerol, J. Rouquerol and K. Sing, "Adsorption by Powers and Porous Solids: Principles Methodology and Applications," Academic Press, San Diego, 1999.
30. T. Sreethawong, S. Chavadej, S. Ngamsinlapasathian and S. Yoshikawa, "A Modified Sol-Gel Process-Derived Highly Nanocrystalline Mesoporous NiO with Narrow Pore Size Distribution," Colloids and Surfaces A: Physicochemical and Engineering Aspects, Vol. 296, No. 1-3, 2007, pp. 222-229. doi:10.1016/j.colsurfa.2006.09.048
31. E. Raymundo-Pinero, V. Khomenko and E. Frackowiak, "Performance of Manganese Oxide/CNTs Composites as Electrode Materials for Electrochemical Capacitors," Journal of The Electrochemical Society, Vol. 152, No. 1, 2005, pp. 229-235.doi:10.1149/1.1834913
32. D. L. Fang, B. C. Wu, A. Q. Mao, Y. Yan and C. H. Zheng, "Supercapacitive Properties of Ultra-Fine MnO_2 Prepared by a Solid-State Coordination Reaction," Journal of Alloys and Compounds, Vol. 507, No. 2, 2010, pp. 526-530. doi:10. 1016/j.jallcom. 2010.08.010
33. M. Toupin, T. Brousse and D. Bélanger, "The Influence of Microstructure on the Charge Storage Properties of Chemically Synthesized Manganese Dioxide," Chemistry of Materials, Vol. 14, No. 9, 2002, pp. 3946-3952. doi:10. 1021/ cm020408q
34. S. J. Bao, B. L. He, Y. Y. Liang, W. J. Zhou and H. L. Li, "Synthesis and Electrochemical Characterization of Amorphous MnO_2 for Electrochemical Capacitor," Materials Science and Engineering: A, Vol. 397, No. 1-2, 2005, pp. 305-309.doi: 10.1016/ j.msea.2005.02.058
35. C. H. Zheng and D. L. Fang, "Preparation of Ultra-Fine Cobalt-Nickel Manganite Powders and Ceramics Derived From Mixed Oxalate," Materials Research Bulletin, Vol. 43, No. 7, 2008, pp. 1877-1882. doi:10.1016/j.materresbull.2007.06.061
36. J. Li, X. Y. Wang, Q. H. Huang, S. Gamboa and P. J. Sebastian, "A New Type of $MnO_2 \cdot xH_2O$/CRF Composite Electrode for Supercapacitors," Journal of Power Sources, Vol. 160, No. 2, 2006, pp. 1501-1505. doi:10.1016/j.jpowsour.2006.02.045

CHAPTER 4

Viscosity Profiles Of Phosphate Glasses Through Combined Quasi-Static and Bob-in-Cup Methods

A.J. Parsons [a,*], N. Sharmin [a], Sharifah I.S. Shaharuddin [a], M.Marshall [b]

[a] *Division of MMS, Faculty of Engineering, University of Nottingham, NG7 2RD, UK*

[b] *Glass Technology Services, 9 Churchill Way, Sheffield, South Yorkshire, S35 2PY, UK*

ABSTRACT

This study used a combined viscosity approach to determine theoretical fibre drawing points for glasses in the series: xP_2O_5, 24MgO, 16CaO, $(60-x)Na_2O$ (x = 40,45,50,55) and yP_2O_5, 24MgO, 16CaO, $(55-y)Na_2O$, $5Fe_2O_3$(y = 40,45,50,55). The points cannot be measured directly since the glasses are only kinetically stable at these points and would crystallise if allowed to equilibrate. Quasi-static and bob-in-cup viscosity data from above and below the range of interest were fitted to the Vogel–Fulcher–Tammann equation and provided good agreement. The theoretical drawing points were taken as the temperature at which the glass has a viscosity of 2 Log Pa·s, based on the known drawing point viscosity of silica glasses. The theoretical drawing points for the glasses ranged from 657 to 839 °C. The viscosity information was also used to assess the fragility of the glasses in comparison with a borosilicate standard by using Doremus and Angell parameters. All of the glasses were of low viscosity and high fragility in comparison to the borosilicate. The fragility improved above 50% content of phosphate in the glass and the addition of iron had little effect on the fragility. Additionally, the limitations of the borosilicate 717a standard glass and the measurement of T_g are discussed.

Keywords: Phosphate glass Viscosity Vogel–Fulcher–Tammann Fragility

1. INTRODUCTION

Phosphate glasses and their fibres are being developed for use in medical applications [1–3]. This is because they possess relatively high strength values (typically 500 MPa to 1.2 GPa [4]) but can also degrade away in an aqueous environment. Ordinarily this is a surface erosion process [5], allowing for the maintenance of properties during degradation [6].

The glasses can be utilised both directly and as reinforcements in composite materials. To make use of the strength properties of the glasses when reinforcing composites, it is necessary to have them in the form of fibres. The production of phosphate optical fibres via pre-form drawing processes is well known [7], but to create reinforcing fibres for composites these processes are not well suited. Typical optical fibres are in the range > 30 μm and often in the hundreds of microns; a desirable diameter for reinforcing fibres is < 25 μm and preferably ~ 9–12 μm [8]. Additionally, reinforcing fibres are created as multi-filament tows that contain hundreds and often thousands of individual filaments all drawn at the same time, making pre-form drawing impractical (although not impossible). To create small diameter, multifilament threads it is necessary to use melt drawing.

Melt drawing is also well known, though overwhelmingly focused on silica-based glasses. Silica fibres (commonly E-glass) have been produced on a commercial scale since the 1960s [9]. However, a commercial process for small diameter phosphate glass fibre production has never been achieved. This is partially due to a lack of perceived applications for phosphate glass fibres until relatively recently and partly due to the properties of the glass.

Silica bases glasses draw very effectively in the commercial melt drawing process because they possess viscosities at the appropriate level for fibre production when the glass is above its liquidus [10]. This means that the molten glass can be held at the correct fibre drawing temperature without a risk of crystallisation and so a continuous process is possible.

Phosphate glasses tend to be fragile and prone to crystallisation, with viscosities much lower than silicates. Critically, the 'drawing point' viscosity for glass fibre production (typically assumed to be ~ 100 Pa·s or 2 Log Pa·s [10]) usually falls below the liquidus point and due to the fragility of the glasses the drawing point temperature window is also relatively narrow. Conventionally this would preclude melt-drawing but a number of groups around the world, including ours, have been producing phosphate glass fibres successfully in the laboratory for many years. This is because glass that is flowing is more resistant to crystallisation, lending a kinetic stability to the process. As such a continuous production process is feasible but to design a long term, reliable process, good information about the viscosity of the glass is necessary. If correct conditions are not achieved and crystals nucleate then gradual growth of these crystals will occur, resulting in reduced quality fibre and eventual blockage of the drawing equipment.

Because the glasses are only usually kinetically stable at their drawing point temperatures it is impossible to take equilibrium viscosity measurements of the drawing point directly. In order to obtain close approximations of these values, this study combines parallel plate quasi-static viscosity measurements using Gent's equation [11] with high temperature bob-in-cup measurements that fall either side of the crystallisation region. Quasi-static measurements provide values in the range of 5.5–8 Log Pa·s, while bob-in-cup measurements provide values in the < 2 Log Pa·s region. The data is then applied to the well-known and generally accepted Vogel–Fulcher–Tammann (VFT) equation (developed in the 1920s) in order to predict a complete viscosity profile and a drawing point temperature. Additionally, the viscosity data is combined with conventional thermal analysis methods to obtain Angell [12,13] and Doremus [14] fragility indices. The data also provides some insights to the effect of additives on the structure of phosphate glasses. The study includes eight glass formulations from the following two systems: xP_2O_5, 24MgO, 16CaO, $(60-x)Na_2O$ (x = 40,45,50,55) and yP_2O_5, 24MgO, 16CaO, $(55-y)Na_2O$, $5Fe_2O_3$(y = 40,45,50,55). The glasses were selected based on previously determined formulations with good cytocompatibility [15].

2. METHODS

2.1. Glass production

Glasses were produced by melt fusing phosphate salts and phosphorus pentoxide in 5%Au/Pt crucibles. NaH_2PO_4, $CaHPO_4$, $MgHPO_4 \cdot 3H_2O$ and $FePO_4 \cdot 2H_2O$ were obtained from Sigma Aldrich UK, P_2O_5 was obtained from Fisher Chemicals. All Reagents were used as received. Each batch was heated for 30 min at 350 °C in a Lenton Furnaces AWF 12/26 to drive off adducted water before being transferred to a Lenton Furnaces UAF 16/10 at 1100 °C for 90 min. The glasses were either cast directly onto a steel plate or poured into 4 mm diameter graphite moulds. Glass compositions and precursor recipes can be found in Table 1.

Table 1. Glass recipes.

Glass code (mol% oxide)	Precursor mixture (mass% ± 0.01 g)				
	NaH_2PO_4	$FePO_4 \cdot 2H_2O$	$CaHPO_4$	$MgHPO_4 \cdot 3H_2O$	P_2O_5
P40Mg24Ca16Na20	43.00	0.00	19.51	37.49	0.00
P45Mg24Ca16Na15	31.63	0.00	19.13	36.77	12.48
P50Mg24Ca16Na10	20.69	0.00	18.77	36.07	24.48
P55Mg24Ca16Na5	10.15	0.00	18.42	35.40	36.03
P40Mg24Ca16Na15Fe5	30.43	15.80	18.40	35.37	0.00
P45Mg24Ca16Na10Fe5	19.92	15.51	18.07	34.73	11.78
P50Mg24Ca16Na5Fe5	9.78	15.23	17.74	34.10	23.14
P55Mg24Ca16Fe5	0.00	14.96	17.43	33.50	34.11

2.2. Glass sample preparation

The glass cast directly was broken up into cullet and used for density and liquidus measurements, and bob-in-cup viscosity. The glass cast into the graphite moulds was annealed at the glass T_g + 10 °C as recommended by Ropp [16], before careful sectioning into parallel faced, ~ 3 mm long rods for quasi-static viscosity by using a South Bay Technologies Inc. Model 650 low speed diamond saw. Pieces of the glasses were also ground to powder for differential scanning calorimetry (DSC), using a Retch Planetary Ball Mill PM 100 with zirconia grinding surfaces. The average particle size was < 45 μm (measured using calibrated testing sieves).

2.3. X-ray fluorescence spectroscopy (XRF)

XRF compositional analysis was carried out on a Bruker S4 Pioneer XRF set. Samples were prepared by re-melting the glass diluted with lithium tetraborate at 1220 °C and casting into platinum dishes. The resulting 40 mm diameter discs are then submitted for analysis, taking care to avoid any contamination from handling. Analysis is performed with the instrument operating in the "standard-less mode" which was developed internally for these types of non soda–lime–silica (SLS) glasses. Results which provide a total sample value outside of the 99.5–100.5% range (a total value of 100% is expected) are rejected and sampling and fusion is repeated.

2.4. Differential scanning calorimetry (DSC)

Thermal analysis of the glasses was performed using a Texas Instruments SDTQ600. Measurements of glass transition temperature were taken in triplicate at 10 °C/min using platinum pans containing ~ 30 mg of powdered glass.

2.5. Liquidus

Liquidus measurements were undertaken according to ASTM C829-81, using a purpose built furnace. Coarse glass cullet (3–5 mm) is placed into a 155 mm × 10 mm × 10 mm 90%Pt/10%Rh tray and melted at 1450 °C for 15 min in order to allow the glass to flow and fill the tray. The tray is then held in the liquidus furnace for 24 h to stabilise before being removed and cooled in air. The extent of crystallisation is determined by using an optical microscope and observing the centre of the resulting glass bar (some nucleation could occur at the walls). The thermal gradient in the furnace is determined by 15 thermocouples with a spacing of 10 mm and confirmed by using a movable thermocouple. The temperature gradient across the length of the sample is typically ~ 80 °C. An approximate temperature can be obtained from the DSC data in order to reduce experimental time.

2.6. Density

Glass density was measured using bubble free cullet in a Micromeritics AccuPyc 1330 gas pycnometer, using helium gas. The pycnometer was calibrated and checked by using steel calibration standards, with 10 measurements made. For the glass, 5 measurements were made for each of three different glass batches of each composition.

2.7. Quasi-static viscosity

The quasi-static viscosity data was obtained via a process described elsewhere [17–19]. Briefly, a small rod of the glass (4 mm diameter × 3 mm length) is subjected to a constant axial force while heated and the rate of change of height of the rod provides a viscosity vs temperature profile. The measurements were taken in triplicate for each composition.

2.8. Bob-in-cup viscosity

Bob-in-cup viscosity measurements were taken essentially in accordance with ISO 7884–2:1987 by using a Theta industries Rheotronic II 1600 °C

Rotating Viscometer. The system uses a 10 mm diameter, 14 mm high platinum double cone bob in a 40 mm high, 30 mm internal diameter platinum crucible with a 1 mm wall thickness. The glass density was used to determine the amount of cullet required to achieve a glass height of ~ 24 mm in the crucible (with the bob inserted). The system measures the shear stress using a Brookfield DV-III Ultra Rheometer and is calibrated by using a multiplication or 'spindle' factor based on a standard material.

Calibration was undertaken by use of the borosilicate glass standard reference material 717a. The glass was heated to 1256 °C and the bob lowered into the crucible to a height of 3 mm above the base. Sufficient spindle RPM was used to achieve > 70% torque (10 rpm). Readings of the viscosity were taken every minute and a minimum determined (see discussion for more explanation). An appropriate spindle factor was then chosen to adjust the minimum viscosity reading to 100,000 cP, or Log Pa·s = 2, as per the values provided for the 717a standard.

To measure a sample, the system is heated to 1100 °C. After 30 min the platinum bob is lowered into the now molten glass so that the base of the bob is 3 mm above the base of the crucible. The system is then left for a further 15 min to come to thermal equilibrium. Viscosity measurements are taken at different RPM values, to probe the viscosity behaviour of the glass. The temperature is lowered in stages, with suitable equilibrium time allowance, until the glass either crystallises or the torque limits of the system are exceeded. Three different batches of each glass type were used. To check that there was no drift in the spring value over the course of the experiments, a fresh batch of the first glass tested was run again at the end. The equipment uses known geometric values and the calibration constant to provide values in cP, which were converted to Log Pa·s.

3. RESULTS

3.1. XRF analysis

XRF analysis of the final glass compositions are provided in Table 2, alongside the expected feed compositions. Values are in mol%. In all cases except for the Na_2O in P40Mg24Ca16Na15Fe5 the results were within 1% of expectation and in the majority of cases within 0.5%, giving good confidence in the composition of the glasses.

Table 2. Glass compositional analysis.

Oxide	Expected	XRF (± 0.5)	Expected	XRF (± 0.5)	Expected	XRF (± 0.5)	Expected	XRF (± 0.5)
	P40Mg24Ca16Na20		P45Mg24Ca16Na15		P50Mg24Ca16Na10		P55Mg24Ca16Na5	
P_2O_5	40	39.8	45	45.1	50	49.4	55	54.0
MgO	24	23.9	24	23.8	24	23.6	24	24.0
CaO	16	16.2	16	16.3	16	16.2	16	16.4
Fe_2O_3	0	0.0	0	0.0	0	0.0	0	0.0
Na_2O	20	20.0	15	14.7	10	10.7	5	5.6
	P40Mg24Ca16Na15Fe5		P45Mg24Ca16Na10Fe5		P50Mg24Ca16Na5Fe5		P55Mg24Ca16Fe5	
P_2O_5	40	39.3	45	44.7	50	49.6	55	54.3
MgO	24	23.5	24	23.7	24	24.1	24	24.4
CaO	16	16.3	16	16.4	16	16.5	16	16.4
Fe_2O_3	5	4.6	5	4.8	5	4.8	5	4.9
Na_2O	15	16.3	10	10.4	5	5.0	0	0.0

3.2. Thermal analysis and density

Glass transition temperature, liquidus and density values for the different glasses are provided in Table 3. Density values drop with increasing phosphate content, and the density increases with iron addition. T_g values show a general trend of increase with phosphate content but with unusual non-linearity. Liquidus values are distinctly non-linear and do not seem to follow a pattern.

Table 3. Thermal data and density.

Glass code (mol% oxide)	T_g (°C)	Liquidus (°C)	Density (kg/m3)
P40Mg24Ca16Na20	443 ± 1	900 ± 5	2689 ± 1
P45Mg24Ca16Na15	450 ± 1	822 ± 5	2641 ± 1
P50Mg24Ca16Na10	465 ± 3	863 ± 5	2588 ± 1

P55Mg24Ca16Na5	460 ± 2	974 ± 5	2536 ± 2
P40Mg24Ca16Na15Fe5	496 ± 2	925 ± 5	2821 ± 1
P45Mg24Ca16Na10Fe5	490 ± 2	957 ± 5	2765 ± 1
P50Mg24Ca16Na5Fe5	501 ± 2	877 ± 5	2698 ± 2
P55Mg24Ca16Fe5	521 ± 2	1000 ± 5	2621 ± 1

3.3. Quasi-static viscosity

Quasi-static viscosity data is provided in Fig. 1. All the glasses showed reasonably linear behaviour in the region of 5.5 to 8 Log Pa·s ($R^2 > 0.99$). Viscosity increased with increasing phosphate content. However, the addition of iron increased the viscosity of the glass greatly, with even the P40Mg24Ca16Na15Fe5 glass demonstrating a higher viscosity than the P55Mg24Ca16Na5 glass. This indicates that iron has a greater effect on the viscosity of these glasses than phosphate content in this viscosity range. Additionally, the magnitude of the effect of increasing phosphorus is different between the glasses containing Fe and those not — there is almost no difference in viscosity between P50Mg24Ca16Na10 and P55Mg 24Ca16Na5 in this region, whereas the jump between P50Mg24 Ca16Na5Fe5 and P55Mg24Ca16Fe5 is the most significant in that series, though the P55Mg24Ca16Fe5 does contain no sodium. Furthermore, a gradual reduction in gradient is observed for the high phosphate content glasses, suggesting an improvement in glass fragility.

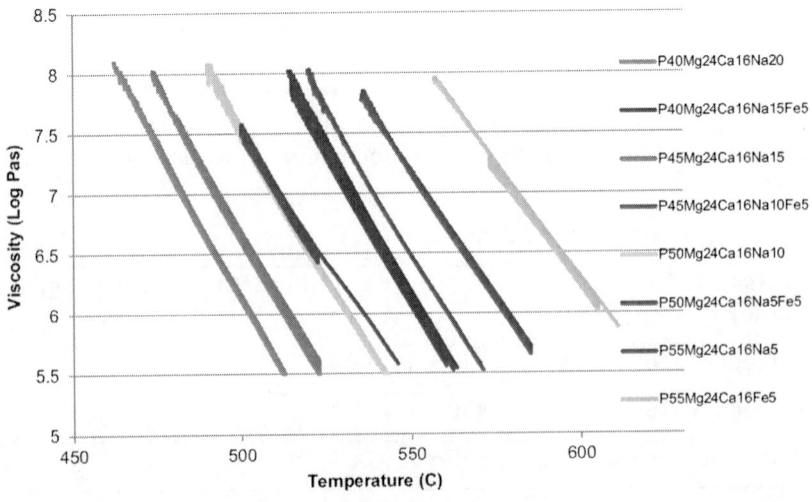

Figure. 1. Viscosity of the glasses in the range 5.5 to 8 Log Pa·s obtained through the quasi-static parallel plate method. The data shown includes three measurements for each glass type, demonstrating the degree of repeatability. Error bars are omitted for clarity, though the assumed error is ± 0.2 Log Pa·s.

3.4. Bob-in-cup viscosity

3.4.1. Newtonian behaviour of phosphate glasses

The viscosity of the glasses was measured for a range of RPM values at each temperature, which in turn affected the shear rate and shear stress. For all of the glasses the viscosity was constant within error for RPM values of 10 or above and the glasses are considered to be Newtonian. For the P50 glasses, crystallisation did not appear to occur until relatively low temperatures were achieved and RPM values of below 10 were required to remain below 100% torque. In these instances there was some divergence in the value of viscosity for different RPM values. This was surmised to be due to slow crystallisation and the results were discounted.

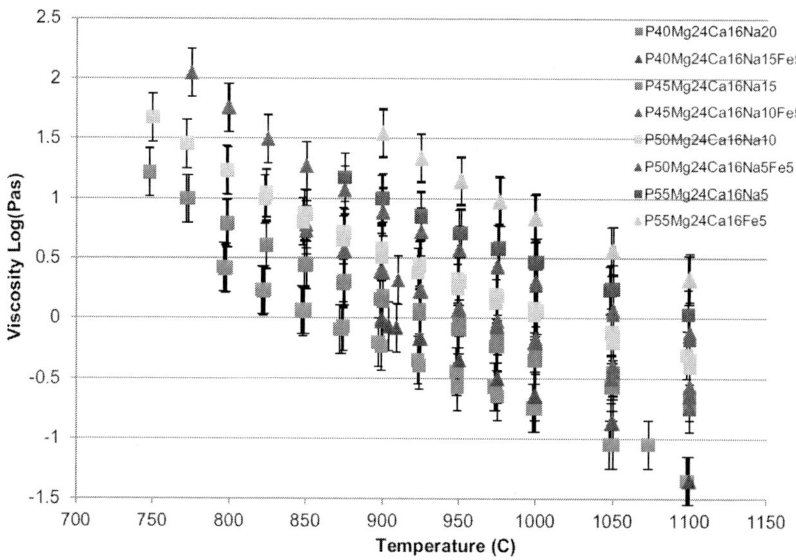

Figure. 2. Viscosity for the glasses in the range – 1.5 to 2 Log Pa·s obtained through the bob in cup method. The data shown includes three measurements for each glass type demonstrating the degree of repeatability. Errors are taken to be ± 0.2 Log Pa·s.

3.4.2. Viscosity data

Bob in cup data is provided in Fig. 2. The glasses showed good linearity over the range of measurement (P40s and P55s $R^2 > 0.99$, P45s $R^2 > 0.98$, P50s $R^2 > 0.97$). Once again, viscosity is seen to increase with increasing phosphate content. However, in this region of viscosity the phosphorus seems to have a much greater effect than the iron. In Fig. 1 the iron containing range of glasses all had higher viscosities than the non-iron containing range. In Fig. 2 the iron and non-iron pairs are quite close together, with a clear split between each level of phosphate content (e.g. P40Mg24Ca16Na20 and P40Mg24Ca16Na15Fe5 lie close together and are below both the P50 glasses). Additionally, the presence of iron seems to have a greater effect on the viscosity at higher phosphate levels.

4. DISCUSSION

4.1. Thermal analysis and density

The T_g values shown in Table 3 generally increase both with increasing phosphate content (due to increasing phosphate chain length) and with increasing iron content (due to cross linking [20]), as would be expected. Above P50 there is a degree of branching, which can perturb the structure and reduce the T_g again, which could explain the peak in T_g at P50 for the non-iron containing glasses. With the iron containing glasses, the effect of iron largely overrides the effect of phosphate in the shorter chain systems and the high value of P55Mg24Ca16Fe5 is likely to be due to the complete absence of sodium as a modifier.

Density also goes up with iron content (as expected [21]) but down with phosphate content. Reduced density occurs in longer chain length glasses due to increasing difficulty in chain packing. The liquidus values do not show an obvious pattern and the reasons for this are not clear. Further analysis of the liquidus surface of these glass compositions would be of benefit in future studies as there may be a eutectic effect.

4.2. Viscosity analysis

4.2.1. Limitations in the measurement of viscosity of phosphate glasses at high temperature and consideration of error

Due to the prevalence of silica based glasses and the relative obscurity of phosphate glasses, the high temperature viscosity equipment and, critically, the related standard materials are intended primarily for the measurement of the former glass type. While the process is applicable directly to phosphate glasses, there are limitations to consider.

When calibrating a viscometer it is advisable to select a standard material that has a comparable viscosity profile, so as to ensure that measurements are made in a similar range of torque. For glass viscosity, there are a very limited number of standard glasses available with certified values for Log Pa·s = 2 and these are all silica based glasses with viscosities typically an order of magnitude or more higher than phosphate glasses at equivalent temperatures. For the fragile phosphate glasses, where measurements are being made at Log Pa·s < 2 this means that torque values can be very low and in some cases less than the recommended minimum for the equipment of 10% torque. As such, error levels for individual readings are likely to be higher than might otherwise be expected. Having said that, triplicate measurements for three different batches of each glass provided very good repeatability and give confidence in the results. Additionally, the final batch run to check for drift in the machine gave a very good match to the first glass, demonstrating good stability in the equipment over several months. Overall error in repeatability of the data is of the order of 0.1 Log Pa·s. This error is in addition to the systematic error provided for the viscosity standard, which is listed as 2.00 ± 0.10 Log Pa·s at 1256 °C. Error in the furnace temperature is comparatively negligible. We will take the overall error to be ± 0.2 Log Pa·s.

The standard reference material provided for the calibration of the equipment was the 717a borosilicate glass. The standard notes that the glass is susceptible to volatilisation at high temperatures but suggest that this is only of real concern above 1400 °C. However, trials in this study determined that the viscosity of this glass is quite variable, even at 1256 °C. This is problematic, since when measuring viscosity the recommendation is to allow a considerable length of time to achieve equilibrium. The standard does not provide guidelines as to how long the material should be held at temperature before taking a measurement and reference to the original papers on the standard glasses does not provide this parameter explicitly either. It can be inferred from the papers that the time before measurement is relatively short [22,23].

After a number of measurements, the authors would recommend that when using this calibration glass a rapid heating rate is used to achieve temperature, followed by a continuous monitoring of the viscosity in order to achieve a minimum, as shown in Fig. 3.

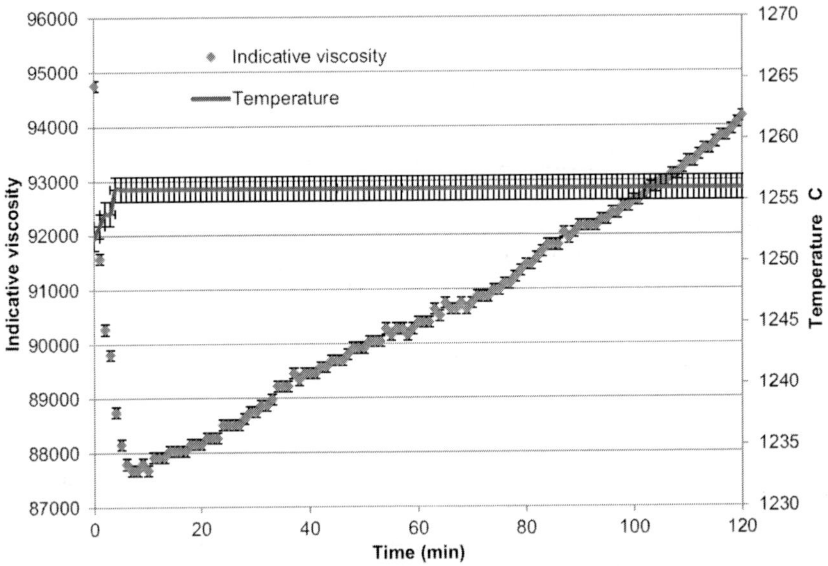

Figure. 3. Variation in the viscosity of the calibration glass with time when held at constant temperature.

This minimum is then taken to be the viscosity at that temperature. Prior to the minimum the glass has not quite achieved the set temperature and so the viscosity is falling. After the minimum the loss of light components (B and possibly Na) becomes the dominant factor and so the viscosity increases. The result is repeatable but the faster the heating rate the better. A slower heating rate results in a convergence with the curve but at a higher minimum, indicating that material loss has already occurred before reaching temperature. A minimum is taken rather than an extrapolation for two reasons. Firstly, the minimum approach is closer to the method used in the papers on the standard glass, i.e. a measurement taken shortly after reaching temperature. Secondly, the temperature of the glass is not measured directly, so it is difficult to determine a '0 min' point. Overall one must consider that the viscosity will be slightly higher than the true value since it is not possible to heat the glass instantaneously and a safe limit for the equipment is considered to be 20 °C/min. Fortunately, the phosphate glasses do not appear to show this instability and it is possible to obtain stable equilibrium measurements, at least below 1100 °C.

Unfortunately 710a soda lime silica and 711 lead glass viscosity standards have been discontinued and so their suitability cannot be considered. It would be greatly beneficial to phosphate glass researchers for a phosphate glass viscosity standard to be established and it is important to highlight the limitations of the 717a calibration material to other users.

Finally, the surface tension of glasses decreases with temperature, providing a practical limit above which the glass will creep up the spindle and cause damage to the supporting components. For the Pt alloy and the glasses used in the study a limit of 1100 °C was imposed. Moving to a Au/Pt low wetting alloy may provide a greater window of measurement.

4.2.2. Combination of viscosity data and application of models

In order to obtain a complete picture of the viscosity profile the quasi static data and bob-in-cup data were combined. There are a number of potential fits for viscosity and Ojovan provides a useful review [24]. This study will be limited to the Vogel–Fulcher–Tammann (VFT) equation (Eq. (1)), which is considered to have the best fit across the full spectrum of viscosity.

$$\text{Log}(\eta) = A + \frac{B}{T - T_0} \tag{1}$$

The Excel Solver addin was used to fit the values of A, B and T_0 by a least squares regression method. An example of the data fit is provided in Fig. 4 and the combined graphs are provided in Fig. 5, with the values of A, B, and T_0, the theoretical drawing point (i.e. the value of temperature at a viscosity of Log Pa·s = 2) (T_D) and the difference between drawing point and liquidus (ΔT) provided in Table 4.

Table 4. VFT parameters and drawing points.

Glass code (mol% oxide)	A (± 0.1)	B (± 100)	To (± 10)	T_D (°C)	Liquidus (°C)	ΔT (°C)
P40Mg24Ca16Na20	- 3.9	2233	277	657 ± 3%	900 ± 5	- 243 ± 8
P45Mg24Ca16Na15	- 3.1	1912	302	680 ± 4%	822 ± 5	- 142 ± 6
P50Mg24Ca16Na10	- 2.6	1811	321	714 ± 5%	863 ± 5	- 149 ± 7
P55Mg24Ca16Na5	- 2.3	1910	305	751 ± 5%	974 ± 5	- 223 ± 11
P40Mg24Ca16Na15Fe5	- 4.1	2366	319	704 ± 3%	925 ± 5	- 221 ± 7
P45Mg24Ca16Na10Fe5	- 3.5	2185	330	730 ± 3%	957 ± 5	- 227 ± 8

P50Mg24Ca16Na5Fe5	– 3.1	2257	329	775 ± 4%	877 ± 5	- 102 ± 4
P55Mg24Ca16Fe5	– 2.8	2462	329	839 ± 4%	1000 ± 5	- 161 ± 7

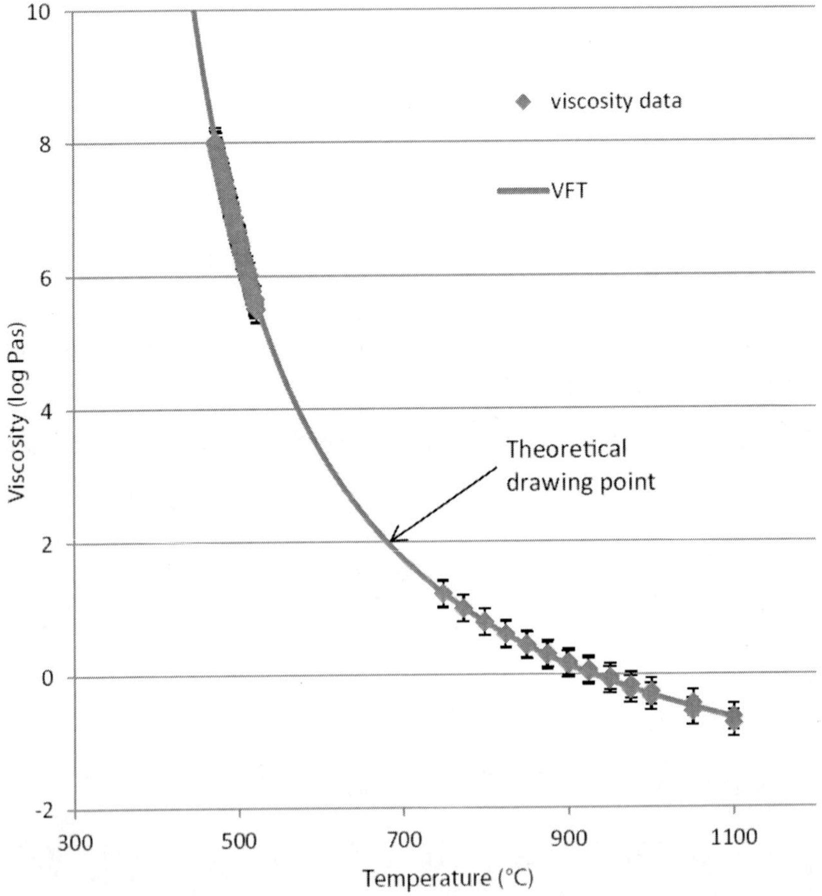

Figure. 4. Example of the fit of the viscosity data to the Vogel–Fulcher–Tammann equation, showing good agreement.

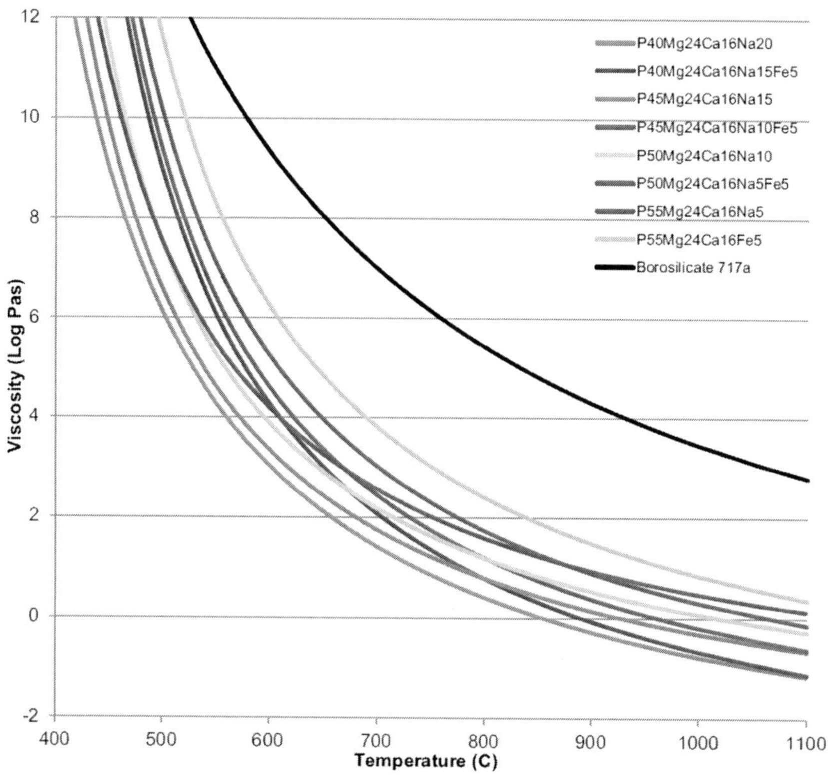

Figure. 5. Combined viscosity profiles fitted using the Vogel–Fulcher–Tammann equation, with the standard glass 717a for comparison. Errors are taken to be ± 0.2 Log Pa·s.

As was considered in Sections 3.3 and 3.4.2, it is apparent from Fig. 5 that the effect of iron on the viscosity of the glass varies. At lower temperature, i.e. higher viscosity, there is a significant increase in viscosity provided by the presence of iron (which is also reflected in the values of T_g in Table 3). At higher temperatures the traces converge, with the presence of iron providing little difference in the viscosity. For greater clarity, the P40 pair are extracted from Fig. 5 and shown in Fig. 6 as an example. This trend appears to be the case for all the glasses from P40 to P55. The reason for this bears further investigation and may be related to the mobility of the Fe^{3+} ion at higher temperature. However, structural analysis of the glasses is beyond the scope of the present study.

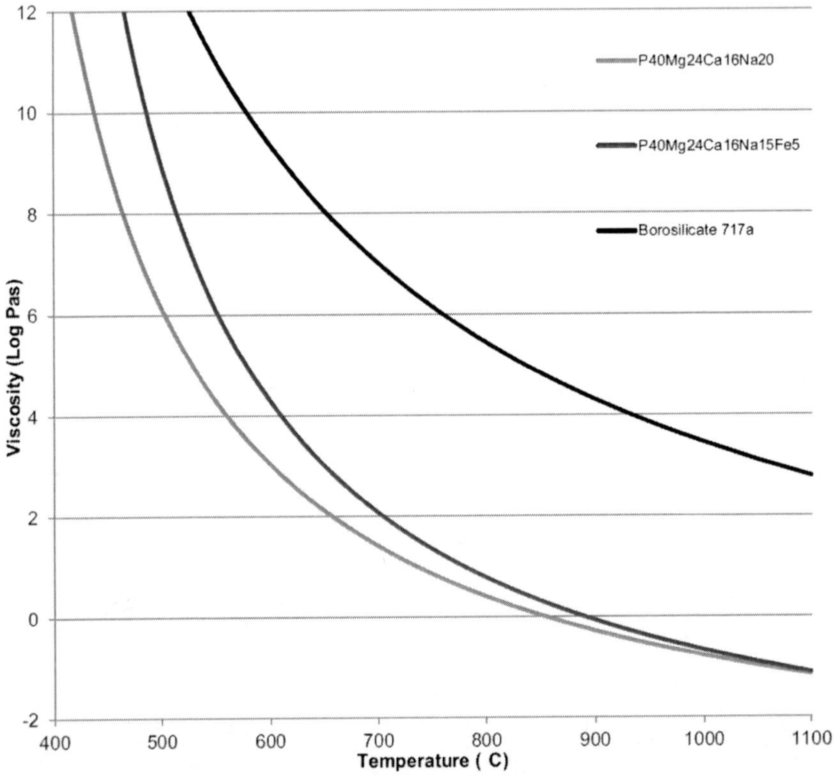

Figure. 6. Combined viscosity profiles fitted using the Vogel–Fulcher–Tammann equation for the P40Mg24Ca16Na20 and P40Mg24Ca16Na15Fe5 glasses, with the standard glass 717a for comparison. Errors are taken to be ± 0.2 Log Pa·s.

There is a small but distinct increase in viscosity of the glasses with increasing phosphate content between 40% and 50%, as might be expected with an increasing phosphate chain length. However the 55% phosphate glass appears to behave in a different fashion, with a more pronounced effect at high temperature but only a small effect or no effect at low temperature (c.f. Figs. 1 and 2). This seems to be the opposite effect to the presence of iron. Again, this bears further investigation but is likely related to the level of branching in the phosphate. It is interesting to note that this effect is not seen in the 55% phosphate containing 5% Fe_2O_3 or, at least, the magnitude of the effect of the presence of the iron masks the effect of the phosphorous.

By plotting the values of drawing point temperature (T_D) against phosphate content of the glass, there is a very good correlation (Fig. 7). The non-iron containing glass and the iron containing glass each appear to follow closely to a polynomial fit, with the iron pushing up the drawing point by a considerable amount. However the fit has no particular basis and is certainly dependent on the modifier component of the glass and so its predictive capacity is limited.

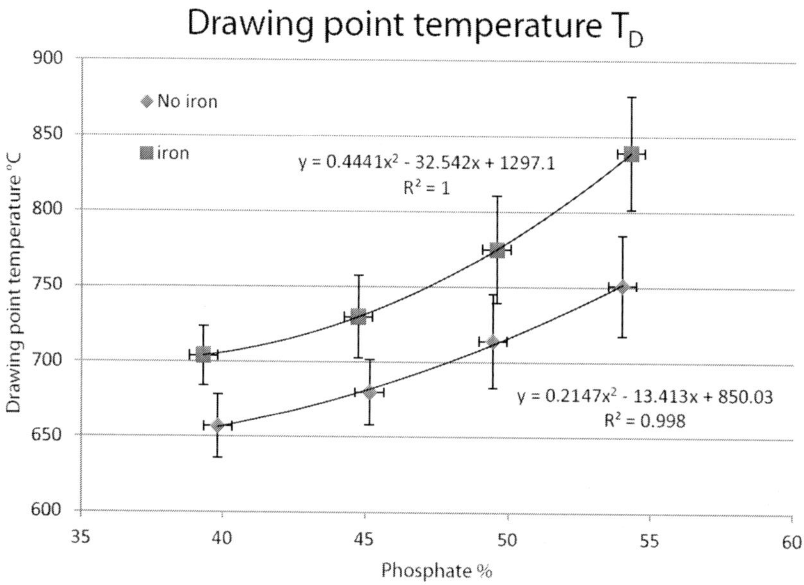

Figure. 7. Drawing point temperatures (T_D) as a function of phosphate content for the iron containing and non-iron containing glasses.

The borosilicate glass is provided for reference. It typically has a higher viscosity than the phosphate glass by several decades, highlighting what was said in Section 4.2.1.

4.3. Fragility

The data can also be used to obtain fragility indices, or relative values of the fragility of the glasses. Fragility gives an indication of the rate of change of the viscosity of the glass with temperature, which is an important parameter in glass drawing — a high fragility requires a much tighter control of processing conditions so that the operation can be held in a much smaller operating window. Fragility can be measured in a number of ways and in this study we will consider three approaches, one by Doremus and two by Angell.

4.3.1. Doremus

At both high and low temperatures, the value of T_0 in the VFT equation (Eq. (1)) approaches 0 and simplified Equations of activation energy can be applied to obtain the high temperature (Q_H) and low temperature (Q_L) activation energies via Eq. (2).

$$\eta = A \cdot \exp\left(\frac{Q}{RT}\right) \tag{4}$$

The ratio of these activation energies provides what is referred to as the Doremus fragility ratio (R_D), with higher values indicating a more fragile glass [14]. If a VFT fit is applied individually to the real data sets (quasi static and bob-in-up) the absolute value of T_0 is < 1 in each case, and so the data has been considered sufficiently linear for the activation energy fit to be used. The activation energies and resulting R_D values are provided in Table 5. The Q_L values fall within a range similar to that observed via fibre elongation tests by Abe [25].

Table 5. Activation energies and Doremus Fragility Index.

Glass code (mol% oxide)	Q_H (kJ/mol)	Q_L (kJ/mol)	R_D
P40Mg24Ca16Na20	570 ± 1	157 ± 1	3.64 ± 0.02
P45Mg24Ca16Na15	577 ± 1	144 ± 1	3.99 ± 0.03
P50Mg24Ca16Na10	583 ± 1	149 ± 1	3.90 ± 0.03
P55Mg24Ca16Na5	513 ± 1	150 ± 1	3.42 ± 0.02
P40Mg24Ca16Na15Fe5	639 ± 1	193 ± 1	3.30 ± 0.02
P45Mg24Ca16Na10Fe5	616 ± 1	164 ± 1	3.75 ± 0.02
P50Mg24Ca16Na5Fe5	572 ± 1	179 ± 1	3.19 ± 0.02
P55Mg24Ca16Fe5	548 ± 1	186 ± 1	2.94 ± 0.02
Borosilicate 717a	453 ± 1	231 ± 1	1.96 ± 0.01

It is quite clear that the phosphate glasses are more fragile than the borosilicate, but this is as expected. It is interesting to note that rather than a gradual decrease in fragility as the phosphate content goes from 40 to 55%, there actually appears to be a peak in fragility at 45% before it drops away again. This is the case for both the iron containing and non-iron containing glasses. The introduction of iron appears to reduce the fragility of the glass in all cases.

4.3.2. Angell

An alternative measurement of fragility is offered by Angell, who used a reduced plot of T_g/T to compare glasses directly in terms of their viscosity gradient near T_g. This causes difficulty in that by its very nature T_g is a rate dependent value and so does not have an equilibrium value. However for practical purposes it is typically the case that a heating rate of 10 °C/min is used to determine T_g. Graphs of Log(η) against T_g/T are provided in Fig. 8.

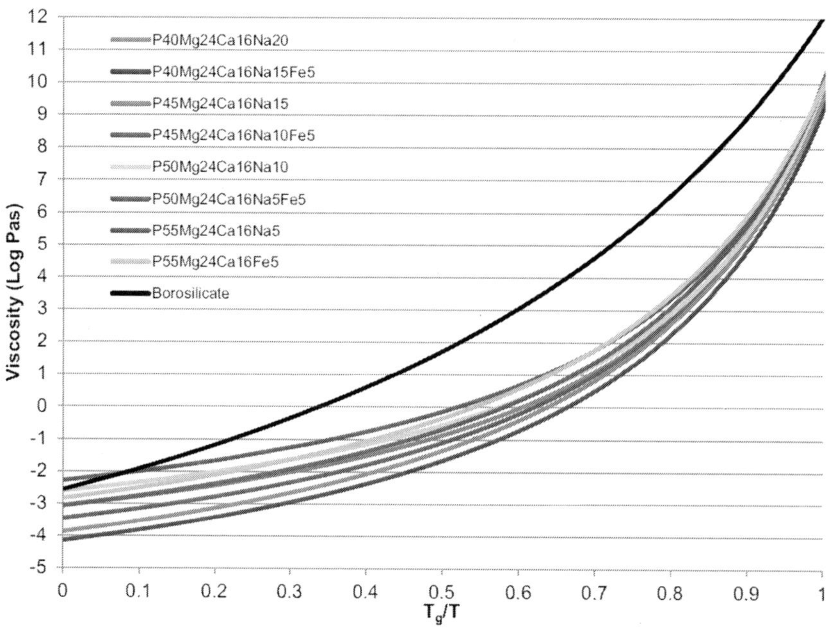

Figure. 8. Viscosity plotted against reduced temperature (T_g/T), with the standard glass for comparison. Errors are taken to be ± 0.2 Log Pa·s.

What is immediately apparent from Fig. 8 is that the values of viscosity at T_g in this reduced plot are in the region of 9–10 Log Pa·s. A general rule for non-fragile glasses is that viscosity is expected to be on the order of 12 Log Pa·s at the T_g[26]. While Moynihan has shown that this does not hold for fragile glasses [27], a value of 9–10 seems very low, particularly in view of the recent results of Muñoz-Senovilla, who observed viscosities of greater than 12 Log Pa·s for metaphosphates at T_g when measured directly via beam bending [28]. The results of Muñoz-Senovilla do fit with the suggestions from Ropp that phosphate glasses need to be annealed 10 °C above T_g[16]. Ultimately this suggests that the 10 °C/min heating rate to obtain the T_g for this plot is not appropriate. Černošek demonstrated that the difference between T_g taken at 10 °C/min and 'equilibrium' T_g can be considerable in chalcogenide glasses [26] and it is not unreasonable to consider that this could be true for phosphate glasses as well.

In order to address this, the VFT parameters in Table 4 were used to determine values of T at a viscosity of 12 Log Pa·s and this was then taken to be the 'equilibrium' T_g. While this is not strictly speaking correct, it provides a useful method of normalisation for comparison of fragility with the standard glass. These 'equilibrium' T_g values are provided in Table 6 and were used to re-normalise the η vs T_g/T plot, which is shown in Fig. 9.

Table 6. Angell Fragility Parameters.

Glass code (mol% oxide)	T_g from VFT	Angel (m – extrapolated)	Angel (m – calculated)	$F_{1/2}$
P40Mg24Ca16Na20	417 ± 2	59 ± 1	47 ± 2	0.440 ± 0.006
P45Mg24Ca16Na15	429 ± 2	63 ± 1	51 ± 3	0.445 ± 0.006
P50Mg24Ca16Na10	445 ± 2	62 ± 1	52 ± 3	0.441 ± 0.007
P55Mg24Ca16Na5	439 ± 2	53 ± 1	47 ± 3	0.381 ± 0.007
P40Mg24Ca16Na15Fe5	466 ± 2	62 ± 1	51 ± 2	0.482 ± 0.006
P45Mg24Ca16Na10Fe5	471 ± 2	59 ± 1	51 ± 3	0.463 ± 0.007
P50Mg24Ca16Na5Fe5	479 ± 2	56 ± 1	48 ± 2	0.423 ± 0.007
P55Mg24Ca16Fe5	495 ± 3	51 ± 1	44 ± 2	0.380 ± 0.007
Borosilicate 717a	526 ± 5	29 ± 1	23 ± 1	0.259 ± 0.007

The first approach used by Angell to consider fragility is the most well-known assessment of fragility, 'm', which is the gradient of the reduced temperature plot at T_g (Eq. (3)[12]).

$$m = \left[\frac{\partial \, \mathrm{Log}_{10}\eta}{\partial\left(\frac{T_g}{T}\right)}\right]_{T=T_g} \tag{3}$$

This can be obtained quite simply from the extrapolated plot, though as Angell notes this measurement can be subjective, particularly in glasses with a rapid change in viscosity at this point, and so the value of 'm' can also be calculated from the VFT parameters by using Eq. (4)[13]. Both these values are provided in Table 6.

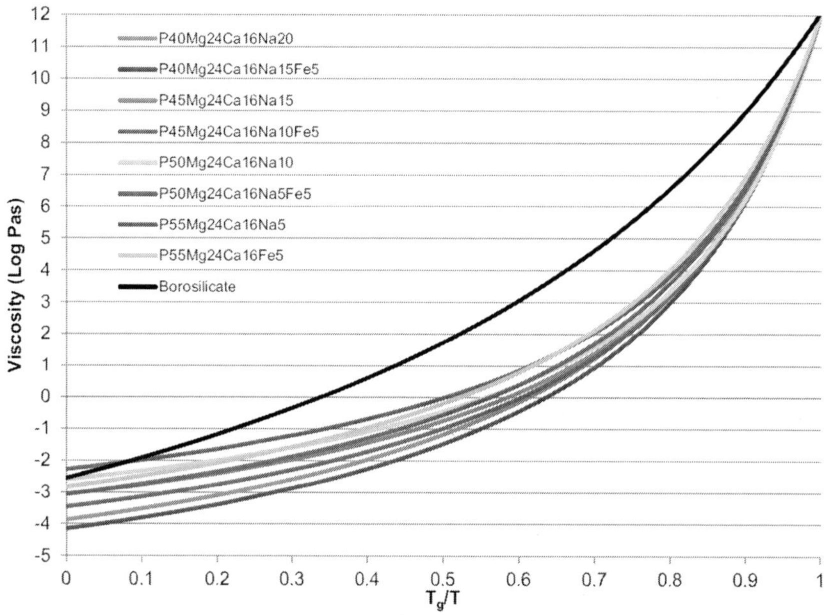

Figure. 9. Viscosity plotted against reduced temperature (T_g/T) using the 'equilibrium' $T_g$values btained from the VFT parameters. Errors are taken to be ± 0.2 Log Pa·s.

$$m = B \cdot T_g \cdot \left(T_g - T_0\right)^{-2} \tag{4}$$

The second approach used by Angell (and Richert) to consider fragility is by using a measure of deviation from Arrhenius behaviour and is referred to as $F_{1/2}$, shown in Eq. (5)[13], where $T_{1/2}$ is the value of T at a viscosity of 3.5 Log Pa·s. In essence it measures the distance of the curve away from the ideal state at a halfway point between 12 Log Pa·s and − 5 Log Pa·s (generally considered a lower common limit of viscosity), with a value of 0 being an ideally strong glass and a value of 1 being an ideally weak glass. The values of $F_{1/2}$are provided in Table 6.

$$F_{\frac{1}{2}} = \frac{2T_g}{T_{\frac{1}{2}}} - 1 \tag{5}$$

Comparison of the fragility parameter from Doremus with those of Angell and Richert shows a largely similar trend. The values are plotted in Fig. 10a and bfor clarity. All the glasses are significantly more fragile than the Borosilicate 717a standard glass.

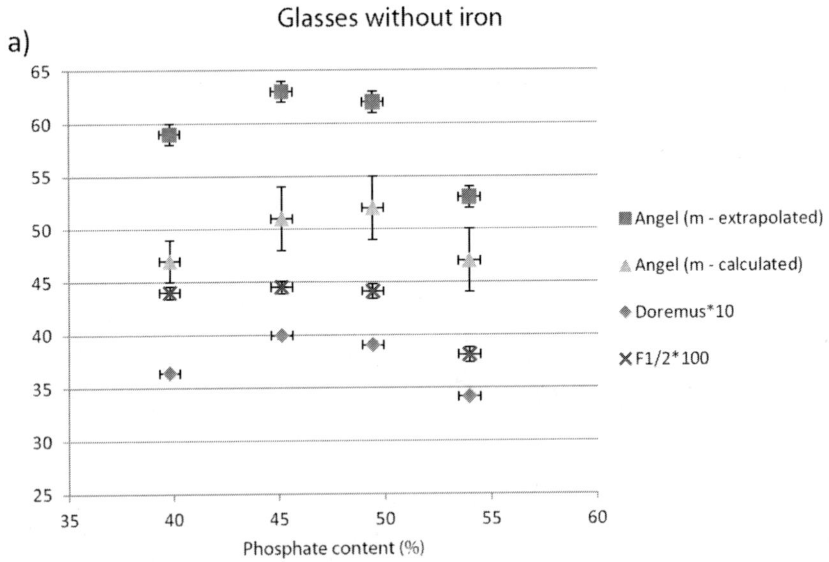

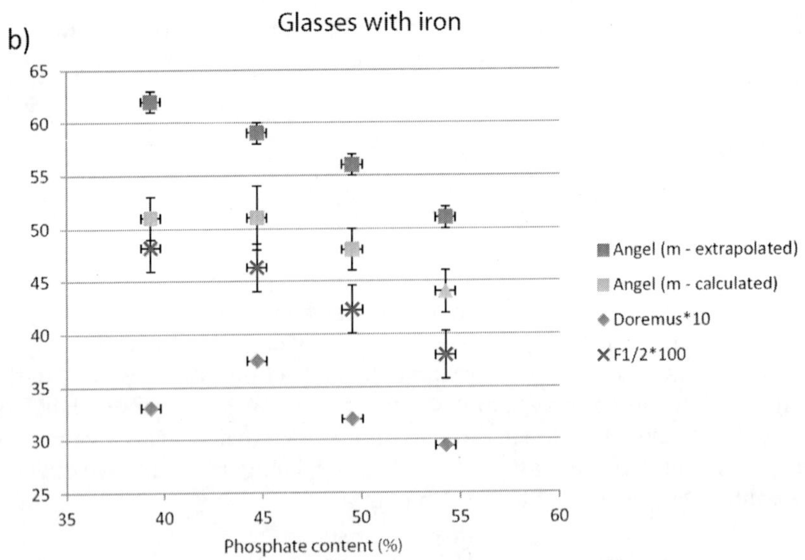

Figure. 10. a. Fragility parameters for the glasses that do not contain iron, as a function of phosphate content. Multiplication factors have been used to allow for easier comparison.

b. Fragility parameters for the glasses that contain iron, as a function of phosphate content. Multiplication factors have been used to allow for easier comparison.

The result in Fig. 10a is somewhat unexpected. An increase in phosphate chain length would be expected to reduce the fragility, since it would take more time for the longer chains to rearrange and relax, and more so with greater branching. This would appear to be the case moving from P45 to P50 to P55 but there is an apparent drop in the fragility at P40. However, it is difficult to assign an uncertainty to these fragility parameters and so the lower value may not be statistically significant. In fact, this lower value may be representative of the level of uncertainty as there is no factor that could (in the author's opinion) cause the fragility of P40 to be lower. Even considering this, the P55 value does appear to be significantly lower than the rest. Since the P50 actually has slightly lower than 50% phosphate content, the P55 is the only glass expected to have very long phosphate chains [29].

The result in Fig. 10b seems to be more clear, with a general reduction in fragility except for the Doremus value for the P45 glass. As was seen in in Fig. 5, iron has a strong effect on the viscosity at lower temperatures, in the range where m is measured. However, the absolute change in the fragility is quite small with the addition of iron, which is surprising. As postulated previously, this may be related to the mobility of iron in the glass and bears further investigation.

5. CONCLUSIONS

The quasi static and bob-in-cup viscosity data fit well to the Vogel Fulcher Tammann equation, providing an estimation of drawing point temperatures of between 657 and 839 °C. The drawing point temperature follows an increasing trend with phosphate content and the presence of iron also increases the drawing point. No glasses achieved a positive ΔT value. Overall, the viscosity of the glasses increases with phosphate content, but the effect may be limited at low temperature. The presence of iron increases the viscosity at low temperature but has only a small effect at high temperature. All the glasses were of considerably lower viscosity than the borosilicate 717a standard glass.

It is difficult to determine an appropriate T_g value suitable for a reduced temperature plot, a DSC heating rate of 10 °C/min appears to be too fast for these glasses. Use of an 'equilibrium' T_g value from the VFT fit enables a more useful comparator. All the glasses were of considerably higher fragility than the borosilicate 717a standard glass. The fragility of the iron free glasses appeared to increase significantly only when the phosphate content is > 50%. Iron has a limited effect on fragility.

Borosilicate 717a is a poor standard material in the conditions that were used, due to a continuous change in viscosity over time.

ACKNOWLEDGEMENTS

The authors would like to thank the Technology Strategy Board (project #101227) for providing funding to support this study, and to Dr David Furniss of the University of Nottingham for his advice.

REFERENCES

1. A.J. Parsons, I. Ahmed, P. Haque, B. Fitzpatrick, M.I.K. Niazi, G.S. Walker, C.D. RuddJ. Bionic Eng., 6 (2009), pp. 318-323, 10.1016/S1672-6529(08)60132-8
2. J.C. KnowlesJ. Mater. Chem., 13 (2003), pp. 2395-2401, 10.1039/b307119g
3. D.S. Brauer, C. Rüssel, S. Vogt, J. Weisser, M. SchnabelrauchJ. Mater. Sci. Mater. Med., 19 (2008), pp. 121-127, 10.1007/s10856-007-3147-x
4. N. Sharmin, A.J. Parsons, C.D. Rudd, I. AhmedJ. Biomater. Appl. (June 17 2014), 10.1177/0885328214539824(published online)
5. B.C. BunkerJ. Non-Cryst. Solids, 64 (1983), pp. 291-316, 10.1016/0022-3093(84)90184-4
6. P. Haque, I. Ahmed, A.J. Parsons, R. Felfel, G.S. Walker, C.D. RuddJ. Non-Cryst. Solids, 375 (2013), pp. 99-109, 10.1016/j.jnoncrysol.2013.05.008
7. D. Hewak (Ed.), Properties, Processing and Applications of Glass and Rare Earth-doped Glasses for Optical Fibres, IEE: INSPEC, London (1998)(ebook)
8. F.T. Wallenberger, J.C. Watson, H. LiD.B. Miracle, S.L. Donaldson (Eds.), ASM Handbook, Vol. 21: Composites, ASM International, Ohio (2001), pp. 27-34
9. J.M. Stickel, M. NagarajanInt. J. Appl. Glass Sci., 3 (2012), pp. 122-136, 10. 1111/j.2041-1294.2012.00090.x
10. F.T. WallenbergerF.T. Wallenberger, P.A. Bingham (Eds.), Fiberglass and Glass Technology, Springer, US (2010), pp. 3-90, 10.1007/978-1-4419-0736-3
11. A.N. GentBr. J. Appl. Phys., 11 (1960), pp. 85-87, 10.1088/0508-3443/11/2/310
12. R. Böhmer, C.A. AngellPhys. Rev. B, 45 (1992), pp. 10091-10994, 10. 1103/PhysRevB.45.10091
13. R. Richert, C.A. AngellJ. Chem. Phys., 108 (1998), pp. 9016-9026, 10.1063/1.476348
14. R.H. DoremusAm. Ceram. Soc. Bull., 82 (2003), pp. 59-63
15. M.S. Hasan, I. Ahmed, A.J. Parsons, G.S. Walker, C.A. ScotchfordJ. Mater. Sci. Mater. Med., 23 (2012), pp. 2531-2541, 10.1007/s10856-012-4708-1
16. R.C. Ropp, **Studies in Inorganic Chemistry 15: Inorganic Polymeric Glasses,** Elsevier Science Publishers B.V., Amsterdam (1992)
17. S.J. Wilson, D. PooleMater. Res. Bull., 25 (1990), pp. 113-118, 10.1016/0025-5408(90)90170-7
18. A.B. Seddon, V.K. Tikhomirov, H. Rowe, D. FurnissJ. Mater. Sci. Mater. Electron., 18 (2007), pp. S145-S151, 10.1007/s10854-007-9190-z
19. S.I.S. Shaharuddin, I. Ahmed, D. Furniss, A.J. Parsons, C.D. RuddGlass Technol. Part A, 53 (2012), pp. 245-251

20. R.K. Brow, C.M. Arens, X. Yu, E. DayPhys. Chem. Glasses, 35 (1994), pp. 132-136
21. A.J. Parsons, C.D. RuddJ. Non-Cryst. Solids, 354 (2008), pp. 4661-4667, 10.1016/j.jnoncrysol.2008.05.050
22. Napolitano, E.G. HawkinsJ. Res. Natl. Bur. Stand. A Phys. Chem., 68A (1964), pp. 439-447, 10.6028/jresv68An5p439_A1b
23. Napolitano, E.G. Hawkins, **Viscosity of a standard borosilicate glass**, National bureau of standards special publication 260-23 (1970)
24. M.I. OjovanAdv. Condens. Matter Phys., 2008 (2008), 10.1155/2008/817829(Article ID 817829, 23 pages)
25. Y. AbeM. Grayson, E.J. Griffith (Eds.), Topics in Phosphorus Chemistry, vol. 11, John Wiley & Sons, New York (1983), pp. 19-67
26. Z. Černošek, J. Holubová, E. Černošková, M. LiškaJ. Optoelectron. Adv. Mater., 4 (2002), pp. 489-503
27. C.T. MoynihanJ. Am. Ceram. Soc., 76 (1993), pp. 1081-1087, 10.1111/j.1151-2916.1993.tb03724.x
28. L. Muñoz-Senovilla, F. MuñozJ. Non-Cryst. Solids, 385 (2014), pp. 9-16, 10.1016/j.jnoncrysol.2013.10.021
29. E.J. Griffith, **Phosphate Fibers**, Plenum Press, New York (1995)

CHAPTER 5

The Structure of Icosahedral Quasicrystals and Seven Blunders in Quasicrystallography

Antony J. Bourdillon

UHRL, San Jose, USA

ABSTRACT

This structural study of quasicrystals is based on extremely dense icosahedral unit cells that are systematically and consistently measured for the first time. The structure and pattern indexation are 3-dimensional. A formula is given for scattering from atoms in hierarchic arrangement and geometric series. The Quasi-Bragg law is a new law in physics, with possible applications beyond crystallography. The structure is compared with previous, unsuccessful, and contradictory, attempts at analysis.

Keywords: Icosahedral Quasicrystals, Logarithmic Periodicity, Quasi-Bragg Law, Metric, Geometric Series, Icosahedral Unit Cell, Blunders

1. INTRODUCTION

The structural solution for icosahedral quasicrystals has been described in several journal publications [1] [2] , monographs [3] -[7] and warmly received at several conferences e.g. [8]. For clarity and simplicity, the solution has been summarized in introductory video records [9]. The theory has been thoroughly debated[1], and all negative reviews are provided and refuted in the work cited. To fix ideas, the many-sided solution is summarized in the next section, and then reinforced by a discussion of errors that have retarded, for a long time, the study of quasicrystals.

The term "quasicrystallography" was coined by its practitioners [5] , though journals have preferred to edit to "quasicrystal crystallography" [2]. The following pages, especially when read in the context of the published reviews, demonstrate that systemic errors disqualify any "quasicrystallographer" from scientific review of this subject.

Quasicrystals were discovered in 1982 and have been long discussed. The first solid, i-Al_6Mn, was supposedly icosahedral in structure [10] , but see Section 3.1, and this is forbidden in crystallography for many reasons, including the fact that none of the 14 Bravais lattices allow the newly observed 5-fold rotational symmetry. Electron microscope images showed that, though the diffraction peaks were sharp, the real structure was not periodic. It was called "quasi-periodic" and the solids were referred to as "quasicrystals". The immediate problem was four-fold: firstly, how does this structure compare with those of crystals or with amorphous materials; secondly, how does the diffraction arise; thirdly how does the pattern relate to structural measurement; and fourthly, what is the driving force that determines these unusual structures? With regard to the second of these, because of fundamental disparities, the new metallic alloys cannot follow Bragg's law—as observed in periodic crystals and in non-periodic amorphous phases—though the scattering was called Bragg diffraction by confusion. The following summary of general features describes a theory that unravels the confusion in many ways, but especially by describing and measuring the metric that is a necessary part of structural analysis and science. We do not ask "What is a quasicrystal?" [11]. "The short answer is no one knows." We find that with the edgesharing unit cells, with 3-dimensional indexation, with the Quasi-Bragg law, the structure is solved and a new law of physics is discovered. This follows from the multiple interplanar spacings in geometric series. The consequent sharp diffraction patterns are systematically simulated. This paper reviews the argument in condensed form.

2. GENERAL FEATURES

The following summary of features suffices for the solution:

2.1. Structure

1) A unique unit cell that is icosahedral and extremely dense:

- With 15 close packed planes through its center;

- With 20 close packed planes on its surface;
- And with a density increase of 17% compared with the equivalent cell in the face-centered cubic phase. This is a crystalline phase that forms the matrix in rapidly quenched Al_6Mn [10] [12] ;

2) With a stoichiometric alloy ratio of 6:1. This occurs on a unit cell with 12 shared surface atoms and one completely contained central atom, in this case Mn.

3) With a significant ratio of solute metal atom diameter/solvent diameter of $\left(\sqrt{\tau^2+1}-1\right)$, where the golden section $\tau=(1+\sqrt{5})/2$. These diameters are consistent with the dense packing[2];

4) With an icosahedral cell that is edge-sharing; not face sharing as in crystals. Notice that many glasses, such as amorphous silica, also have dense, edge-sharing cells; but the diffraction patterns are diffuse and circular because the sharing is by single, uncorrelated edges. By contrast, the quasicrystal cells share two or more adjacent edges, and these result in unique orientation with sharp diffraction patterns by a mechanism that is simulated.

5) With an agglomeration of cells that is represented in an ideal hierarchic structure that is uniquely icosahedral, uniquely aligned and infinitely extensive;

6) With a structure that is open to simulation.

2.2. Diffraction Pattern

1) With structural and diffraction pattern sequences that are geometric (as below);

2) With complete, simple and systematic 3-dimensional indexation of icosahedral axes and diffraction planes [1] ;

3) With 3-dimensional pattern indexation that is simple, consistent and complete;

4) With an explanation for the diffraction that is systematically, completely, and simply described by means of the Quasi-Bragg law that has been derived and is now enhanced in appendix A.1 ~ A.2.

2.3. Measurement

1) With simulated quasi-structure factors that match experimental values extraordinarily well [5] [7] ;

2) With a structure that is simulated and systematically measured without guessing [cf. [5] [13]], and which matches both phase contrast microscopic images, and atomic dimensions;

3) With a simple demonstration of the measured metric [4] [5] , including serious consequences for image analysis (see below) and for structural analysis.

2.4. Driving Force

1) As derived above consistently, especially in sections 2.1.1 to 2.1.3;

2) With a logarithmic electron energy band structure [2] having high density of states at long wavelengths that stabilize the structure, etc.

Notice that the 'logarithmically periodic solid' described is the only proposed solution with a single unit cell and obvious driving force, and is the only solution that has been systematically and consistently measured. The structure is illustrated in Figure 1 viewed close to the direction of the 5-fold axis. No other method of solution is free of one or more of the blunders to be described in the next section.

3. SEVEN BLUNDERS

3.1. The Data of Shechtman et al. [10] Are Not Icosahedral

Their two-fold patterns are inconsistent (Figure 2). Spatial symmetries constitute the foundation of crystallography. If Shechtman's data were correct, the inconsistency would have extremely significant structural implications. The fact of inconsistency was (1) overlooked by quasicrystallographers for 25 years; (2) denied by many reviewers e.g. [5] ; but when occasionally admitted, claimed to be trivial [7] , in spite of their false simulations. In fact we have found, firstly, that his data are excluded by our simulations on the ideal icosahedral, hierarchic structure; and secondly that they do not match icosahedral systems other than Shechtman's Al_6Mn [citations in ref. 1]. Two assumptions are therefore fairly made: his data are mistakenly transcribed; and a fundamental feature of crystallography has passed unnoticed. The pity is that the experimental discovery is so novel.

3.2. The Diffraction Pattern Sequence Is Not Properly Fibonacci

This sequence has the following properties: each term is the sum of its two preceding terms $f_{n+1}= f_n + f_{n-1}$ and the ratio between consecutive terms in a series, f_{n+1}/f, tends to the golden section τ (Appendix A.2) as order $n \to \infty$. However in the quasicrystal diffraction pattern, this ratio is the constant τ, and so the series is the geometric series, $g_n=a\tau^n$. This fact has profound consequences for the description of the diffraction pattern whose indices are properly described in integral powers of τ, as shown for poles and diffraction planes in Figure 2. Detailed illustrations of these facts are provided elsewhere [1] [2] , but nowhere else in the review literature are those details described, discussed or even cited.

Figure 1. Hierarchic icosahedral cluster viewed from near a 5-fold axis of rotational symmetry. Individual icosahedra are represented by golden triads constructed from golden rectangles, each having the significant length/width ratio τ :1. In i-Al_6Mn, the center of each golden triad locates a Mn atom surrounded, at corners, by Al atoms. The unit cell is extremely dense; the ideal structure is infinitely extensive, uniquely aligned, and uniquely icosahedral. The stretching factor is τ^2.

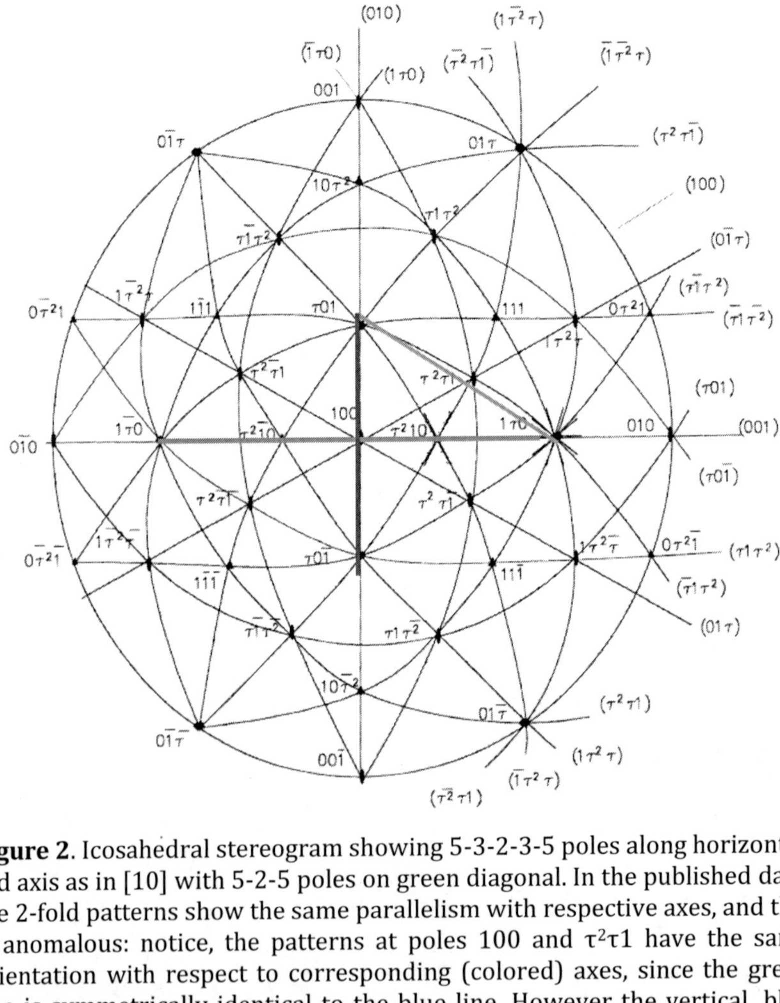

Figure 2. Icosahedral stereogram showing 5-3-2-3-5 poles along horizontal, red axis as in [10] with 5-2-5 poles on green diagonal. In the published data, the 2-fold patterns show the same parallelism with respective axes, and this is anomalous: notice, the patterns at poles 100 and $\tau^2\tau 1$ have the same orientation with respect to corresponding (colored) axes, since the green line is symmetrically identical to the blue line. However the vertical, blue axis with 5-2-5 poles shows that the crossed, two-fold diffraction patterns for the icosahedral structure are mutually normal, Simulations match data from other "icosahedral" alloys [1] and so confirm a transcription error; not a systematic structural anomaly in Al_6Mn. Shechtman's data are not icosahedral (Original stereogram courtesy Elsevier, [1]).

It is interesting to note, nevertheless, that the geometric series on the ratio τ has some interesting properties in common with the Fibonacci sequence. For example, while free electron energy bands have slope 2 when plotted on logarithmic scales, quasi-Brillouin scattered electron energy bands in a quasicrystal have a negative slope half that value, −1 [2]. This is a consequence of the first of the equalities formulated in the preceding paragraph. Even so a Fibonacci series in which, by definition, the first two terms can be any variables, $f_1 = a$; $f_2 = b$, is too loose and inaccurate a description of the diffraction patterns in quasicrystals. In this application, f_2 is strictly constrained: $b=a\tau$, and the series ratios $f_{n+1}/f_n=\tau$ for all n. The diffraction pattern spacings are then geometric: a, $a\tau$, $a\tau^2$, $a\tau^3$,··· This geometric pattern is moreover common to both the hierarchic model, and to its set of interplanar spacings. Illustrations of the last are given elsewhere [5] [7] for the ideal hierarchic structure and they are commonly seen in electron micrographs.

3.3. Quasicrystals Are Not 6-Dimensional

In real space, the structure is unambiguously 3-dimensional. However for reasons given in the previous section, neither the structure nor the diffraction pattern are 6-dimensional in any sense that can survive Occam's razor: dimensions "should not be multiplied without necessity". Unfortunately, unphysical dimensionality is the most strongly held tenet of quasicrystallography [5] [13] [14].

A reader will easily find a simple understanding for their widespread belief in 6-dimensional quasicrystals: indexation, for quasicrystallographers, is based on a double Fibonacci series construction, though the series is, in fact, simply three dimensional and geometric – as in real life. An illustration can be given for the relationship between those Fibonacci sequences and the single geometric series, base τ, that is proper in quasicrystals. This geometric series G can be written as the sum of two Fibonacci series, F_1, F_2, one of them multiplied by τ [5]. As an example, it is shown that the second term of the geometric series is written $g_2= f_1+ \tau f_2= \tau^2$, f_1 and $f_2$being the first two terms in the same Fibonacci series. Furthermore, the golden section is an irrational number, which is why G is not linearly harmonic (as quoted in Section 3.6 below, but typical only in crystals).

As in the stereogram, geometric, 3-dimensional indexation is easily and completely represented, without obfuscation [1] [2] [5] [7] [15]. What has to be understood, as is done through simulation, is the diffractive effect that is due to sets of planes that are spaced by orders τ^m, $-\infty < m < \infty$, m integral and 3-dimensional. The doubled sequence on F has led to confusion with 6 dimensions.

3.4. The Diffraction Is Not Bragg Diffraction

This law applies to periodic crystals, as it does to amorphous materials. In each case, the scattering is specular and is commonly represented on 2-

dimensional ray diagrams. The scattering is constrained by the Ewald sphere construction, and also by absorption, crystal orientation and other features. Such features vary according to applications in X-ray optics or in electron or neutron optics etc. In particular, electron diffraction relaxes the Ewald sphere constraint [16] supplying instead a deviation parameter that partly determines scattering and absorption. Simulations of scattering from crystals are typically 3-dimensional. In quasicrystals, however, the differences become so radical, that the diffraction ceases to be describable as Bragg diffraction except by extreme use of the Humpty-Dumpty principle[3]. In the first place, quasicrystals contain multiple interplanar spacings that are all active in each diffractive event. Simulations from an extensive, ideal, hierarchic, icosahedral quasicrystal show that Bragg's law is transmuted from $n\lambda = 2d \sin(\theta)$ for scattering of X-rays of wavelength λ from a crystal with interplanar spacing d at Bragg angle θ. The orders are linear in n. In quasicrystals, structure factors calculated by Bragg's law are all zero. The new Quasi-Bragg law become4s $\lambda\tau^m = d'\sin(\theta')$, where d' is a corrected quasi interplanar spacing $d' = d/c_S$; where the compromise spacing effect c_S is discovered to be constant for all quasi-structure factors; and where the Quasi-Bragg angle θ' differs from the Bragg angle in crystals, chiefly because of the multiple interplanar spacings. The order m is now geometric. Every variable except wavelength is different in the Quasi-Bragg law and even the constant 2 becomes unity. This occurs because of the approximately half-integral nature of the interplanar sequences [5] , especially at low m. The simulated quasi-structure factors match experimental values extraordinarily well [5] [7]. A simple illustration of the special diffraction, that is described in terms of calculated quasi-Bloch waves, is available elsewhere [3] [5] [9].

Many consequences follow an inadequate understanding of the special diffraction; two (in imaging and in structural analysis) will be described below as illustrations. Quasicrystallographers have failed to understand or explain the special diffraction in quasicrystals that relates to the multiple interplanar spacings: multi-dimensionality multiplies only confusion; the interplanar spacings are indeed evident in phase-contrast electron microscopy e.g. [17]. The practitioners resort to over-economical and erroneous guesses (or "choices"), without measurement [5]. The multiple interplanar spacings in geometric series in quasicrystals cause a new law in physics: the Quasi-Bragg law. Where Bragg's law transformed a crystal lattice onto a 3-dimensional reciprocal lattice; the Quasi-Bragg law transforms a 3-dimensional geometric pattern onto an hierarchic structure.

3.5. HRTEM Does Not Image Atoms

High resolution transmission electron microscopy (HRTEM) is complicated science that is typically ambiguous and that is easily misrepresented [16]. The contrast results from interference of Bloch waves generated during transmission through a crystalline foil in an electron microscope. The contrast is strongly dependent on specimen thickness, specimen orientation, defocus, estimates of absorption rates of individual Bloch waves, etc. When it can be proved that the Bloch waves peak on atomic columns, as is common in crystals, then the interference can be interpreted as an image of the columns. Even so, significant interpretation requires systematic simulation of through-focal series in multiple images, etc. However, without a proper theory of diffraction, no HR image is of any significance. In fact diffraction simulations, including the compromise spacing effect [7] that results from the multiple interplanar spacings and the measured metric, demonstrate that the Bloch waves do not peak on atomic columns. HRTEM does not therefore image atoms in quasicrystals. Any interpretation that is based on structure from HR images is misrepresented.

3.6. Fourier Transforms Are Misleading

How should a crystallographer interpret real structure from a diffraction pattern that he does not understand? An example is given in the following review [5] :

"We know that for every reflection with index **H** and its higher harmonics with indices n**H**, there exists a periodic average structure (PAS) with period d_H which has to be used in Bragg's equation. Then it can be applied as usual. The scaling symmetry by powers of τ of the diffraction pattern, which is a Fourier module of rank n ($n > 3$) does not need to enter Bragg's equation; we just have to choose d_H of the respective scales PAS properly."

The trouble is, he doesn't know how to make the proper choice, so his simulation is without measurement, without consistency and without significance. Moreover, what mental confusion allows a "periodic average structure" in a "non-periodic material"? Furthermore, his "harmonics" are not observed. If there are harmonics, they are in geometric space, but only sound visual inspection proves it. Any interpretation that lacks the measured metric is unphysical.

Worse still, is the misallocation of effort. Fourier transform analysis hides defects where these are of the greatest significance for structure, especially in rapidly quenched material, like Shechtman's. That is why sterogram's such as that in Figure 2 is of greater significance: in providing a perspective for the diffraction they enable dark field imaging of the defects.

3.7. Multiple Cells Higgledy Piggledy

Mathematically, it is as easy to multiply cells as to multiply dimensions, but neither contributes to understanding why and how quasicrystals form. Amorphous silica can be constructed from an infinite number of unit cells. Quasicrystallographers typically contrive cells that could contribute to icosahedral structure [18] ; but they are not able to measure them and narrow down to a solution. In crystallography, only one cell is allowed, and this in periodic array with a precise symmetry. Typically, chemical conditions during formation determine the symmetry, the unit cell, and therefore the structure. The conditions include cooling rate.

Multiplicity of cells fails many requirements of acceptable theory: it does not explain stoichiometry in this type of alloy; it does not explain peculiar atomic sizes; it does not reveal the driving force for these unusual materials. Only the ideal of logarithmic periodicity satisfies these requirements. Crystallography abhors multiple unit cells!

4. CONCLUSION

Antony J. Bourdillon The structure of icosahedral quasicrystals is solved. It depends on an icosahedral unit cell that is extremely dense and that is edge-sharing. By contrast, seven quasi-conventional blunders are described: Shechtman's data are not icosahedral; the diffraction pattern is not Fibonacci; quasicrystals are not 6-dimensional; there is no Bragg diffraction; HRTEM does not image atoms; Fourier transform analysis, without the measured metric, is misleading; and crystallography abhors multiple unit cells. The Quasi-Bragg law is a new law in physics.

REFERENCES

1. Bourdillon, A.J. (2013) Micron, 51, 21-25. http://dx.doi.org/10. 1016/j. micron. 2013.06.004
2. Bourdillon, A.J. (2009) Solid State Communication, 149, 1221-1225. http://dx. doi.org/10.1016/j.ssc.2009.04.032
3. Bourdillon, A.J. (2011) Logarithmically Periodic Solids, Nova Science.
4. Bourdillon, A.J. (2011) Logarithmically Periodic Solids—Properties, Evidence and Uncertainties. In: Puckermann, B.E., Ed., Quasicrystals: Types, Systems, and Techniques, Nova Science.
5. Bourdillon, A.J. (2012) Metric, Myth and Quasicrystals.

6. Bourdillon, A.J. (2010) Quasicrystals' 2D Tiles in 3D Superclusters.
7. Bourdillon, A.J. (2009) Quasicrystals and Quasi Drivers. [Citation Time(s):7]
8. (2013) American Physical Society March Meeting, Baltimore, 18-23 March 2013, Session W43. In Response to Journals, the Meeting Instructed the Speaker: "Be aggressive;" "Remove the kid-gloves," "Add insult".
9. (2013) Quasicrystals at the American Physical Society March Meeting. http://www.youtube.com/watch?v=xD30KF93qio
10. Shechtman, D., Blech, I., Gratias, D. and Cahn, J.W. (1984) Physical Review Letters, 53, 1951-1953. http://dx.doi.org/10.1103/PhysRevLett.53.1951
11. Senechal, M. (2006) Notices to the American Mathematical Society, 3, 886-997.
12. Bourdillon, A.J. (1987) Philosophical Magazine Letters, 55, 21-26. http://dx.doi.org/10.1080/09500838708210435
13. Steurer, W. (2004) Zeitschrift für Kristallographie, 219, 391-446. http://dx.doi.org/10.1524/zkri.219.7.391.35643
14. Steurer, W. and Deloudi, S. (2008) Acta Crystallographica, A64, 1-11.
15. Bourdillon, A.J. (2013) Quasicrystallography. www.UHRL.net
16. Hirsch, P., Howie, A., Nicholson, R.B., Pashley, D.W. and Whelan, M.J. (1977) Electron Microscopy of Thin Films. 2nd Edition, Krieger, New York.
17. Bursill, L.A. and Peng, J.L. (1985) Nature, 316, 50-51. http://dx.doi.org/10.1038/316050a0
18. Takakura, H., Gomez, C.P., Yamamoto, A., De Boisieu, M. and Tsai, A.P. (2006) Nature Materials, 61, 58-63.
19. Huntley, H.E. (1970) The Divine Proportion, Dover.

APPENDIX

A.1. Fourier Summations on Geometric Series In grating optics and Fraunhofer diffraction, Fourier transforms of the grating are used to describe diffraction patterns observed in the image plane. In crystal optics, it is often possible to approximate atoms to point scatterers and so apply Fourier series to the analysis. We did this before in simulations of quasi-structure factors. The calculations were performed in 3-dimensions over a truncated and ideal hierarchic model that is, in principle, infinitely extensive. Now we show how the method can be used to understand quasi-Bragg diffraction on atoms in linear geometric series.

Start with the Quasi-Bragg law described earlier, $\lambda\tau^m = d'\sin(\theta')$. Proceed to simulate quasi structure factors in the centrosymmetric hierarchic model by following the crystalline case, where a structure factor F_{hkl} for a reflection with Miller indices h, k, l, is written:

$$F_{hkl} = \sum_j f_j^{\varsigma} \cos\left(2\pi\left(hu_j + kv_j + lw_j\right)c_s\right) \tag{A1}$$

summed over atoms j, each with atomic scattering amplitude f_j^{ς}, on coordinates u_j, v_j and w_j, where the superscript ς identifies the chemical species on the site: either Al or Mn. Notice the compromise spacing effect metric c_s that has unit value in crystals, but that has a smaller measured value in quasicrystals because the multiple interplanar spacings are here non-uniform and geometric. Then the quasi-Miller indices h', k',l' have typical values in individual powers of τ^m; while the atomic locations u'_j v'_j w'_j are expressed in values belonging to powers τ^n, e.g. $u_j = \sum_n a_n \tau^n$ where the a_n factors are integral values that express the number of times a particular interplanar space is contained in u_j. The factors are inherent in the simulations cited earlier. Quasi structure factors, calculated for superclusters up to order 6, with 10^8 atomic sites, match experimental values.

Now we introduce a further feature that is the evident result of many calculations. We have found the line width of quasi-structure factors to be dependent on sample sizes used in simulations. For example in a cluster of ~100 atoms the line width is ~5% of the quasi-Bragg angle; reducing to 0.005% in the supercluster order 6. The most probable reason for the systematic trend is the oscillatory behavior in the cosine term in Equation (A1).

We now proceed to further details in the crystallography of quasicrystals. Begin by noticing the octahedral subgroup within the icosahedral point group symmetry. This is enabling, both for the 3-dimensional indexation, while it also allows us to write for an interplanar spacing $d'c_s = 1/\sqrt{h'^2 + k'^2 + l'^2}$ since $d' = d/c_s$ with the measured metric, and where the unit of length is the edge length of the icosahedral unit cell. Simplifying to one dimension by choosing $k' = l' = 0$, expand equation A1 in series:

$$F_{hkl} = \sum_j f_j^\varsigma \sum_p \left[1 + (i4\pi h u_j)^{2p} / (2p)!\right],$$
$$= \sum_j f_j^\varsigma \sum_{m,n,p} \left[1 + (i4\pi \tau^m c_s^m a_n \tau^n)^{2p} / (2p)!\right], \tag{A2}$$

with, $-\infty < m < \infty$, $-\infty < n < \infty$, $0 < p < \infty$, m, n, p all integral. As before, the factor 4π has replaced 2π in (A1) because of the approximately halfintegral values of τ^m at low powers [5]. Notice that the second term in the square brackets is oscillatory. This explains the narrowing of quasi-structure factor line widths previously remarked. In large specimens, the second term therefore sums approximately to zero, and we are left with a structure factor in Fourier series that depends on the number of atoms, including their scattering powers f_j^ς. The result is typical, and the metric is consistently measured already in the 3-dimensional quasi-structure factor calculations. The metric correlates with the quasi-Bragg angle and corresponds to the variety of interplanar spacings found in the ideal hierarchic model.

In quasicrystals, linearly periodic electromagnetic waves are scattered by atoms in an approximately hierarchic structure having logarithmic periodicity, in such a way that the resulting diffraction pattern is geometric as simulated.

A.2. Geometric and Fibonacci Series Compared

Table A.1 illustrates the source of confusion for a 6-dimensional construction of a three dimensional structure with logarithmic periodicity. Start with the simplest Fibonacci sequence based on 0 and 1 (column 1) then calculate the ratios between consecutive terms. Notice that they vary about τ (column 2); that they tend to τ at high order; and that the

Table A.1. Comparison between Fibonacci series (not observed) and Geometric series (as observed).

Simplest Fibonacci series	Ratios between consecutive terms	Fibonacci with integers and reals	Ratios	Geometric series
$f_{n+1}=f_n+f_{n-1}$	f_n/f_{n-1}			g_n
				τ^-infinity
				.
				.
				τ^−2
				τ^−1
0		1		τ^0
1	infinity	τ	τ	τ^1
1	1	1 + τ	τ	τ^2
2	2	1 + 2τ	τ	τ^3
3	1.5	2 + 3τ	τ	τ^4
5	1.6666...	3 + 5τ	τ	τ^5
8	1.6	5 + 8τ	τ	τ^6
13	1.625	8 + 13τ	τ	.
21	1.615	13 + 21τ	τ	.
.	.	.	.	.
.	.	.	.	.
	.		.	.
	τ = 1.618034....		τ	τ^infinity
	Not observed		**As observed in quasicrystals**	
		Redundant 6D		**Definitely 3D**

quasicrystal diffraction series that is observed is not Fibonacci because it lacks the property of varying ratios [19]. Next set up a Fibonacci series that starts with the ratio τ by means of initial terms 1 and τ (column 3). Calculate sequential ratios and notice they are constant τ, the same as the first term. The observed constant ratios in quasicrystal diffraction patterns are a property of the geometric series. This is firmly three dimensional. It is also the sum of two Fibonacci series on integral bases (column 2), with one of them having the irrational number τ as multiplier. The diffraction series can then be indexed using 6 numbers, though three of them are in fact redundant. The quasicrystal structure is firmly three dimensional and is simply comprehensible in terms of classical crystallography with the extremely dense, edge-sharing, unit cell. In this context, 3-dimensional indexation is important because it opens analysis to the range of techniques established in classical crystallography. The first of these techniques is the quasi-structure factors and measured metric.

There are further differences between the Fibonacci and geometric series. The geometric series is both infinite and infinitesimal; whereas the Fibonacci series truncates at zero. The two series can be made equivalent if the unit in the Fibonacci series is made indefinitely small. Quasicrystallographers can do this because their unit is not measured; but they deny measurement as a core requirement in science, and are content with guess-work.

NOTES

[1]see especially the response to the section editor of Acta. Cryst. in ref 5?

[2]Along with layered types of quasicrystal having reduced symmetry, more complex types have been found with modified structures, but they can all be understood by relation to the more symmetric icosahedral structure that is described here.

[3]"When I use a word... it means what I choose it to mean—neither more nor less" Dodgson CL, Alice through the Looking Glass.

CHAPTER 6

Diamond as a Solid State Quantum Computer with a Linear Chain of Nuclear Spins System

G. V. López

Departamento de Fsica de la Universidad de Guadalajara, Guadalajara, México

ABSTRACT

By removing a ^{12}C atom from the tetrahedral configuration of the diamond, replacing it by a ^{13}C atom, and repeating this in a linear direction, it is possible to have a linear chain of nuclear spins one half and to build a solid state quantum computer. One qubit rotation, controlled-not (CNOT) and controlled-controlled-not (CCNOT) quantum gates are obtained immediately from this configuration. CNOT and CCNOT quantum gates are used to determined the design parameters of this quantum computer.

Keywords: Quantum Computer; Controlled-Not Gate; Diamond

1. INTRODUCTION

So far, the idea of having a working quantum computer with enough number of qubits (at least 1000) has faced two main problems: the decoherence [1-8] due the interaction of the environment with the quantum system, and technological limitations (pick up signal from NMR quantum computer [9,10], laser control capability in ion trap quantum computer [11,12], physical build up for more than two qubits like in photons cavities [13], atoms traps [14,15], Josephson's joint ions [16],

Aronov-Bhom devices [17], diamond NV device [18], or high field and high field gradients in linear chain of paramagnetic atoms with spin one half [19]). In particular, the linear chain of paramagnetic atoms of spin one half became a good mathematical model to make studies of quantum gates [20], quantum algorithms [21], and decoherence [22] which could be applied to other quantum computers. In this paper, one put together the ideas of using the diamond stable structure and the linear chain of spin one half nucleus. To do this, on the tetrahedral ^{12}C (with nuclear spin zero) configuration of the diamond main structure, one removes a ^{12}C element of this configuration and replace it by a ^{13}C (with nuclear spin one half) atom, and one repeats this replacement along a linear direction of the crystal. By doing this replacement, one obtains a linear chain of atoms of nuclear spin one half which is protected from the environment by the crystal structure and the electrons cloud. Therefore, one could have a quantum computer highly tolerant to environment interaction and maybe not so difficult to build it, from the technological point of view.

2. ^{12}C-^{13}C DIAMOND AND SPIN-SPIN INTERACTION

The above idea is represented in Figure 1, where the ^{13}C atoms are place on the position of some ^{12}C atoms. This replacement could be done using the same technics used to construct the diamond NV structure [23], or using ion implantation technics [24] and neutralization of ^{13}C in the diamond [25]. It is assumed in this paper that this configuration can be built somehow.

Now, as one can see, the important interaction on this configuration is the spin-spin interaction between the nucleus of the ^{13}C atoms. This interaction is well known [26] and is given by

$$U = \frac{\mu_o}{4\pi} \frac{(\boldsymbol{m}_1 \cdot \boldsymbol{x})(\boldsymbol{m}_2 \cdot \boldsymbol{x}) - \boldsymbol{m}_1 \cdot \boldsymbol{m}_2}{|\boldsymbol{x}|^3}, \tag{1}$$

where the magnetic moment $\boldsymbol{m}_i, i = 1, 2$ of ^{13}C's is related with the nuclear spin as

$$\boldsymbol{m}_i = \gamma \boldsymbol{S}_i, \tag{2}$$

being γ the proton gyromagnetic ratio $(\gamma \approx 2.675 \times 10^8\ \text{rad/T}\cdot\text{s})$. Without loosing the main idea, it will be assumed here that ^{13}C magnetic moment is

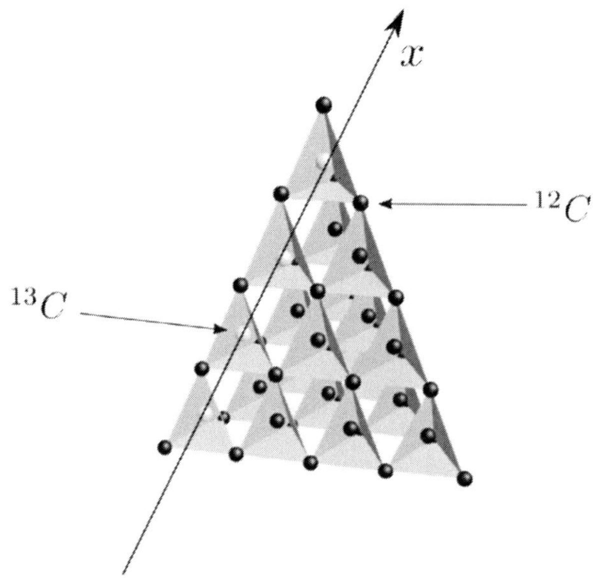

Figure 1. Diamond 12C -13C.

due to proton. The variable $\boldsymbol{x}$ indicates the separation vector between two ^{13}C nucleus, which has magnitude $a = |\boldsymbol{x}| \sim 10^{-10}$. Aligning the chain of ^{13}C nucleus along the x-axis of the reference system and assuming Ising interaction between ^{13}C nucleus, this energy can be written as

$$U = \frac{J}{\hbar} S_1^z S_2^z, \tag{3}$$

where the coupling constant J has been defined as

$$J = \frac{\mu_o \gamma^2 \hbar}{4\pi a^3}. \tag{4}$$

3. HAMILTONIAN OF THE SYSTEM

Consider a magnetic field of the form

$$\boldsymbol{B}(x,t) = \left(b\cos(\omega t + \varphi), -b\sin(\omega t + \varphi), B_0(x)\right), \tag{5}$$

where b , φ , and ω are the magnitude, the phase, and the frequency of the transverse rf-field. The z-component of the magnetic field has a gradient on the x-axis, determined by the difference on Larmore's frequencies of the ^{13}C's nuclear magnetic moments,

$$\left(\frac{\Delta B_0}{\Delta x}\right) = \frac{\Delta\omega}{\gamma\Delta x}. \tag{6}$$

The magnetic field at the location of the ith-^{13}C atom is $\boldsymbol{B}_i(t) = \boldsymbol{B}(x_i, t)$, and the interaction energy of the magnetic moments of the ^{13}C atoms with the magnetic field is

$$U = -\sum_{i=1}^{N} \boldsymbol{m}_i \cdot \boldsymbol{B}_i(t), \tag{7}$$

where N is the number of ^{13}C atoms aligned along the x-axis. This energy can be written as

$$U = -\sum_{j=1}^{N} \omega_j S_j^z - \frac{\Omega}{2} \sum_{k=1}^{N-1} \left(e^{i\theta} S_k^- + e^{-i\theta} S_k^+ \right), \tag{8}$$

where ω_j is the Larmore's frequency of the ith-^{13}C,

$$\omega_j = \gamma B_0(x_j), \tag{9}$$

Ω is the Rabi's frequency,

$$\Omega = \gamma b, \tag{10}$$

S_j^- and S_j^+ are the ascent and descent spin operators, $S_j^{\pm} = S_j^x \mp i S_j^y$, and θ has been defined as

$$\theta = \omega t + \varphi. \tag{11}$$

Let us consider first and second neighbor interactions among ^{13}C nuclear spins, and assuming equidistant separation between any pair of spins, the Hamiltonian of the system is

$$H = -\sum_{j=1}^{N} \omega_j S_j^z + \frac{J}{\hbar} \sum_{k=1}^{N-1} S_j^z S_{j+1}^z$$

$$+\frac{J'}{\hbar}\sum_{l=1}^{N-2} S_l^z S_{l+2}^z - \frac{\Omega}{2}\sum_{j=1}^{N}\left(\mathrm{e}^{i\theta} S_j^- + \mathrm{e}^{-i\theta} S_j^+\right), \tag{12}$$

where J is the coupling constant of first neighbor ^{13}C atoms, and J' is the coupling constant of second neighbor ^{13}C atoms which must be about one order of magnitude lower than J. One can write this Hamiltonian as $H = H_0 + W(t)$, where H_0 and W are defined as

$$H_0 = -\sum_{j=1}^{N}\omega_j S_j^z + \frac{J}{\hbar}\sum_{k=1}^{N-1} S_j^z S_{j+1}^z + \frac{J'}{\hbar}\sum_{l=1}^{N-2} S_l^z S_{l+2}^z, \tag{13}$$

And

$$W(t) = -\frac{\Omega}{2}\sum_{j=1}^{N}\left(\mathrm{e}^{i\theta} S_j^- + \mathrm{e}^{-i\theta} S_j^+\right). \tag{14}$$

The operator H_0 is diagonal on the basis $\left\{|\xi\rangle = |\xi_N \mathrm{L}\ \xi_1\rangle\right\}_{\xi_k=0,1}$ of the Hilbert space of 2^N dimensionality. Its eigenvalues defines the spectrum of the system,

$$E_\xi = \frac{\hbar}{2}\left\{-\sum_{j=1}^{N}(-1)^{\xi_j}\,\omega_j + \frac{J}{2}\sum_{k=1}^{N-1}(-1)^{\xi_k+\xi_{k+1}} + \frac{J'}{2}\sum_{l=1}^{N-2}(-1)^{\xi_l+\xi_{l+2}}\right\} \tag{15}$$

Since $J' < J \ll \omega_j$ for $j = 1,\cdots,N$, this spectrum is not degenerated with $E_{|00\cdots0\rangle}$ as the energy of ground state, and $E_{|11\mathrm{L}\,1\rangle}$ as the energy of the most exited state. To calculate the spectrum, one has used the following action of S_j^z operator

$$S_j^z|\xi\rangle = \frac{\hbar}{2}(-1)^{\xi_j}|\xi\rangle. \tag{16}$$

The Schrödinger's equation,

$$i\hbar\frac{\partial|\Psi\rangle}{\partial t}=H|\Psi\rangle, \tag{17}$$

is solved by proposing a solution of the form

$$|\Psi\rangle=\sum_{\xi}C_{\xi}(t)|\xi\rangle, \tag{18}$$

which brings about the following system of first order differential equations on the interaction representation

$$i\hbar\dot{a}_{\delta}=\sum_{\xi}a_{\xi}\mathrm{e}^{i(E_{\delta}-E_{\xi})t/\hbar}W_{\delta,\xi}(t), \tag{19}$$

where a_{δ} and $W_{\delta,\xi}$ are defined as

$$a_{\delta}(t)=C_{\delta}(t)\mathrm{e}^{-iE_{\delta}t/\hbar} \tag{20}$$

And

$$W_{\delta,\xi}(t)=\langle\delta|W(t)|\xi\rangle. \tag{21}$$

This is very well known procedure to solve time dependent Schrödinger's equation, and the solution of Equation (19) brings about he unitary evolution of the system (given the initial condition $|\Psi_{o}\rangle$).

Defining the evolution parameter τ through the change of variable $t=\omega_{o}\tau\,(\omega_{o}=2\pi\ \mathrm{MHz})$, the parameters ω_{j}, Ω, J and J' are real numbers given in units of ω_{o}. This evolution parameter will be used below in the analysis of the CNOT quantum gate.

4. ANALYSIS OF THE SYSTEM

In order to get an operating quantum computer, one needs to show that, at least, one qubit rotation gate $(N=1)$ and two qubits CNOT gate $(N=2)$ or three qubits controlled-controlled-not (CCNOT) gate $(N=3)$ can be constructed from this quantum system. Because this quantum system is homeomorphic [27] to the linear chain of paramagnetic atoms with spin one half system [28], it is clear from the point of view of mathematical models that the above gates can be constructed with this ^{12}C-^{13}C diamond system. However, one needs to assign realistic workable parameters for the real design of a ^{12}C-^{13}C diamond quantum computer. To do this, one studies in this section the behavior of a quantum CNOT and CCNOT gates as a function of several parameters. One neglect one qubit rotation $(N=1, J=J'=0)$ because it is obvious that one can get it through an arbitrary pulse on the rf-field with the frequency given by the Larmore's frequency of the qubit $(\omega=\omega_1)$, for a single ^{13}C atom in the diamond structure. In particular, the NOT quantum gate is obtained using a π-pulse duration $(\tau=\pi/\Omega)$ with this frequency. Therefore, the study of the CNOT $(N=2, J\neq 0, J'=0)$ and CCNOT $(N=3, J\neq 0, J'\neq 0)$ quantum gates is of the most interest. The equations for the two and three qubits dynamics are shown on the appendix. CNOT quantum gate corresponds to the transition $|10\rangle \leftrightarrow |11\rangle$, and CCNOT quantum gate corresponds to the transition $|110\rangle \leftrightarrow |111\rangle$. The first and second transitions are obtained through the resonant frequencies

$$\omega=(E_{11}-E_{10})/\hbar, \quad \text{and} \quad \omega=(E_{111}-E_{110})/\hbar. \tag{22}$$

Larmore's frequencies are denoted by ω_1 and ω_2, and ω_3 is parametrized as

$$\omega_2=\omega_1(1+f), \quad \omega_3=\omega_1(1+2f) \tag{23}$$

where f measures the relative change of the frequencies of qubits. The separation of the ^{13}C nucleus, a, is parametrized as

$$a=\xi\cdot 10^{-10}\ \text{m}. \tag{24}$$

For the CNOT quantum gate, one has the initial conditions $C_{00}(0)=C_{01}(0)=C_{11}(0)=0$ and $C_{10}(0)=1$. The time is allowed to last a π-pulse $(\tau=\pi/\Omega)$, and one takes the coupling constant as a fixed paramenter,

$$J=0.12, \quad \xi=1. \tag{25}$$

Figure 2 shows the fidelity parameter,

$$F = \left| \langle \Psi_{ideal} | \Psi_{real} \rangle \right|^2, \tag{26}$$

at the end of the - π pulse, as a function of the Rabi's frequency, where $|\Psi_{real}\rangle$ is the state obtained with the simulation, and $|\Psi_{ideal}\rangle$ is the expected state $(|11\rangle)$. The simulation was done for two different weak magnetic fields and for

$$f = 0.01 \quad (1),$$

$$f = 0.05 \quad (2),$$

$$f = 0.1 \quad (3),$$

and

$$f = 0.2 \quad (4).$$

The oscillations seen on this picture are due to the low and high contribution of the non resonant states ($(|00\rangle$ and $|01\rangle)$) to the dynamics of the system, which depends on Rabi's frequency and they are explained by the $2\pi k$-method [19]. As one can see from this picture , the CNOT gate is very well produced either with $B_{01} = 0.1\text{ T}$ and $f = 0.2$ or with $B_{01} = 0.5\text{ T}$ and $f = 0.05$.

Figure 3 shows the gradient of magnetic field along the x-axis, the coupling constant J, and the fidelity F of the CNOT quantum gate as a function of the two qubits separation (characterized by the parameter ξ, Equation (24)), having $f = \mathbf{0.05}$. As one can see, the fidelity is not sensitive for relatively wide variation of ξ, meanwhile the gradient and coupling constant have the strong variation deduce from Equation (6) and Equation (4). Considering the separation of the two ^{13}C atoms about the the length of the diamond unit cell, one can select $\xi = 3$, corresponding to a coupling constant of $J = 0.00445$, and a magnetic field gradient of $\Delta B_0 / a \approx 0.83 \times 10^6$ T/m.

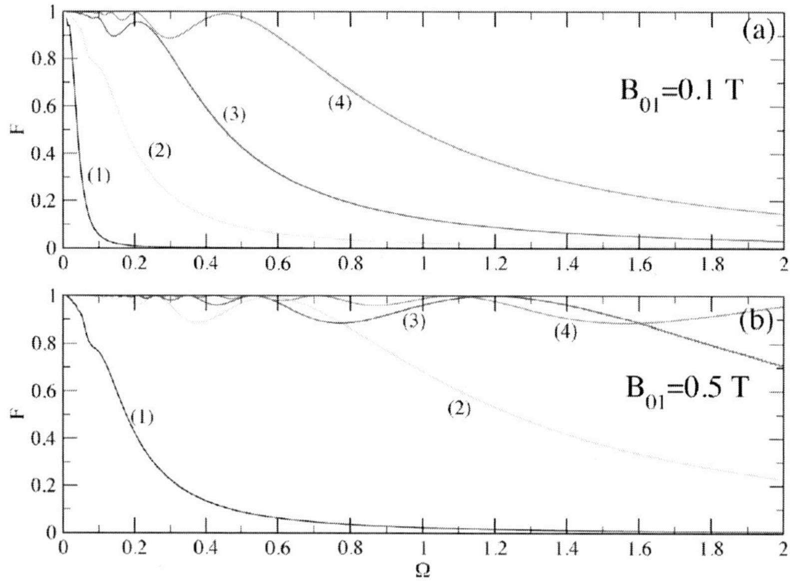

Figure 2. Fidelity at the end of the π-pulse for CNOT.

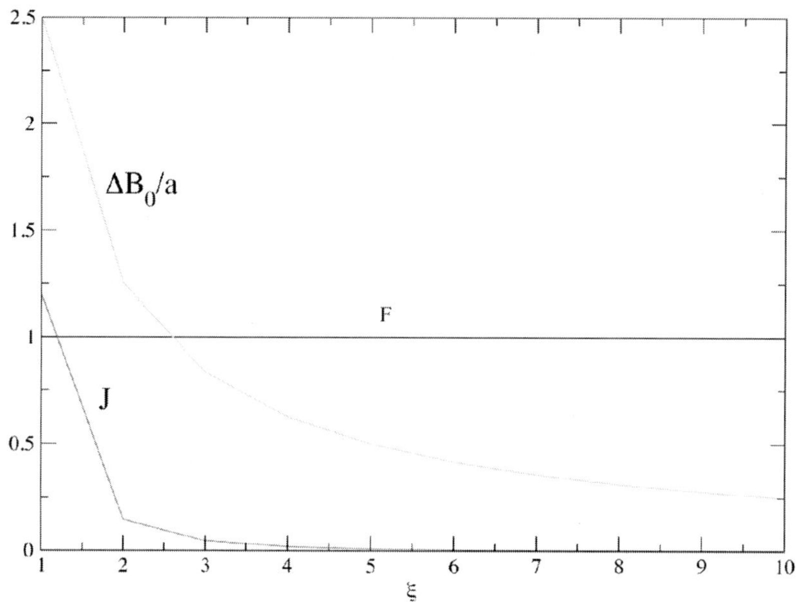

Figure 3. Effect of 13C-13C separation.

One needs to mention that in the case the alignment of the ^{13}C atoms be along the z-axis (the same direction of the longitudinal magnetic field), the coupling constant deduced from Equation (1) would be given by $-2J$, with J given by Equation (4), and basically the results are the same as the presented here.

According to these results, one has now an idea of the value of the parameters for the design of a quantum computer with the ^{12}C - ^{13}C diamond quantum system: (a) Separation between ^{13}C atoms is $a = 3\times 10^{-10}$ m which can be aligned along the x-axis, (b) coupling constant is $J = 0.00445(2\pi\ \mathrm{MHz})$, (c) longitudinal magnetic field is $B_{01} = 0.05\ \mathrm{T}$, (d) gradient of this longitudinal magnetic field along the x-axis is $\Delta B_0/a = 0.83\times 10^6\ \mathrm{T/m}$, and (e) magnitude of the rf-magnetic field on the plane x-y is $b = 0.00608\ \mathrm{T}$ (Rabi's frequency $\Omega = 0.259(2\pi\ \mathrm{Mhz})$).

For the CCNOT quantum gate, one has the initial conditions $C_{kji}(0) = 0$ for $k = j \neq 1$ and $C_{110}(0) = 1$. The time is allowed to last a π-pulse $(\tau = \pi/\Omega)$. The coupling parameter J is taken as before, and second neighbor coupling parameter $J' = J/10$. Figure 4 shows the fidelity parameter as a function of Rabi's frequency for magnetic field intensities (1) $B_o = 0.01\ \mathrm{T}$, (2)

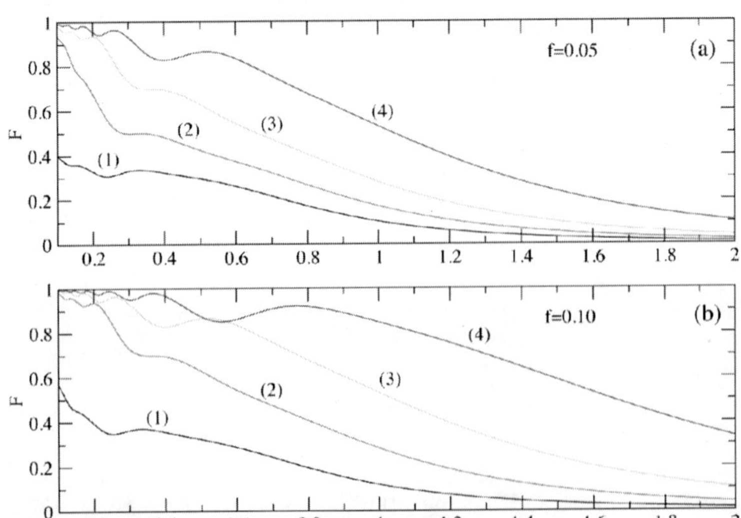

Figure 4. Fidelity at the end of the π-pulse for CCNOT.

$B_o = 0.05\,\mathrm{T}$, (3) $B_o = 0.1\,\mathrm{T}$, and (4) $B_o = 0.2\,\mathrm{T}$, for $f = 0.05$ (a), and for $f = 0.1$ (b). As one can see fro these plots, a good CCNOT quantum gate can be obtained by choosing $f = 0.1$ (implying and encresing of the gradient of the magnetic field by a factor of two), and with the other parameters given as defined with the CNOT quantum gate.

Although the gradient of the magnetic field might be a concern, the magnitude of the longitudinal magnetic field is low enough to think that this gradient can be achieved. The scalability of the system is clear, the read out system could be based on single spin measurement technics [29], and studies on decoherence remains to be done on this system. This quantum computer resembles a solid state NMR system [30].

5. CONCLUSION AND DISCUSSION

G. V. López (0.5 T). In principle, it is possible to replace a ^{12}C atom by any other spin one half atom. However, an unclose configuration of electrons in the lattice makes necessarily to take into account the interaction of electrons with this atom (as it is the case of diamond NV configuration) which makes the analysis and the quantum computer much more complicated and sensitive to environment interaction. The misplacement of the ^{13}C atom along the x-axis produces different coupling constant in the interaction, but according to **Figure 4**, the fidelity of the CNOT quantum gate does not change, and one would expect the same result for quantum algorithms. The displacement of ^{13}C atoms off x-axis changes the coupling constant and the interaction itself, which has to be studied. In addition, it still remains to study the decoherence on this system. Finally, one recalls that the carbon isotopesatoms ^{12}C, ^{13}C and ^{14}C occur naturally on Earth with a percent of about 99%, 1% and 10^{-4}%, ^{14}C being a radioactive isotope with a half life of about 5730 years and having spin zero. Therefore, ^{14}C is not an important composition at all in the diamond structure.

REFERENCES

1. H.-P. Breuer and F. Petruccione, "The Theory of Open Quantum Systems," Oxford University Press, Oxford, 2006.
2. O. Caldeira and A. T. Legget, Physica A: Statistical Mechanics and its Applications, Vol. 121, 1983, pp. 587- 616. http://dx.doi.org/10.1016/0378-4371(83)90013-4
3. W. G. Unruh and W. H. Zurek, Physical Review D, 1989, Vol. 40, pp. 1071-1094http://dx.doi.org/10.1103/PhysRevD.40.1071

4. B. L. Hu, J. P. Paz and Y. Zhang, Physical Review D, Vol. 45, 1992, pp. 2843-2861.http://dx.doi.org/10.1103/PhysRevD.45.2843
5. Venugopalan, Physical Review A, Vol. 56, 1997, pp. 4307-4310. http://dx.doi.org/10.1103/PhysRevA.56.4307
6. H. D. Zeh, Foundations of Physics, Vol. 3, 1973, pp. 109-116. http://dx.doi.org/10.1007/BF00708603
7. J. P. Paz and W. H. Zurek, Proc. Les Houches, Vol. 111A, 1997, p. 409.
8. G. Lindblad, Communications in Mathematical Physics, Vol. 48, 1976, pp. 119-130.http://dx.doi.org/10.1007/BF01608499
9. W. S. Warren, Science, Vol. 277, 1997, pp. 1688-1690.http:// dx.doi.org /10.1126/science.277.5332.1688
10. L. M. L. Vandersypen, M. Steffen, G. Breyta, C. S. Yannoni, M. H. Sherwood and I. L. Chuang, Nature, Vol. 414, 2001, p. 883. http://dx.doi.org/10.1038/414883a
11. M. H. Holzschelter, Los Alamos Science, Vol. 27, 2002, p. 264.
12. C. Monroe and J. Kim, Science, 2013, Vol. 339, p. 1164.http://dx.doi.org/10.1126/science.1231298
13. H. Walter, B. T. H. Varcoe, B. G. Englert and T. Becker, Reports on Progress in Physics, Vol. 69, 2006, p. 1325. http://dx.doi.org/10.1088/0034-4885/69/5/R02
14. D. Jaksch, J. I. Cirac, P. Zoller, S. L. Rolston, R. Côté and M. D. Lukin, Physical Review Letters, Vol. 85, 2000, pp. 2208-2211. http://dx.doi.org/ 10.1103/PhysRevLett.85.2208
15. K. C. Younge, B. Knuffman, S. E. Anderson and G. Raithel, Physical Review Letters, Vol. 104, 2010, Article ID: 173001. http://dx.doi.org/10.1103/PhysRevLett.104.173001
16. Chiorescu, Y. Nakamura, C. J. P. M. Harmans and J. E. Mooij, Science, Vol. 299, 2003, pp. 1869-1871. http://dx.doi.org/10.1126/science.1081045
17. Yu. Kitaev, Annals of Physics, Vol. 303, 2003, pp. 2-30. http://dx.doi.org/10.1016/S0003-4916(02)00018-0
18. L. Childress and R. Hanson, MRS Bulletin, Vol. 38, 2013, pp. 134-138. http://dx.doi.org/10.1557/mrs.2013.20
19. G. P. Berman, D. I. Kamenev, D. D. Doolen, G. V. López and V. I. Tsifrinovich, Contemporary Mathematics, Vol. 305, 2002, p. 13.http://dx.doi.org/10.1090/conm/305/05213
20. G. V. López and L. Lara, Journal of Physics B: Atomic, Molecular and Optical Physics, Vol. 39, 2006, p. 3897. http://dx.doi.org/10.1088/0953-4075/39/18/019
21. G. V. López, T. Gorin and L. Lara, Journal of Physics B: Atomic, Molecular and Optical Physics, Vol. 41, 2008, Article ID: 055504. http://dx.doi.org/10.1088/0953-4075/41/5/055504
22. G. V. López and P. López, Journal of Modern Physics, Vol. 3, 2012, pp. 85-101.http://dx.doi.org/10.4236/jmp.2012.31013
23. K. Lakoubovskii and G. J. Adriaenssens, Journal of Physics: Condensed Matter, Vol. 13, 2001, p. 6015.
24. R. W. Hamm and M. E. Hamm, "Industrial Acceleretors and Their Applications," World Scientific, Singapore, 2012.
25. M. A. Cazalilla, N. Lorente, R. D. Muño, J. P. Gauyacq, D. Teillet-Billy and P. M. Echenique, Physical Review B, Vol. 58, 1998, pp. 13991-14006.http:// dx.doi.org/10.1103/PhysRevB.58.13991
26. J. D. Jackson, "Classical Electrodynamics," 3rd Edition, Chapter 5.6, John Wiley and Sons, Inc., Hoboken, 1999.
27. N. Kolmogorov and S. V. Fomin, "Introductory Real Analysis," Dover Publications, Inc., Mineola, 1970.
28. M. A. Nielsen and I. L. Chuang, "Quantum Computation and Quantum Information," Cambridge University Press, Cambridge, 2000.
29. D. Rugar, R. Budakian, H. J. Mamin and B. W. Chui, Nature, Vol. 430, 2004, p. 329.http://dx.doi.org/10.1038/nature02658
30. M. J. Duer, "Introduction to Solid-State NMR Spectroscopy," Blackwell, Oxford, 2004.

Appendix

Two qubits dynamics is obtained from Equations (13), (14), and (19), resulting the equations

$$i\dot{a}_{00} = -\frac{\Omega}{2}\left(\mathrm{e}^{-i\left(\omega t+\varphi+\left(E_1-E_0\right)t/\hbar\right)}a_{01} + \mathrm{e}^{-i\left(\omega t+\varphi+\left(E_2-E_0\right)t/\hbar\right)}a_{10}\right) \tag{27}$$

$$i\dot{a}_{01} = -\frac{\Omega}{2}\left(\mathrm{e}^{+i\left(\omega t+\varphi+\left(E_1-E_0\right)t/\hbar\right)}a_{00} + \mathrm{e}^{-i\left(\omega t+\varphi+\left(E_3-E_1\right)t/\hbar\right)}a_{11}\right) \tag{28}$$

$$i\dot{a}_{10} = -\frac{\Omega}{2}\left(\mathrm{e}^{+i\left(\omega t+\varphi+\left(E_2-E_0\right)t/\hbar\right)}a_{00} + \mathrm{e}^{-i\left(\omega t+\varphi+\left(E_3-E_2\right)t/\hbar\right)}a_{11}\right) \tag{29}$$

$$i\dot{a}_{11} = -\frac{\Omega}{2}\left(\mathrm{e}^{+i\left(\omega t+\varphi+\left(E_1-E_3\right)t/\hbar\right)}a_{01} + \mathrm{e}^{+i\left(\omega t+\varphi+\left(E_3-E_2\right)t/\hbar\right)}a_{10}\right) \tag{30}$$

where the complex variables a_{ij} for $i, j = 0,1$ correspond to the amplitude of probability to find the system on the states $|00\rangle, |01\rangle, |10\rangle$ and $|11\rangle$, and one has $a_{ij}(0) = C_{ij}(0)$ and $|a_{ij}(t)|^2 = |C_{ij}(t)|^2$. Decimal notation on the energies i corresponds to its binary elements $\beta_i\alpha_i$ for $i = 0,1,2,3$.

As before, three qubits dynamics is described by the equations

$$i\dot{a}_{000} = -\frac{\Omega}{2}\left(\mathrm{e}^{i\phi_{01}}a_{001} + \mathrm{e}^{i\phi_{02}}a_{010} + \mathrm{e}^{i\phi_{04}}a_{100}\right) \tag{31}$$

$$i\dot{a}_{001} = -\frac{\Omega}{2}\left(\mathrm{e}^{i\phi_{10}}a_{000} + \mathrm{e}^{i\phi_{13}}a_{011} + \mathrm{e}^{i\phi_{15}}a_{101}\right) \tag{32}$$

$$i\dot{a}_{010} = -\frac{\Omega}{2}\left(\mathrm{e}^{i\phi_{23}}a_{011} + \mathrm{e}^{i\phi_{21}}a_{001} + \mathrm{e}^{i\phi_{25}}a_{101}\right) \tag{33}$$

$$i\dot{a}_{011} = -\frac{\Omega}{2}\left(e^{i\phi_{32}} a_{010} + e^{i\phi_{31}} a_{001} + e^{i\phi_{37}} a_{111}\right) \tag{34}$$

$$i\dot{a}_{100} = -\frac{\Omega}{2}\left(e^{i\phi_{45}} a_{101} + e^{i\phi_{46}} a_{110} + e^{i\phi_{43}} a_{011}\right) \tag{35}$$

$$i\dot{a}_{101} = -\frac{\Omega}{2}\left(e^{i\phi_{54}} a_{100} + e^{i\phi_{57}} a_{111} + e^{i\phi_{51}} a_{001}\right) \tag{36}$$

$$i\dot{a}_{110} = -\frac{\Omega}{2}\left(e^{i\phi_{67}} a_{111} + e^{i\phi_{64}} a_{100} + e^{i\phi_{62}} a_{010}\right) \tag{37}$$

$$i\dot{a}_{111} = -\frac{\Omega}{2}\left(e^{i\phi_{76}} a_{011} + e^{i\phi_{75}} a_{101} + e^{i\phi_{73}} a_{011}\right) \tag{38}$$

with $a_{ijk}(0) = C_{ijk}(0)$ and $\left|a_{ijk}(t)\right|^2 = \left|C_{ijk}(t)\right|^2$,

$i, j, k = 0,1$. The phases ϕ_{ij} are defined as

$\phi_{ij} = \pm\omega t + \varphi + \left(E_i - E_j\right)t/\hbar$ where decimal i corresponds to its binary elements $\gamma_i \beta_i \alpha_i$ for $i = 0, \ldots, 7$.

For both cases, the energies E_i are deduced from Equation (15).

CHAPTER 7

Novel Screen-Printed All-Solid-State Copper(II)-Selective Electrode for Mobile Environmental Analysis

Johannes Schwarz[1*], Kathrin Trommer[1], Michael Mertig[1,2]

[1]*Kurt-Schwabe Institut für Mess- und Sensortechnik e.V. Meinsberg, Waldheim, Germany*

[2]*Institut für Physikalische Chemie, Technische Universität Dresden, Dresden, Germany*

ABSTRACT

Based on poly(vinyl chloride) membranes, a novel miniaturized screen-printed all-solid-state copper(II)-selective electrode has been developed for applications in environmental monitoring. Per- formance and applicability of the ion-selective electrode (ISE) have been proved by potentiometric investigations. Conducting polymers were used as intermediate layers and as solid contacts between the ion-selective membrane and the graphite transducer. The ion-complexing reagent 2-mercapto-benzoxazole was incorporated into poly(vinyl chloride) membranes. In the concentration range 10^{-6} - 10^{-2} mol/L, the ISE exhibited a linear Nernstian potential response to copper(II) with an average slope value of 28 mV/decade. The detection limit was 3×10^{-7} mol/L. The electrode exhibits a short response time (<10 s) and can be used in the range of pH = 3 - 7. Selectivity coefficents against certain interfering ions are investigated. The life time of the electrode under laboratory conditions was approximately 12-month. The electrode was applied in the investigation of different aqueous environmental samples and the electrode characteristics were described.

The copper(II) ASS electrode has also successfully been used in potentiometric, complexometric titrations with ethylenediaminetetraacetic acid.

Keywords: All-Solid-State Copper(II)-Selective Electrode, Conducting Polymers, Potentiometric Titration, Environmental Samples

1. INTRODUCTION

Contamination of ecosystems by heavy metals is of worldwide concern and represents a major problem in environmental pollution of today. Heavy metals are non-biodegradable and can accumulate in human organism. Copper is one of the most widely distributed and essential trace elements. It is of great importance for industrial applications and plays a central role in many biochemical processes. Its determination in various media is, therefore, of great interest. According to the German Drinking Water Act, the concentration of copper in drinking water has to be lower than 2 mg/L [1]. According to the United States Environmental Agency (EPA) and the World Health Organization (WHO), the maximum containment level (MCL) of copper is even lower: 1.3 mg/L [2] [3].

Many instrumental analytical methods such as atomic absorption spectrometry (AAS), inductively-coupled- plasma optical emission spectrometry (ICP-OES) and inductively-coupled-plasma mass spectrometry (ICP-MS) are used routinely for copper analysis [4] - [6]. These laboratory-based methods have good precision, high sensitivity and low detection limits. Disadvantages of these methods are complex analytical operations as well as the need of qualified personnel and high cost equipment. They are also not suitable for field analysis.

Therefore, new analytical tools are required for economical real-time monitoring of environmental pollutants. During the last years, there has been an increasing demand and a high interest for miniaturized electrochemical devices for field investigations of heavy metals. Due to their portability, low cost, compactness, simplicity, easy- to-use capability and robustness, electrochemical devices are promising alternatives compared to costly instrumental laboratory equipment [7] - [12]. The search for rapid, sensitive, simple, easy-to-handle and robust analyzers is still going on.

Ion-selective electrodes (ISEs) are a class of electrochemical sensors with increasing importance for on-site analysis of environmentally relevant ions. Potentiometric ion detection with ion-selective all-solid-state (ASS) electrodes has been widely developed in recent years and will be broadly applicable due to their simplicity and fast analysis. Potentiometric detection requires minimum sample pretreatment and preparation and has great potential for field analysis. In literature, the use of different ion-complexing compounds like calixarenes, Schiff's bases and thiocrown ethers is described for potentiometric copper(II) detection applying ISEs of different construction forms [13] - [25].

There are recent reports on ASS copper(II) electrodes [26] - [29]. These electrodes are mainly based on graphite pastes and expensive nanomaterials. Furthermore, the use of new synthesized ionophores is often required which results in time-consuming electrode preparation steps.

Conducting polymers (CPs) have been used as solid contact materials and intermediate layers in ASS ion-se- lective electrodes [30] - [36]. They are of interest because of their mixed ionic and electronic conductivity which allows a definite transfer of ion to electrical charge without using an internal liquid electrolyte. The incorporation of conducting polymers as solid contact offers new possibilities for simple and reproducible electrode construction. The internal electrolyte can therefore be replaced.

Screen-printing is a well-established technique for manufacturing of portable and disposable electrochemical electrodes. It allows the design of electrodes with small dimensions, good reproducibility and variable electrode arrangements [9] [37].

Here we report on the design, preparation, potentiometric characterization and analytical application of screen- printed, solid-contacted ASS copper(II)-selective electrodes consisting of different functional polymeric layers. As neutral carrier and chelating agent, 2-mercapto-benzoxazole (2-MBA) has been used for the first time in ASS ion-selective electrodes. 2-MBA forms stable complexes with copper(II) ions [38].

2. EXPERIMENTAL

2.1. Reagents

All reagents were analytical grade and used without any further purification. Solutions were prepared with deionized water. All inorganic salts used were purchased from Sigma-Aldrich Co. LCC.. Sample solutions of different ion concentration were prepared by diluting the stock solutions. Graphite and silver pastes used for screen-printed electrodes were obtained from DuPont de Nemours (Germany) GmbH. High molecular weight poly (vinyl chloride) (PVC), ortho-nitrophenyl octyl ether (o-NPOE), pyrrole, poly (3,4-ethylenedioxythiophene)- poly(styrene-sulfonate) (PEDOT:PSS), tetrahydrofuran (THF), 2-MBA and ethylenediaminetetraacetic acid (EDTA) were purchased from Sigma-Aldrich Co. LCC. and used as received.

2.2. Apparatus and Electrodes

Potentiometric measurements were carried out at room temperature, using the pH/ion analyzers pMX 3000 (WTW Weilheim, Germany) and LM 2000 (Sensortechnik Meinsberg GmbH, Germany). Ag/AgCl reference electrodes with aqueous saturated KCl as the electrolyte solution have been used.

A half-automatic screen printer (EKRA, FEI Germany) has been used to manufacture the screen-printed ASS electrodes.

Atomic adsorption spectrometry (novAA®, Analytik Jena, Germany) was used as reference method.

3. ELECTRODE PREPARATION

Copper(II)-selective electrodes were fabricated with different functional layers on screen-printed graphite electrodes based on Al_2O_3-ceramic substrates. The manufacture of the electrodes consists of 3 steps:

- Screen-printing of graphite on a ceramic substrate (layer 1),
- Application of CP (layer 2),
- Drop-casting of ion-selective membrane cocktails (layer 3).

Reproducible layer configurations were obtained. Commercially available graphite ink is screen-printed on ceramic substrates within an area of 5 mm × 30 mm. The thickness of the graphite layer is ~500 μm. The graphite discs have diameters of 2.5 mm. Polypyrrole (PPy) was prepared by in-situ electropolymerization of 0.1 mol/L pyrrole in KCl-containing solution. PEDOT:PSS was applied by drop casting of an aqueous solution of the polymer directly onto the graphite layer. Both polymers form stable layers and show good adhesion to the graphite substrates. After applying the CP on the graphite surface, the electrodes were rinsed with H_2O and air dried overnight. The thickness of the polymer layer is between 5 μm and 10 μm.

The mixture of the membrane components was dissolved in 2 ml THF and allowed to evaporate at room temperature till a viscous concentrate was obtained. After the deposition of the CPs, the ion-selective cocktail was drop casted by pipetting about 5 μl onto the CP. The thickness of the resulting membrane is about 100 μm. The composition of the ion-selective membrane components has been optimized. Best results with average slopes of >28 mV/decade were obtained using the following membrane composition:

- PVC (32 wt.%),
- 2-nitrophenyl octyl ether (65 wt.%),
- 2-MBA (3 wt.%).

Before the first measurement, the electrodes were conditioned in 10^{-3} mol/L copper(II) salt solutions for at least 5 hours.

In **Figure 1**, the chemical structure of the 2-MBA is shown. **Figure 2** shows a schematic representation of the ASS electrodes. A photograph of the ASS electrode is shown in **Figure 3**.

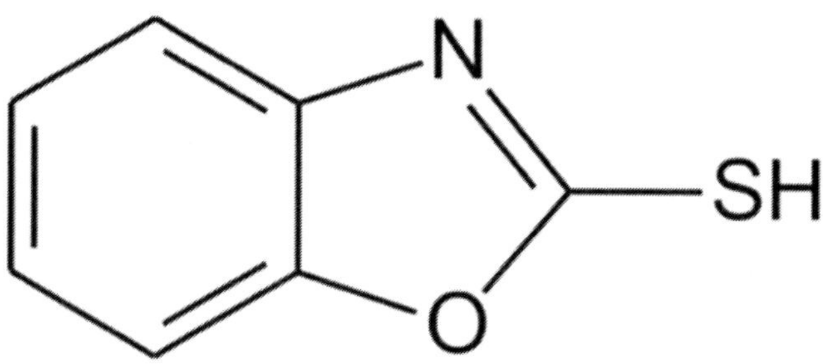

Figure 1. Chemical structure of the neutral ion-carrier 2-mercapto-benzoxazole (2-MBA).

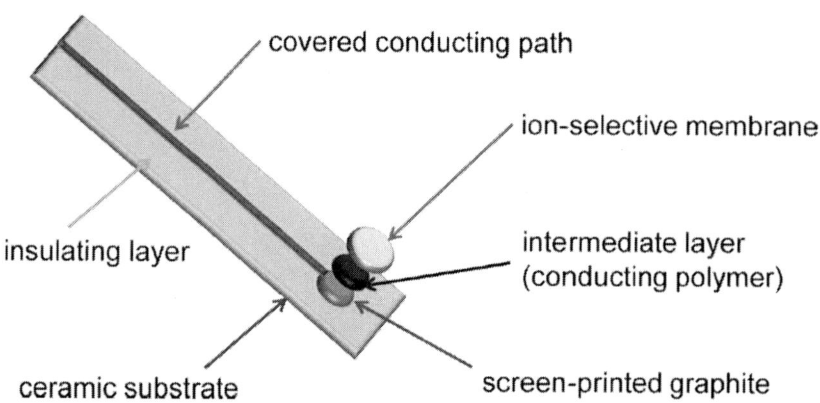

Figure 2. Schematic representation of the ASS electrode.

4. RESULT AND DISCUSSION

4.1. Potentiometric Investigations

The PVC-based membrane electrodes, containing 2-MBA as neutral ion carrier, generated stable responses in $CuSO_4$ solutions. The dynamic response behaviour of a PEDOT: PSS-modified ASS electrode and the corresponding calibration plot are shown in Figure 4. The linear dynamic range extends from 10^{-6} mol/L to 10^{-1} mol/L.

To evaluate the high reversibility and the fast response of the electrode, high-to-low repetitive sample concentrations were tested as shown in Figure 5. The response behaviour confirms the reversibility of the measured potentials and the fast response. The response time is <10 s. The calculated slope is on average 28 mV/decade. The value of the regression coefficient is $R^2 = 0.99$. The relative standard deviations of the potentials for 10^{-4} mol/L and 10^{-3} mol/L copper(II) salt solutions are in both cases less than 4% of 10 replicate measurements indicating good reproducibility of measurements. The working pH-range of the electrode is between 3 and 7 and remained constant in this range. Over 12 month the electrodes showed no significant divergence in slope potential and response time. During non-usage, the electrodes were stored in air and before use they were conditioned by dipping into 10^{-3} mol/L copper(II) salt solutions for 2 - 3 hours. Consequently, the electrodes generate stable potentials when in contact with copper(II) containing solutions.

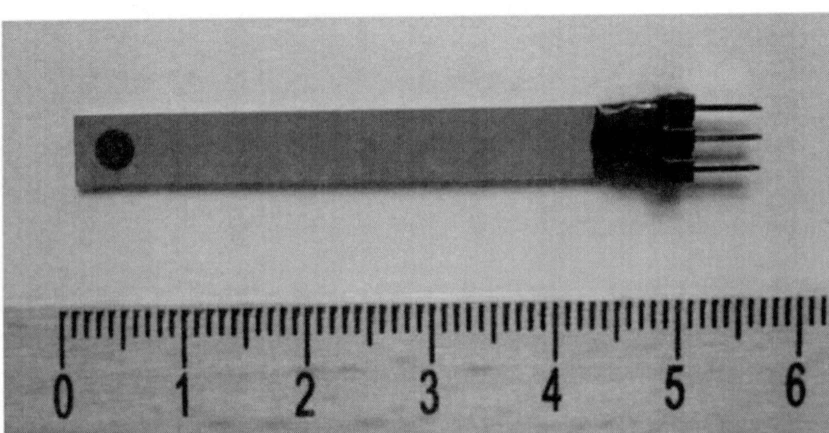

Figure 3. Photograph of a screen-printed ASS ion-selective electrode.

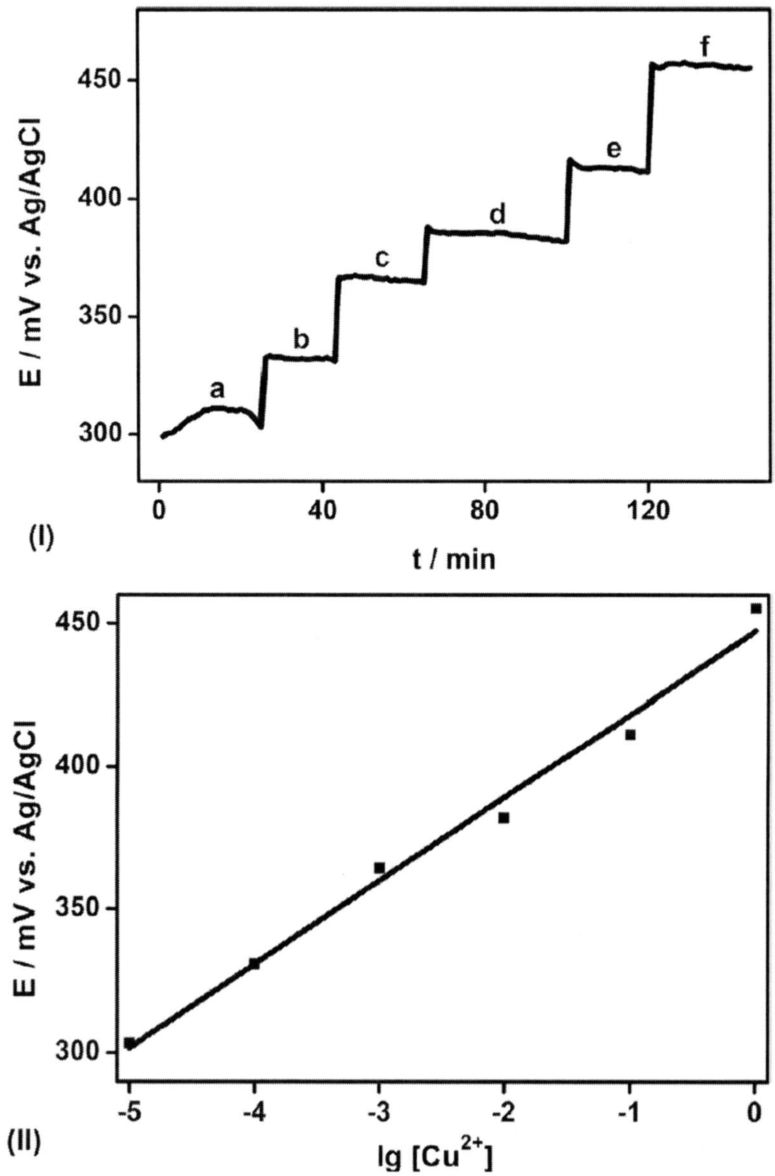

Figure 4. Dynamic potentiometric response behaviour (I) and calibration plot (II) of a PEDOT:PSS-modified ASS copper(II)-selective electrode. $CuSO_4$ concentrations—a: 10^{-5}mol/L, b: 10^{-4} mol/L, c: 10^{-3} mol/L, d: 10^{-2} mol/L, e: 10^{-1} mol/L, f: 1 mol/L.

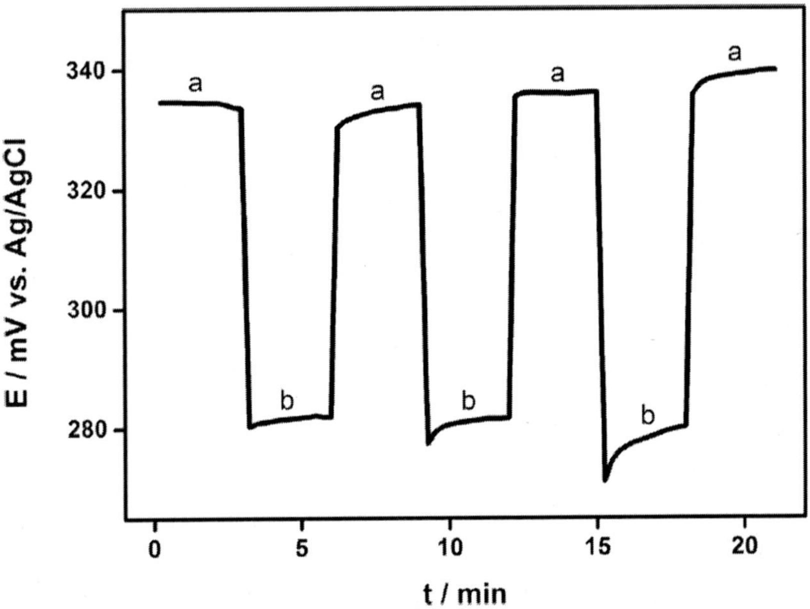

Figure 5. Dynamic response characteristics of a PEDOT: PSS-modified ASS copper(II)-selective electrode for several high- to-low and low-to-high concentration cycles. $CuSO_4$ concentrations: a: 10^{-2} mol/L, b: 10^{-4}mol/L.

The potentiometric selectivity coefficients K^{pot} have been evaluated after the separate solution method. The resulting selectivity-coefficients are given in **Table 1**, indicating that the proposed electrodes are selective towards the tested interfering ions.

The advantages of the new electrode, compared to other known ISEs, are the short response time, as well as a reasonable long-term stability with life times of about one year [13] [17] [20]. Due to the thin functional layer configuration, only few chemicals and materials in relatively small amounts are needed for a reproducible electrode fabrication. This allows the fabrication of low-cost electrodes in large quantities. All used chemicals, including the ion-complexing compound, are commercially available. Therefore, no time-consuming synthesis steps are necessary as described for most solid-contacted copper(II)-selective electrodes [21] - [23]. The use of screen-printing technology for the manufacture of the graphite transducers is also advantageous with regard to a possible mass production compared to devices produced by conventional methods.

4.2. Analytical Applications

The disposable ISEs were successfully applied as indicator electrodes in combination with Ag/AgCl-reference electrodes for complexometric titrations of copper(II) ions with EDTA solutions as titrant in standard and in spiked environmental samples. The results show that the amount of copper(II) ions can be accurately determined by potentiometric titrations using the electrodes described here. The titration plots including the first derivatives with respect to the added volume of the EDTA solution are shown in Figure 6 (standard solution) and Figure 7 (spiked river sample). PPy-modified ASS electrodes are used. The inflection point on the graph agreed well with the calculated equivalent point. The samples were prepared as follows: 40 ml solutions of copper(II) (10^{-3} mol/L) (Figure 6) and 50 ml of a spiked real water sample from the Zschopau River, Germany (Figure 7) were titrated with 10^{-2} mol/L EDTA solution. The pH was adjusted to pH 5. Comparative analyses by colorimetric titrations were carried out using the color indicators 1-(2-Pyridylazo)-2-naphthole and 4-(2-Pyridylazo) resorcinole and by AAS as a routinely used copper detection method. The results of all three methods agree well.

The titration curves are characterized by sharp equivalent points. The copper(II) content can be analyzed with high accuracy in spiked and standard samples. The potential jump near the equivalent point is nearly 50 mV. The copper(II) content in environmental samples can be evaluated with high precision as has been demonstrated in potentiometric titrations using the novel electrode. The good accordance with the colorimetric titration must be emphasized and proves the practical utility and applicability of the novel copper(II) ASS electrode.

Environmental samples spiked with defined amounts of copper(II) have been analyzed by potentiometry using the novel electrodes and compared with AAS. The results are shown in **Table 2**.

Table 1. Selectivity coefficients.

Interfering ion	lg K^{pot}_{MI}
Cd^{2+}	−2.6
Ca^{2+}	−3.4
Mg^{2+}	−4.5
Pb^{2+}	−2.4

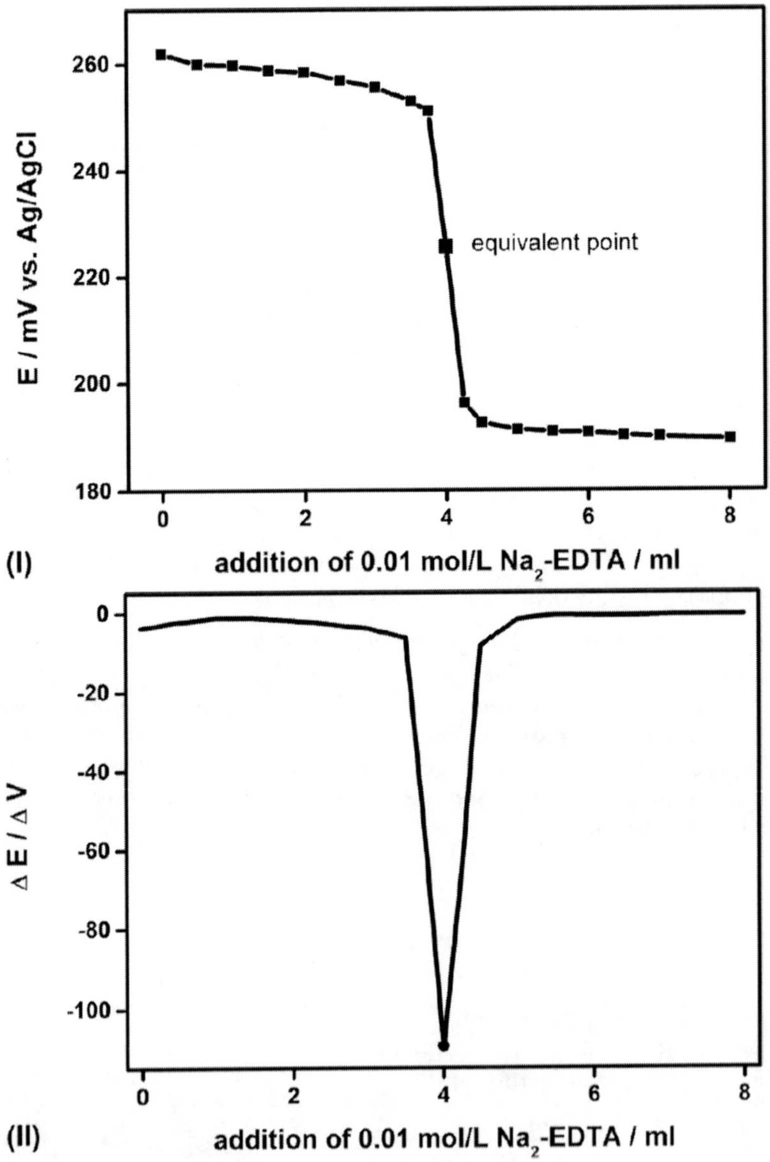

Figure 6. Potentiometric titration curve of copper(II) with EDTA using a PPy-modified ASS electrode. (I) Titration plot. (II) First derivative of the potential with respect to the added volume of the EDTA solution (ΔE/ΔV). Sample concentration: 10^{-3} mol/L $Cu(NO_3)_2$.

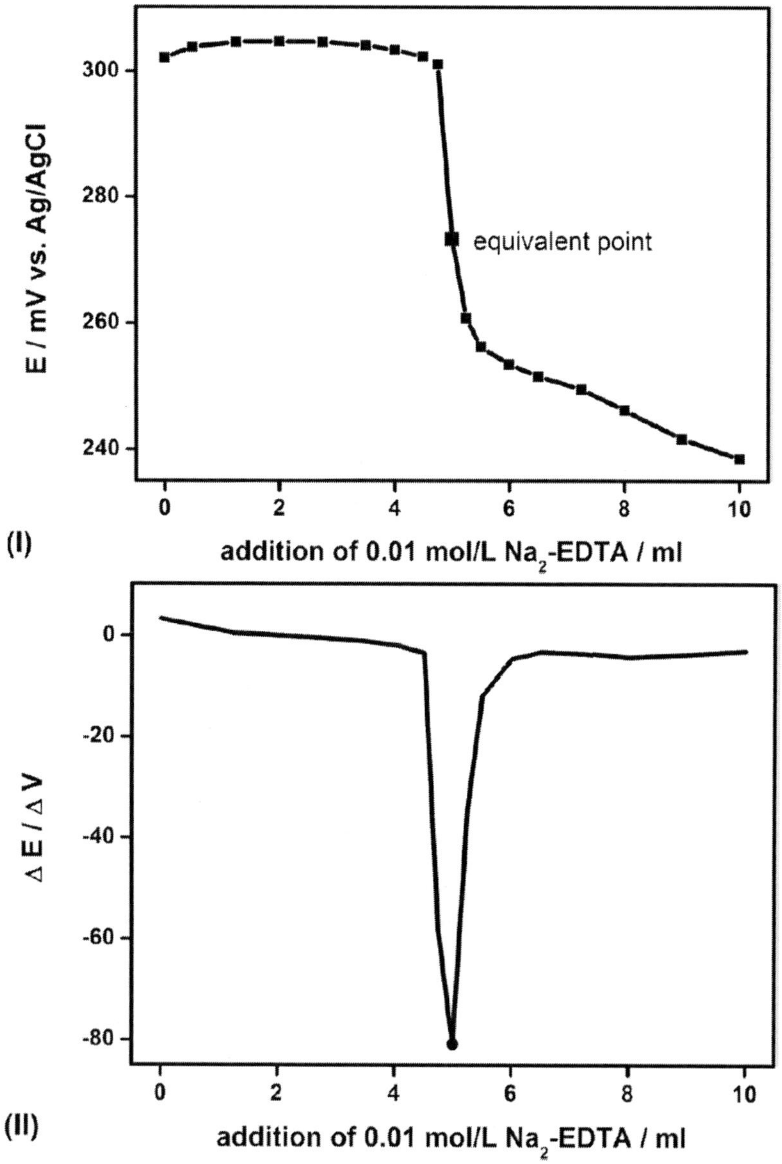

Figure 7. Potentiometric titration curve of copper(II) with EDTA using a PPy-modified ASS electrode. (I) Titration plot. (II) First derivative of the potential with respect to the added volume of the EDTA solution (ΔE/ΔV). Sample: spiked river water (Zschopau River, Germany).

Table 2. Determination of copper(II) ions in real water samples.

Spiked real samples	Copper(II) concentration		
	of spiked sample [mg/L]	determined by ISE [mg/L]	determined by AAS [mg/L]
River water (Zschopau, Germany)	6.4	7.7	5.8
Sea water (Baltic sea, Kolobrzeg, Poland)	63.5	66.1	57.1
River water (Rhein, Germany)	63.5	62.7	53.1
Sea water (Baltic sea, Glowe, Germany)	63.5	57.5	53.9

5. CONCLUSION

A novel liquid-free copper(II)-selective electrode was developed and tested in standard and real samples. The electrode is based on an intermediate layer of conducting polymer deposited between the graphite substrate and the ion selective membrane. The solid contact realized by conducting polymer ensures stable and reproducible potentials. Thus, the integration of different functional polymer layers on screen-printed graphite substrates enables reproducible electrode constructs. The developed ASS copper(II)-selective electrode has a simple and stable design, can be simply prepared, provides easy handling and exhibits a long life-time. The developed electrode has good operating characteristics (sensitivity, response time, detection limit, linear range, selectivity) and high mechanical stability. Successful applications and good performance have been shown by complexometric titrations and by direct potentiometric analysis. The results confirm the analytical utility and applicability of the developed electrodes. The electrode can be successfully employed for copper(II) detection in water samples. It requires no extensive, cost intensive equipment or time consuming sample preparation, and it allows a real-time in-situ monitoring of copper(II). The novel electrode shows competitive characteristics to commercially available electrodes enabling sensitive mobile and decentralized copper(II) analysis. The presented screen-printed electrodes are small and easy to handle, and thus, attractive for field studies and environmental monitoring.

ACKNOWLEDGEMENTS

The authors gratefully thank the Bundesministerium für Bildung und Forschung in the context with the project BioNEWS (project number 03WKCL03G) for financial support. We also thank F. Gerlach for preparing the screen-printed electrodes and for helpful discussions in optimizing electrode layout and arrangement.

REFERENCES

1. TrinkwV (2001) Deutsche Trinkwasserverordnung.
2. National Primary Drinking Water Standards, Water Research Center. www.water-research.net/standards.htm
3. US Environmental Protection Agency (EPA) (2003) National Primary Drinking Water Standards. www.epa.gov/safewater
4. Gong, R., Zhang, D., Zhong, K., Feng, M. and Liu, X. (2008) Determination of Trace Copper in Water Samples by Flame Atomic Absorption Spectrometry after Preconcentration on a Phosphoric Acid Functionaliced Cotton Chelator. Journal of the Serbian Chemical Society, 73, 249-258. http://dx.doi.org/10.2298/JSC0802249G
5. Chwastowska, J., Skwara, W., Sterbinska, E., Dudek, J., Dubrowska, M. and Pszonicki, L. (2008) GF AAS Determination of Cadmium, Lead and Copper in Environmental Materials and Food Products after Separation on Dithizone Sorbent. Chemia Analityczna (Warsaw), 53, 887-894.
6. Townsend, A.T., Miller, K.A., Mclean, S. and Aldrous, S. (1998) The Determination of Copper, Zinc, Cadmium and Lead in Urine by High Resolution ICP-MS. Journal of Analytical Atomic Spectrometry, 13, 1213-1219. http://dx.doi.org /10.1039/ a805021j
7. Pujol, L., Evrard, D., Groenene-Serrano, K., Freyssinier, M., Ruffien-Cizsak, A. and Gros, P. (2014) Electrochemical Sensors and Devices for Heavy Metals Assay in Water: The French Groups Contribution. Frontiers in Electrochemistry, 19, 1-14. http://dx.doi.org/10.3389/fchem.2014.00019
8. Bobacka, J., Ivaska, A. and Lewenstam, A. (2008) Potentiometric Ion Sensors. Chemical Reviews, 108, 329-351. http://dx.doi.org/10.1021/cr068100w
9. Tymecki, L., Glab, S. and Koncki, R. (2006) Miniaturized, Planar Ion-Selective Electrodes Fabricated by Means of Thick-Film Technology. Sensors, 6, 390-396. http://dx.doi.org/10.3390/s6040390
10. Honeychurch, K.C. (2012) Screen-Printed Electrochemical Sensors and Biosensors for Monitoring Metal Pollutants. Insciences Journal, 2, 1-51.
11. Hayat, A. and Marty, J.-L. (2014) Disposable Screen Printed Electrochemical Sensors: Tools for Environmental Monitoring. Sensors, 14, 10432-10453. http://dx.doi.org/10.3390/s140610432
12. Renedo, O.D., Alonso-Lomillo, M.A. and Arcos Martinez, M.J. (2007) Recent Developments in the Field of Screen- Printed Electrodes and Their Related Applications. Talanta, 73, 202-219. http://dx.doi.org/10.1016/j.talanta.2007.03.050
13. Emadi, D., Yaftian, M.R. and Rayati, S. (2007) N,N-bis(1-hydrooxy-2-acetonaphthone) Propylenediamine: Synthesis, Extractive Properties and Use as an Ionophore in a Cu(II)-Selective Potentiometric Sensor. Turkish Journal of Chemistry, 31, 423-433.
14. Ansari, R., Mosayebzadeh, Z., Arvand, M. and Mohammad-Khah, A. (2013) A Potentiometric Solid State Copper Electrode Based on Nanostructure Polypyrrole Conducting Polymer Film Doped with 5-Sulfosalicyl Acid. Journal of Nanostructure in Chemistry, 3, 33-42. http://dx.doi.org/10.1186/2193-8865-3-33
15. Ganjali, M.R., Poursaberi, T., Babaei, L.H., Rouhani, S., Yousefi, M., Kargar-Razi, M., Moghimi, A., Aghabozorg, H. and Shamsipur, M. (2001) Highly Selective and Sensitive Copper(II) Membrane Coated Graphite Electrode Based on a Recently Synthesized Schiff's Base. Analytica Chimica Acta, 440, 81-87. http://dx.doi.org/10.1016/S0003-2670(01)01051-0
16. Mahajan, R.K. and Sood, P. (2007) Novel Copper(II)-Selective Electrode Based on 2,2':5',2"-Terthiophene in PVC- Matrix. International Journal of Electrochemical Science, 2, 832-847.

17. Ardakani, M.M., Mirhoseini, S.H. and Salavati-Niasari, M. (2006) Copper Selective Electrode Based on 1,8-bis(2-Hy- droxynaphthaldiminato)3,6-Dioxaoctane. Acta Chimica Slovenica, 53, 197-303.

18. Firooz, A.R., Mazloum, M., Safari, J. and Amini, M.K. (2002) Coated-Wire Copper(II)-Selective Electrode Based on Phenylglyoxal-α-Monoxime Ionophore. Analytical and Bioanalytical Chemistry, 372, 718-722. http://dx.doi.org/10.1007/s00216-001-1205-7

19. Isa, I.M., Saidin, M.I., Ahmad, M., Hashim, N., Ghani, S.A. and Si, S.M. (2013) Development of New Copper(II) Ion-Selective Poly(vinyl chloride) Membrane Electrode Based on 2,6-Diacetylpyridine-(1R)-(-)-Fenchone Diazine Ligand. International Journal of Electrochemical Science, 8, 11175-11185.

20. Ganjali, M.R., Ghafarloo, A., Faridbod, F. and Norouzi, P. (2012) Copper-Selective PVC Membrane Sensor. International Journal of Electrochemical Science, 7, 3706-3726.

21. Yoshimoto, S., Mukai, H., Kitano, T. and Sohrin, Y. (2003) Copper(II)-Selective Membrane Electrode Based on Hydrotris(3-isopropylpyrazolyl)methane in a Poly(vinyl chloride) Matrix. Analytica Chimica Acta, 494, 207-213. http://dx.doi.org/10.1016/S0003-2670(03)01012-2

22. Vlascici, D., Popa, J., Fagadar-Cosma, G., Popvici, H. and Fagadar-Cosma, E. (2013) Potentiometric Detection and Removal of Copper Using Porphyrins. Chemistry Central Journal, 7, 111-117. http://dx.doi.org/10.1186/1752-153X-7-111

23. Norouzi, P., Ganjali, M.R., Faridbod, F. and Salavati-Nisari, M. (2006) Determination of Copper in Black, Red Pepper and the Waste Water Samples by a Highly Selective Sensitive Cu(II) Microelectrode Based on a New Hexadentates Schiff's Base. Bulletin of the Korean Chemical Society, 27, 1439-1444. http://dx.doi.org /10. 5012/bkcs.2006.27.9.1439

24. Ganjali, M.R., Emani, M. and Salavati-Niasari, M. (2002) Novel Copper(II)-Selective Sensor Based on a New Hexadentates Schiff's Base. Bulletin of the Korean Chemical Society, 23, 1393-1398.

25. Yu, S.-Y., Li, Y.-C., Xiong, T., Yuan, Q., Liu, Y.-M., Yuan, Z.-Y. and Xiao, Y. (2014) A Ladder Conjugated Polymer Transducer for Solid-Contact Cu2+-Selective Electrodes. Chinese Chemical Letters, 25, 364-366. http://dx.doi.org/ 10.1016/ j.cclet. 2013.11.015

26. Liang, R., Yin, T. and Qin, W. (2015) A Simple Approach for Fabricating Solid-Contact Ion-Selective Electrodes Using Nanomaterials as Transducers. Analytica Chimica Acta, 853, 291-296. http://dx.doi.org/10.1016/j.aca.2014.10.033

27. Tutulea-Anastasiu, M.D., Wilson, D., del Valle, M., Schreiner, C.M. and Cretescu, I. (2013) A Solid-Contact Ion Selective Electrode for Copper(II) Using a Succinimide Derivative as Ionophore. Sensors, 13, 4367-4377. http://dx.doi.org /10.3390/s130404367

28. Ganjali, M.R., Rezapour, M., Pirali-Hamedani, M. and Rashedi, H. (2015) Cu(II)-All Solid State Sensor Ion Selective Electrode (ASS-ISE) with a Nano-Molar Detection Limit and Its Use for the Analysis of Waste Water Samples. International Journal of Electrochemical Science, 10, 6924-6934.

29. Ali, T.A., Mohamed, G.G. and Othman, A.R. (2015) Design and Construction of New Potentiometric Sensors for Determination of Copper(II) Ion Based on Copper Oxide Nanoparticles. International Journal of Electrochemical Science, 10, 8041-8057.

30. Faridbod, F., Ganjali, M.R., Dinarvand, R. and Norouzi, P. (2008) Development in the Field of Conducting and Non-Conducting Polymer Based Potentiometric Membrane Sensors for Ions over the Past Decade. Sensors, 8, 2331-2412. http://dx.doi.org/10.3390/s8042331

31. Zachara, J.E., Toczylowska, R., Prokrop, R., Zagorska, M., Dybko, A. and Wroblewski, W. (2004) Miniaturized All-Solid-State Potentiometric Ion Sensors Based on PVC-Membranes Containing Conducting Polymers. Sensors and Actuators B: Chemical, 101, 207-212. http://dx.doi.org/10.1016/j.snb.2004.02.052

32. Bobacka, J. (2005) Conducting Polymer-Based Solid-State-Ion-Selective Electrodes. Electroanalysis, 18, 7-18. http://dx.doi.org/10.1002/elan.200503384

33. Rahman, M.A., Kumar, P., Park, D.-S. and Shim, Y.-B. (2008) Electrochemical Sensors Based on Organic Conjugated Polymers. Sensors, 8, 118-141. http://dx.doi.org/10.3390/s8010118
34. Ramanavicius, A., Ramanaviciene, A. and Malinauskas, A. (2006) Electrochemical Sensors Based on Conducting Polymer—Polypyrrole. Electrochimica Acta, 51, 6025-6037. http://dx.doi.org/10.1016/j.electacta.2005.11.052
35. Gupta, N., Sharma, S. and Mir, I.A. (2006) Advances in Sensors Based on Conducting Polymers. Journal of Scientific & Industrial Research, 65, 549-557.
36. Kirchmeyer, S. and Reuter, K. (2005) Scientific Importance, Properties and Growing Applications of Poly(3,4-ethylenedioxythiophene). Journal of Materials Chemistry, 15, 2077-2088. http://dx.doi.org/10.1039/b417803n
37. Menil, F., Debeda, H. and Lucat, C. (2005) Screen-Printed Thick-Films: From Materials to Functional Devices. Journal of the European Ceramic Society, 25, 2105-2113. http://dx.doi.org/10.1016/j.jeurceramsoc.2005.03.017
38. Akhond, M., Ghaedi, M. and Tashkhourian, J. (2005) Development of a New Copper(II) Ion-Selective Poly(vinyl chloride) Membrane Electrode Based on 2-Mercaptobenzoxazole. Bulletin of the Korean Chemical Society, 26, 882-886. http://dx.doi.org/10.5012/bkcs.2005.26.6.882

CHAPTER 8

Direct Solid-State Fermentation of Soybean Processing Residues for the Production of Fungal Chitosan by Mucor rouxii

Andro Mondala*, Ramea Al-Mubarak, James Atkinson, Shaun Shields, Brian Young, Yurguen Dos Santos Senger, Jan Pekarovic

Department of Chemical and Paper Engineering, Western Michigan University, Kalamazoo, USA

ABSTRACT

The feasibility of utilizing soybean-processing residues such as soybean meal and hulls as substrates for chitosan production by the fungus Mucor rouxii ATCC 24905 via solid-state fermentation (SSF) was investigated. The effects of the type of soybean-based substrate, length of cultivation period, substrate moisture content, substrate pH, incubation temperature and extraction conditions on chitosan yield were determined. The results showed that a maximum fungal chitosan yield of up to 3.44% by dry substrate weight (34.4 g/kg) could be achieved using a pure soybean meal substrate with an initial moisture content of 50% (w/w) and pH of 5 - 6 incubated for six days at 25°C. A more severe heat treatment (autoclaving vs. refluxing) resulted in higher chitosan extraction yields regardless of the strength of extraction reagents. Fourier transform infrared (FTIR) analysis of the fungal chitosan revealed its degree of deacetylation (DDA) to be between 55% and 60%.

Keywords: Solid-State Fermentation, Fungal Chitosan, Biopolymers, Bioconversion, Agro-Industrial Residues

1. INTRODUCTION

Chitosan is a natural biopolymer that has gained increased attention in recent years due to its interesting and beneficial biological and molecular characteristics, which enable its application in a wide variety of industrial and biotechnological fields. These fields range from agriculture, environmental remediation, pulp and paper coatings and barrier films, food preservation, and cosmetics to more advanced technology emerging fields such as functional materials in flexible printed organic electronic devices, protein separations, drug delivery, and tissue engineering [1]. Chitosan is a heterogeneous polysaccharide containing β-(1-4)-linked 2-amino-2-deoxy-D- glucose (glucosamine) and *N*-acetyl-2-amino-2-deoxy-D-glucose (N-acetyl glucosamine) units, with the former consisting of more than 60 % of the copolymer [2]. It is obtained through deacetylation of chitin, which contains mostly *N*-acetyl glucosamine units. Its unique and beneficial properties enabling its numerous applications include its cationic nature, chelation and ion-binding ability, protein immobilization, film- and gel-forming characteristics, chemical reactivity, amenability to modification, biocompatibility, and antimicrobial activity [3]. These properties are a consequence of its heterogeneous molecular structure and the presence of highly reactive amine groups at the glucosamine residues [4]. Fields and applications that can be served by utilizing chitosan and exploiting its properties include biomedical engineering, food and agriculture, environmental remediation, and polymers, films, fibers and coatings [1] - [3].

Chitosan is traditionally produced by subjecting crustacean shells to severe alkali, acid, and/or enzymatic treatments to isolate chitin, by removing impurities such as proteins, minerals and pigments, and by deacetylating of chitin to chitosan—that is, the removal of the acetyl groups in the N-acetyl glucosamine units comprising the majority of chitin to increase the proportion of glucosamine units in the polymer. Harsh reaction conditions are required due to the complex organization of chitin with other components in the crustacean exoskeleton [5]. Thus, the conventional process results in high labor and processing costs. Additionally, these raw materials are seasonally and geographically limited and have highly variable characteristics, which can lead to inconsistent product yield and molecular characteristics. This is undesirable for high purity applications such as in the pharmaceutical and biomedical fields [3] [6]. On the other hand, chitosan is also found as a major fungal cell wall component particularly of fungal strains belonging to the order Mucorales of the Zygomycetes class [6] [7]. These fungal species produce chitosan directly unlike crustacean-derived chitosan, which has to be produced via isolation and chemical conversion of chitin from the exoskeletons of these organisms. Fungal chitosan is more advantageous than crustacean-based chitosan, because the former can be produced in a controlled environment all year round independent of the seasonal seafood industry [8]. Fungal growth and chitosan productivity and properties can be controlled through the manipulation of fermentation and extraction conditions to give chitosan more consistent and desired physico-chemical properties compared with chitosan obtained from crustacean sources [9]. Moreover, the extraction process for isolating chitosan from fungal cell walls is relatively simpler and milder than obtaining chitosan from crustacean shells, thus producing less waste [9]. In terms of chitosan properties, chitosan produced by most of the identified producer fungal strains has roughly the same degree of deacetylation, and relatively lower viscosities and molecular weight than crustacean-derived chitosan, which improves its potential for processing and modification for its numerous applications particularly in agriculture and biomedical engineering [10].

Agro-industrial crops, by-products and residues can be used as inexpensive carbon sources for growing the fungi, which can help alleviate some environmental concerns on the disposal of these waste materials [11]. These materials are particularly suited for use as substrates and microbial growth support in solid-state fermentation (SSF) processes, which exploit the natural growth conditions of filamentous fungi and their capabilities to biosynthesize and excrete hydrolytic enzymes to directly ferment solid biomass substrates. SSF methods have been applied previously for the production of fungal chitosan using a variety of agro-industrial residues and fungal strains, though not as prevalent as using submerged liquid fermentation (SLF) cultures. Potato chip processing wastes [12] , cottonseed hulls and corn residues [13] , sweet potato [14] , and residues from the manufacture of soymilk and mungbean noodles [15] were utilized as fermentation substrates for growth and chitosan production of various filamentous fungal strains. The filamentous fungus Mucor rouxii, a Zygomycete strain, has been well known to produce high amounts of chitosan using SLF [8] [16] [17]. The possibility of utilizing this fungus for SSF of agro-residues for chitosan production has not been tested previously.

Soybean processing residues, such as soybean meal and soy hulls, are excellent options for use as fermentation media/substrates for the growth of fungi that can produce chitosan due to their abundance and high nutrient content. According to the American Soybean Association, one bushel (60 lbs) of soybeans contain on a mass basis approximately 18.3% oil, 80%

meal, and 1.7% of other residues such as the hulls and mill runs [18].With a total annual soybean production of 89.5 million metric tons, the amount of soymeal production is approximately 71.6 million metric tons while that of soybean hulls is estimated to be 1.5 million metric tons [19]. The conventional use of soybean meal includes supplementation of animal feed due to its high carbohydrate and protein content. However, such use is limited due to the presence of oligosaccharides that are indigestible by non-rumi- nant livestock and poultry [20]. To some extent, soy hulls can be used as animal feed fillers. But due to their relatively low protein and high fiber content, they are not as widely used as soymeal for this application. Instead, they are used as a boiler fuel additive. These characteristics, however, can make both soybean meal and hulls an ideal solid-state fermentation (SSF) substrate for the growth and cultivation of fungi that can be used to produce chitosan. The fungal chitosan product, with all its interesting molecular properties and applications, can be used as a renewable and "green" biobased alternative not only for crustacean chitosan, but also for petroleum based polymers.

The goal of this study is to investigate fungal chitosan production via solid-state fermentation of soybean processing residues using Mucor rouxii. Experiments were conducted to determine the optimum cultivation and extraction conditions for maximum chitosan yield. The physicochemical characteristics of the produced chitosan were then determined using analytical methods.

2. EXPERIMENTAL

2.1. Materials

Mucor (Amylomyces) rouxii ATCC 24905, the fungal strain used in this study, was purchased from the American Type Culture Collection (ATCC). The freeze-dried culture was rehydrated with sterile deionized (DI) water and stock cultures were generated by cultivation of the rehydrated fungal cells on potato dextrose agar (PDA) slants at 25°C for five days. The stock cultures were then stored and maintained at 4°C until use. Soybean meal and hulls provided by three local soybean processors was tested as a solid-state fermentation substrate. These were then stored in airtight plastic containers in a refrigerator at 4°C until use. Representative samples of the soybean meal and hulls from the different sources were sent to Northland Labs, Inc. (Northbrook, IL) for analysis of carbohydrates, proteins, moisture, and ash using standard AOAC methods. The soymeal were analyzed and used as received while whole soy hulls were homogenized to granular form using a Waring blender. Composite samples consisting of equal amounts from each of the three suppliers were used for the fermentation process. Commercial grade sodium hydroxide pellets and Optima grade glacial acetic acid were purchased from Fisher Scientific (Pittsburgh, PA, USA). Low molecular weight chitosan purchased from Sigma Aldrich (St. Louis, MO, USA) was used as the reference chitosan for fungal chitosan characterization. The chemicals and standards were used as received.

2.2. Fungal SSF Cultivation and Chitosan Production

The fungal spore suspension inocula used in the SSF cultures were prepared as follows: ten milliliters of sterile DI water was transferred aseptically into a M. rouxii ATCC 24905 stock PDA slant. The slant tube was then gently inverted ten times to suspend the fungal spores in water. After this, 2-mL of the spore suspension was immediately transferred aseptically into a new PDA plate. The spore suspension was distributed onto the PDA surface using a sterile disposable L-spreader. The inoculated plates were incubated at 30°C for five days. After the incubation period, the entire mycelial growth on the agar surface was scraped using a sterile metal spatula and transferred into a 250-mL flask containing 100 mL of sterile DI water. The mixture was gently shaken to suspend the fungal spores in the water. After this, the suspension was filtered through a sterile coarse filter paper into a new sterile collecting flask to remove the large agar and fungal debris.

The soybean residue substrates were prepared and inoculated as a single batch in a sterile autoclavable plastic bag. Sterile DI water was added to the filtered fungal spore suspension to attain the desired initial moisture content setting (30%, 50%, or 70% w/w). The pH of the sterile culture dilution water was previously set according to the desired initial pH setting of the substrate (pH 4, 5, or 6). The wet soymeal mash was shaken and mixed thoroughly inside the bag and then distributed in 10-gram portions (wet basis) in sterile polystyrene deep petri dishes. The petri dishes containing the soybean meal fungal cultures were then incubated at the desired temperature setting (25°C, 30°C, or 35°C) for the set cultivation period (3, 6, 9, 12, or 15 days). All experimental treatments were tested in triplicate petri dish cultures.

2.3. Measurement of Fungal Growth

The extent of growth of M. rouxii in the soybean meal cultures was determined indirectly by spectrophotometric measurement of glucosamine in samples of the fermented soymeal substrate according to a method published elsewhere [21].

2.4. Chitosan Extraction

Fungal chitosan produced in the soybean residue cultures was extracted and isolated using an approach based on previously published methods [12] [22] [23]. The entire contents of each petri dish soy residues culture were processed first by extraction of residual proteins and other alkali soluble materials with NaOH solution with the desired concentration (1 M or 46% w/v) and volume-to-solids ratio (30:1 or 10:1, mL/g dry mass). Two different heated extraction methods were also tested: autoclaving at 121°C, 15 psi for 15 min or refluxing at 46°C for 8 h. Alkali insoluble materials (AIMs) were recovered by vacuum filtration and washed to neutral pH with DI water. The washed AIMs were transferred and weighed into tared glass petri dishes and dried in an oven at 60°C overnight or until a constant weight was achieved. The dry weight and moisture content of the AIMs were then calculated using the wet and dry weights measured. Chitosan was extracted from the AIMs with 2% (v/v) aqueous acetic acid. Two different volume-to-solids ratios (40:1 or 10:1, mL/g dry mass AIMs) and heat treatments (autoclaving at 121°C, 15 psi for 15 min or refluxing at 95°C for 8 h) were tested. After the extraction period, the slurries were transferred into tared centrifuge tubes and centrifuged at 3000 RPM for 20 minutes, after which the insoluble fraction was discarded. The pH of the supernatant was adjusted to pH 10 by the dropwise addition of 4 M NaOH in order to precipitate the chitosan out of solution. The suspension was then centrifuged at 3000 RPM for 20 minutes, after which the supernatant was discarded. The chitosan pellets were then washed successively with DI water, 95% ethanol, and acetone; frozen at −20°C overnight; freeze-dried; and weighed to calculate the chitosan yield (% w/w dry mass of initial soymeal substrate).

2.5. Characterization of Fungal Chitosan by FTIR Spectroscopy

The Fourier transfer infrared (FTIR) spectra of the product fungal chitosan from soymeal fermentation were obtained by analyzing solid powder samples with an Alpha-E FTIR spectrometer (Bruker Optics Inc., Billerica, MA, USA). Sixty-four scans were conducted per sample between wavelength number of 4000 - 700 cm^{-1} at a resolution of 2 cm^{-1}. The degree of deacetylation (DDA) of the chitosan samples was estimated according to the method of [24]. The absorbance peak at wavelength number ≈ 1560 cm^{-1} (amide II) was used as the characteristic signal while the peak at wavelength number ≈ 1070 cm^{-1} (C-O bond stretching) was used as the reference signal. DDA values were calculated using the absorbance ratio

A_{1560}/A_{1070} and the corresponding calibration curve relating the absorbance ratio to DDA.

3. RESULTS AND DISCUSSION

3.1. Characterization of Soybean Processing Residues

Following analysis, the soybean meals were found to contain almost 50% by weight proteins, which is roughly five times that of the soy hulls. In terms of the carbohydrate content, soy hulls had around 79% by weight of carbohydrates, which is almost twice that of the carbohydrates in soymeal. Both meal and hulls contain similar residual fats while the meal has slightly higher moisture and ash contents than the hulls. The high carbohydrates and protein content of these residues could serve as carbon and nitrogen sources for direct fungal growth under solid-state cultivation and support the rationale behind the selection of these agricultural residues as fermentation substrates for chitosan production.

3.2. Effect of Substrate Type

Pure soybean meal, soy hulls, and 1:1 meal-hull mixtures were tested as SSF substrates for growth and chitosan production by M. rouxii ATCC 24905. The results (Figure 1) show that fungal growth (as measured by glucosamine in the culture) under solid-state cultivation conditions in a pure soymeal culture or in a 1:1 meal/hull mixture were not significantly different, while growth in pure soy hull was the lowest. This result was expected as the substrates containing the soymeal contained higher protein (nitrogen) contents than pure soy hulls, resulting in the increased biosynthesis of fungal mycelial materials. On the other hand, chitosan yield was highest from the pure soymeal cultures (1.63% ± 0.16% w/w dry substrate) while those from the pure soy hull (1.26% ± 0.15%) and 1:1 meal/hull mixtures (0.97 ± 0.15) were not significantly different. The chitosan yield displayed a positive correlation with fungal growth levels for both the pure soymeal and the 1:1 meal:hull mixtures. Being a fungal cell wall component, the production of chitosan is considered to be growth-associated (i.e., production rates correlate with growth rate) under carbon- and nutrient-rich condition. However, under a condition of high carbohydrate-to-protein proportion such as in soy hull cultures, the chitosan yield displayed an opposite trend relative to fungal growth between the pure soy hull and 1:1 meal/hull cultures. It appears that despite the lower fungal growth level in the soy hull culture, the fungal cell wall mass generated contained a relatively higher composition of chitosan.

Based on these results, pure soymeal substrates were utilized in the succeeding fermentation tests.

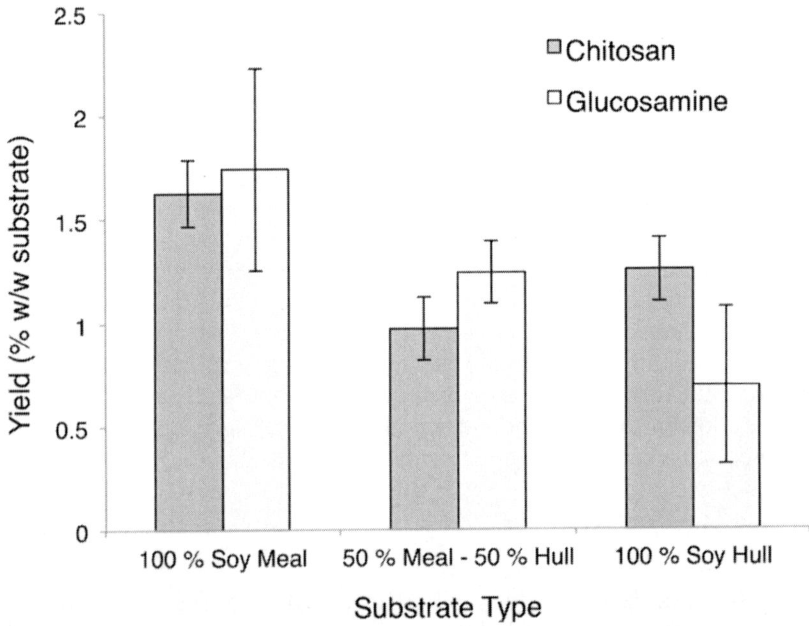

Figure 1. Growth and chitosan production by Mucor rouxii ATCC 24905 on different soybean processing residues. Incubation temperature—30°C, initial pH unadjusted, initial mo- isture content—50% (w/w), and cultivation time—six days.

3.3. Effect of Incubation Period

The effect of the length of the cultivation period on fungal growth and chitosan yield using pure soymeal SSF cultures is shown in Figure 2. All other fermentation variables were kept constant at the following levels: 50% (w/w) initial substrate moisture content, 30°C incubation temperature, and unadjusted initial substrate pH. The fungal growth pattern (as glucosamine) showed a possible diauxic pattern, with a first stationary period attained between 6 to 12 days of cultivation followed by a sharp increase at day 15. It is possible that during the first stationary period, the fungal cells utilized the readily available adhered carbon and nutrients in the soymeal particle surfaces for partial growth and assembly of hydrolytic enzymes (cellulase, proteases, etc.). During the stationary period, the synthesized enzymes were utilized by the fungal cells for the degradation of the polymeric carbohydrates and proteins into more readily assimilable carbon and nutrients, which were then made available for a succeeding exponential growth phase. Despite the observed increase in growth, the results show that maximum chitosan yield was achieved after only six days of cultivation after which chitosan yields fluctuated (not significantly different). Previous studies have indicated that during the attainment of the stationary growth phase, the fungal cell walls may have begun to lose their fluidity and potential for chitosan formation due to consolidation by chitin crystallization and bond formation with other cell wall components, which could lead to reduced extraction yields [22] [25]. Thus, six days was considered to be the optimum length of cultivation period for maximum chitosan yield and was maintained in the succeeding experiments.

3.4. Effect of Initial Moisture Content

Figure 3 shows the effect of the initial substrate moisture content on growth and chitosan yield of M. rouxii ATCC 24905 in SSF soymeal cultures. In these experiments, the incubation period was set at six days based on the optimum level obtained in the prior experiments, the incubation temperature at 30°C, and initial substrate pH was not adjusted. The lowest moisture level tested (30% by wt.) resulted in the lowest fungal biomass growth and chitosan production. However, fungal growth and chitosan yields were not significantly different between the 50% and 70% initial moisture content treatments (chitosan yield is between 1% and 2% w/w dry substrate). The moisture content of the substrate is a critical parameter related to the water activity of the microorganisms in SSF processes where moisture levels are relatively low. The fungal cells need an appropriate level of water in which to exercise the various biochemical reactions for the enzymatic breakdown and uptake of the biomass substrate. The results in this study indicated that M. rouxii ATCC 24905 was able to tolerate a lower substrate moisture content of 50% for growth and chitosan production in SS unlike other fungal strains previously studied for chitosan production via SSF, that were grown in 60% - 70% moisture substrates [12] [22] [26]. Using less water to moisten the soybean meal substrate could mean less dilution of nutrients and growth factors due to water addition in the culture. Furthermore, using soymeal cultures with 70% initial moisture content appeared to produce high variability in the results. Thus, the initial soymeal substrate moisture content of 50% was considered to be the optimum level for maximizing fungal growth and chitosan production using the strain M. rouxii ATCC 24905.

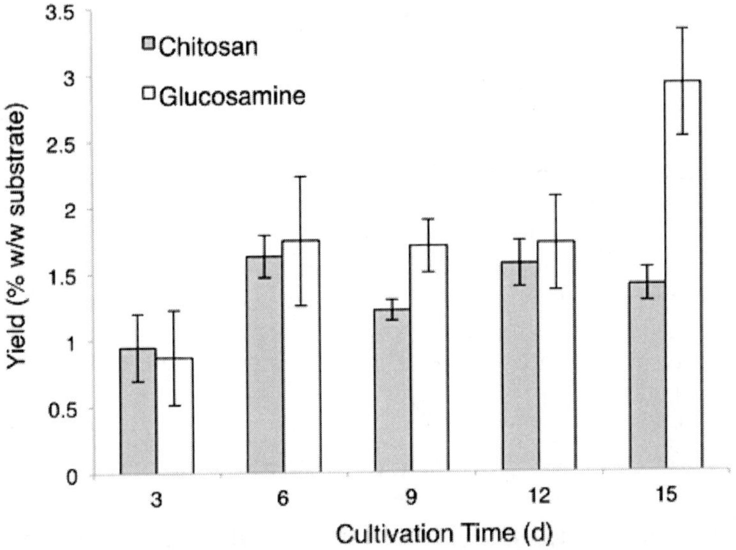

Figure 2. Effect of cultivation time on growth and chitosan yield by Mucor rouxii ATCC 24905 on pure soymeal substrate. Incubation temperature—30°C, initial pH unadjusted, initial moisture content—50% (w/w), and cultivation time—six days.

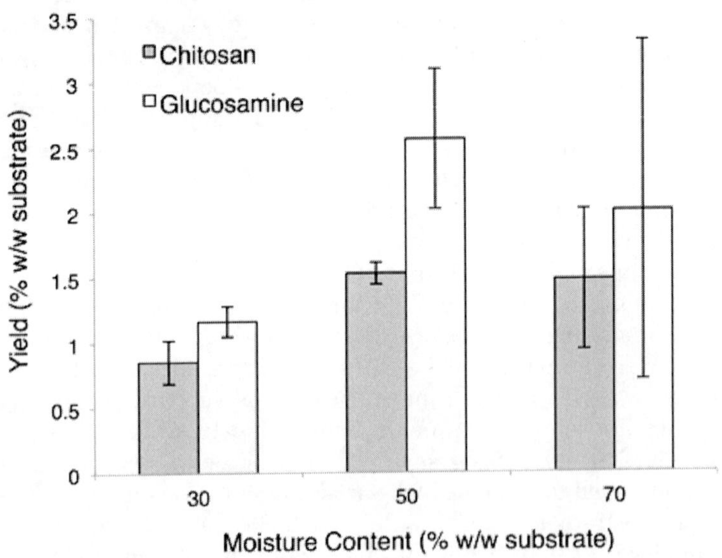

Figure 3. Effect of initial substrate moisture content on growth and chitosan production by Mucor rouxii ATCC 24905 on pure soymeal substrate. Incubation temperature—30°C, initial pH unadjusted, and cultivation time—six days.

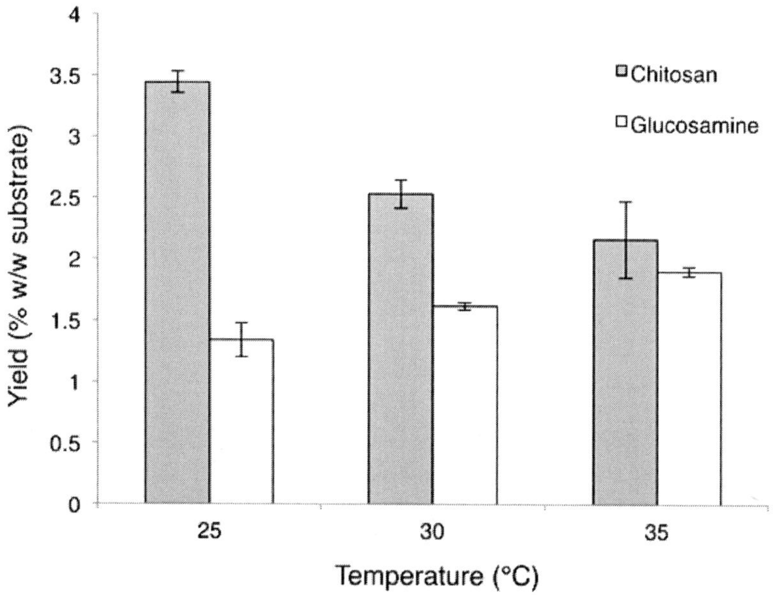

Figure 4. Effect of incubation temperature on growth and chitosan production by Mucor rouxii ATCC 24905 on pure soymeal substrate. Initial pH unadjusted, initial moisture content—50% (w/w), and cultivation time—six days.

3.5. Effect of Incubation Temperature

Figure 4 shows the effect of incubation temperature on fungal growth and chitosan production in solid-state cultures of M. rouxii ATCC 24905. In these experiments, the incubation period was set at six days and the initial moisture content of the soymeal substrate was 50% (w/w) based on the optimum levels obtained in the prior experiments. The initial substrate pH was not adjusted. The experiments demonstrated opposite trends of growth and chitosan yield with increasing incubation temperature. The highest chitosan yield of around 3.44% ± 0.09% (w/w of total solid substrate) was achieved at 25°C and decreased considerably at incubation temperatures of 30 and 35°C. On the other hand, total fungal growth was shown to increase as the incubation temperature increased. A higher incubation temperature appeared to favor fungal growth in the solid soymeal culture but results in lower chitosan yields. A possible explanation is that the fungal cell walls might be losing their fluidity in favor of a denser and more consolidated structure dominated by chitin, which results in reduction of chitosan composition and/or extractability at incubation temperatures higher than 25°C. This could be a defense mechanism by the fungal cells to counteract the potentially detrimental effects of high temperatures on fungal growth and metabolism. On the operational standpoint, incubation at lower temperatures could lead to lower costs attributed to heating and humidification of fermentation chambers but could imply the necessary fermentation bed cooling mechanisms to counteract the buildup of temperature gradients across the substrate bed thickness.

3.6. Effect of Initial Substrate pH

Figure 5 shows the effect of the initial pH of the moistened soybean meal substrate on fungal growth and chitosan production in solid-state cultures of M. rouxii ATCC 24905. In these experiments, the incubation period was set at six days, the incubation temperature at 30°C, and initial substrate moisture content was at 50% (w/w) according to the results of the prior experiments. The results obtained demonstrate that there are no significant differences in terms of fungal growth and chitosan yields among the initial substrate pH levels investigated. It seems logical to select the unadjusted pH as the best condition from the operational cost standpoint, which would not require prior treatment of the soymeal substrate with acid and/or alkali for pH adjustment. However, an initial pH setting between 5 and 6 might be ideal for helping to minimize the variability of the product chitosan yields and characteristics. Additional tests will be conducted to confirm this hypothesis.

3.7. Effect of Extraction Conditions

In addition to testing the effect of the fermentation variables on fungal growth and chitosan yield, different extraction methods were also tested to determine their effects on chitosan recoveries. Different methods were selected from literature. The results of these tests are summarized in Figure 6. In these experiments, the fermentation conditions were kept constant at six days cultivation period, 50% initial substrate moisture content, unad justed initial substrate pH, and 30°C incubation temperature. Figure 6 shows that chitosan recoveries differed significantly among the different methods investigated with regards to the concentration of reagents used as well as the severity of the heat treatment method used. For instance, Methods 1 and 2 utilized the same reagent concentrations but with differing volume ratios, while Method 2 used a lower volume of reagent. Although a higher volume-to-mass ratio for both the base and acid reagents (Method 1) resulted in a slightly higher chitosan recovery, the results were not significantly different. The same case was observed with Methods 3 and 4, which used a higher base concentration but with heating and refluxing instead of autoclaving. Compared with Methods 1 and 2, Methods 3 and 4 resulted in a significantly lower chitosan recovery. This finding implies the importance of the base-assisted deproteinization step for maximizing the chitosan yield. However, Figure 5 also shows that when a more severe reaction condition (i.e., autoclaving) was employed for the acetic acid extraction of chitosan step, the chitosan recovery was significantly higher than all the other extraction methods tested. Despite this, a more thorough economic and cost-benefit study should be made to justify the use of more severe extraction conditions and correlations must be made with the effect of fermentation conditions.

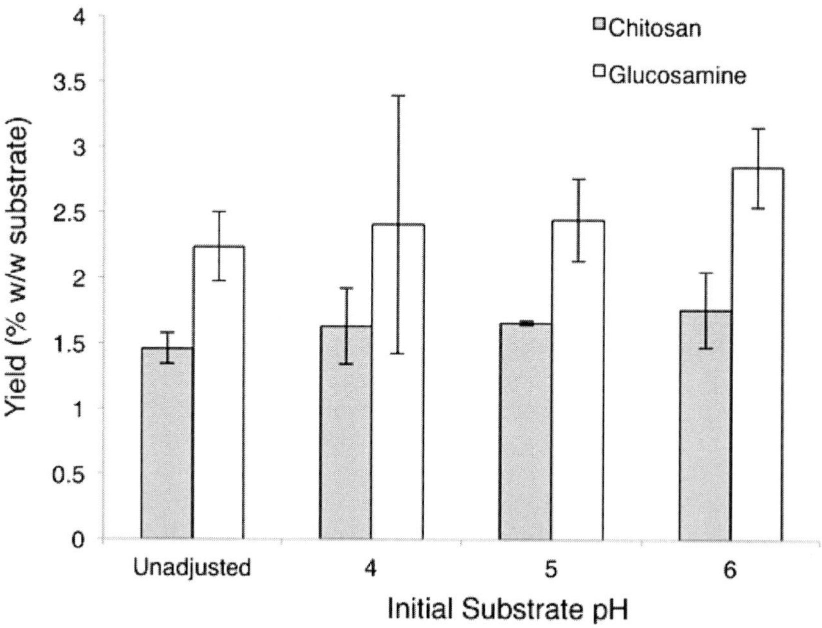

Figure 5. Effect of initial substrate pH on growth and chitosan production by Mucor rouxii ATCC 24905 on pure soymeal substrate. Incubation temperature?30°C, initial moisture content—50% (w/w), and cultivation time—six days.

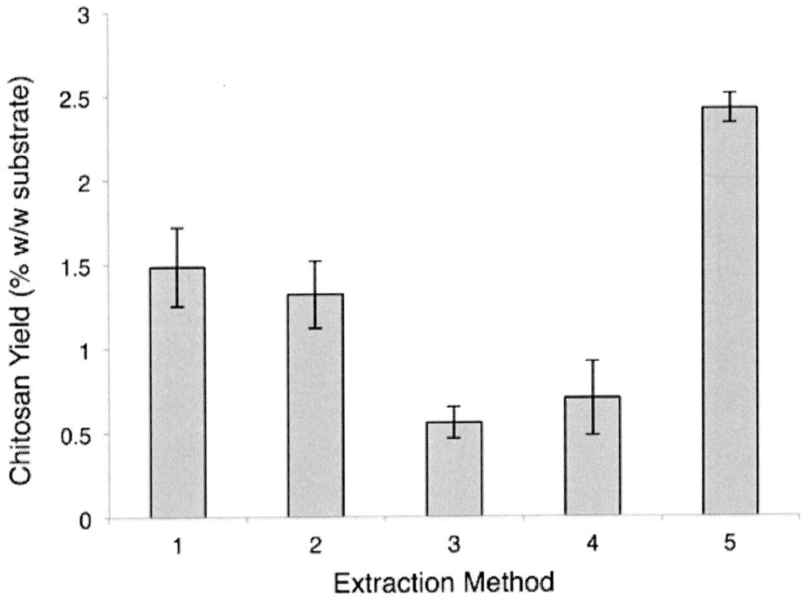

Figure 6. Effect of extraction method on growth and chitosan yield by Mucor rouxii ATCC 24905. Method 1—1 M NaOH at 121°C for 20 min (30 mL per g substrate) followed by 2% (v/v) acetic acid (40 mL per g substrate) at 95°C with refluxing for 8 h. Method 2—1 M NaOH at 121°C for 20 min (10 mL per g substrate) followed by 2% (v/v) acetic acid (10 mL per g substrate) at 95°C with refluxing for 8 h. Method 3—46% (w/w) NaOH at 46°C for 13 h (30 mL per g substrate) followed by 2% (v/v) acetic acid (40 mL per g substrate) at 95°C with refluxing for 8 h. Method 4—46% (w/w) NaOH at 46°C for 13 h (10 mL per g substrate) followed by 2% (v/v) acetic acid (10 mL per g substrate) at 95°C with refluxing for 8 h. Method 5—1 M NaOH at 121°C for 20 min (30 mL per g substrate) followed by 2 % (v/v) acetic acid (40 mL per g substrate) at 121°C for 20 min.

3.8. Characterization of Fungal Chitosan Using FTIR

The chitosan extracted from Mucor rouxii ATCC 24905 fungal cultures in soybean meal was characterized by FTIR spectroscopy and compared with commercially available chitosan. The purpose of this analysis was to confirm the identity of the extracted chitosan pellets and to estimate the degree of deacetylation (DDA) of the fungal chitosan product, an important parameter that influences the activity of the chitosan polymer for its intended applications. The DDA of the fungal chitosan produced from solid-state fermentation of soybean meal was estimated using the absorbances of the FTIR signals at approximate wavelength numbers of 1560 and 1070

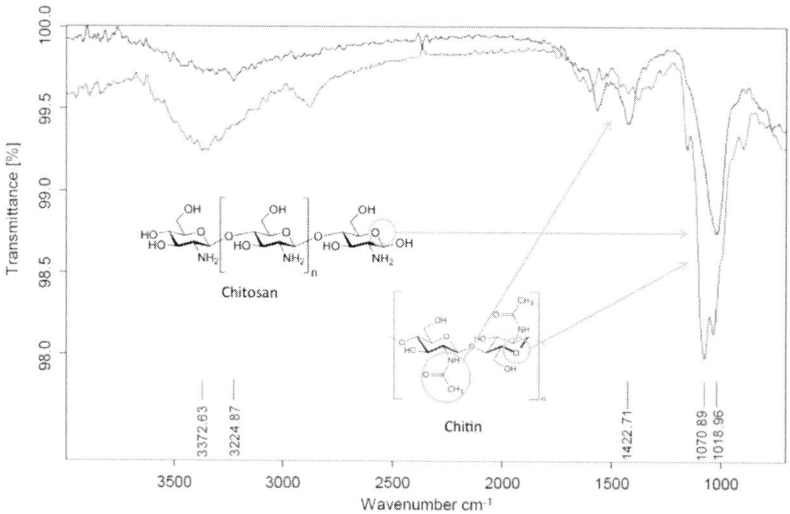

Figure 7. FTIR spectra of commercial chitosan (red) and Mucor rouxii ATCC 24905 chitosan from fermentationofpure soymeal substrate (blue).

cm^{-1}(Shigemasa et al., 1996)·As shown in Figure 7, the signal at 1560 cm^{-1} corresponds to the characteristic band of the amide group (Amide II) in the N-acetylglucosamine residue of chitin, which accounts for more than 90% of the monomer residues in the polymer chain. Although definitions vary among different sources, the chitin chain is identified as chitosan if the polymer chain contains a significant proportion of the deacetylated glucosamine residue. Thus the intensity of the band at 1560 cm^{-1} is expected to decrease as the DDA increases. The signal at 1070 cm^{-1}corresponding to C-O bond stretching in the chitin/chitosan monomers was used as the reference band. Using the calibration curve developed by Shigemasa et al., the DDA of representative samples of fungal chitosan from different fermentation and extraction conditions was estimated to be between 55% - 60%.

4. CONCLUSION

Solid-state fermentation (SSF) of soybean processing residues using the fungus Mucor rouxii ATCC 24905 for chitosan production was for the first time investigated. The optimum SSF conditions for maximum chitosan yields were determined. FTIR analysis of the fungal chitosan derived from soybean meal fermentation confirmed its molecular identity and estimated its degree of deacetylation to be around 55% - 60%. Extraction conditions such as the concentration of acid and base reagents and severity of heat treatment were also shown to influence the chitosan yield.

ACKNOWLEDGEMENTS

Andro Mondala, Ramea Al-Mubarak, James Atkinson, Shaun Shields, Brian Young, Yurguen Dos Santos Senger, Jan Pekarovic (USB) for providing the funding for this research (USB Project # 1440-512-5283): Archer Daniels Midland, Zeeland Farms, and CHS for supplying the soybean meal and hull samples; and Dr. Linda Kim-Habermehl of USB/Omni Tech International, Ltd. for her valuable inputs on the project.

REFERENCES

1. Castro, S.P.M. and Paulín, E.G.L. (2012) Is Chitosan a New Panacea? Areas of Application. In: Karunaratne, D.N., Ed., The Complex World of Polysaccharides, ISBN: 978-953-51-0819-1, InTech. http://www.intechopen.com/books/the-complex-world-of-polysaccharides/is-chitosan-a-new-panacea-areas-of-application
2. Kaur, S., and Dhillon, G.S. (2014) The Versatile Biopolymer Chitosan: Potential Sources, Evaluation of Extraction Methods and Applications. Critical Reviews in Microbiology, 2, 155-175. http://dx.doi.org/10.3109/1040841X.2013.770385
3. Badawy, M.E.I. and Rabea, E.I. (2011) A Biopolymer Chitosan and Its Derivatives as Promising Antimicrobial Agents against Plant Pathogens and Their Applications in Crop Protection. International Journal of Carbohydrate Chemistry, 2011, Article ID: 460381, 29 p. http://dx.doi.org/10.1155/2011/460381
4. Dutta, P.K., Dutta, J., and Tripathi, V.S. (2004) Chitin and Chitosan : Chemistry, Properties and Applications. Journal of Scientific & Industrial Research, 63, 20-31. http://nopr.niscair.res.in/handle/123456789/5397
5. Bruck, W.M., Slater, J.W., and Carney, B.F. (2011) Chitin and Chitosan from Marine Organisms. In: Kim, S.K., Ed., Chitin, Chitosan, Oligosaccharides, and Their Derivatives: Biological Activities and Applications, CRC Press, Boca Raton, 11-24.
6. Knezevic-Jugovic, Z., Petronijevic, Z., and Smelcerovic, A. (2011) Chitin and Chitosan from Microorganisms. In: Kim, S.K., Ed., Chitin, Chitosan, Oligosaccharides, and Their Derivatives: Biological Activities and Applications, CRC Press, Boca Raton, 25-36.
7. Peter, M.G. (2002) Chitin and Chitosan in Fungi. Biopolymers Online, 6, 123-157. http://dx.doi.org/10.1002/3527600035.bpol6005
8. White, S.A., Farina, P.R. and Fulton, I. (1979) Production and Isolation of Chitosan from Mucor rouxii. Applied and Environmental Microbiology, 38, 323-328.
9. Tan, S.C., Tan, T.K., Wong, S.M. and Khorb, E. (1996) The Chitosan Yield of Zygomycetes at Their Optimum Harvesting Time. Carbohydrate Polymers, 30, 239-242. http://dx.doi.org/10.1016/S0144-8617(96)00052-5
10. Pochanavanich, P. and Suntornsuk, W. (2002) Fungal Chitosan Production and Its Characterization. Letters in Applied Microbiology, 35, 17-21. http://dx.doi.org/10.1046/j.1472-765X.2002.01118.x
11. McGahren, W.J., Perkinson, G.A., Growich, J.A., Leese, R.A. and Ellestad, G.A. (1984) Chitosan by Fermentation. Process Biochemistry, 19, 88-90.

12. Kleekayai, T. and Suntornsuk, W. (2010) Production and Characterization of Chitosan Obtained from Rhizopus oryzae Grown on Potato Chip Processing Waste. World Journal of Microbiology and Biotechnology, 27, 1145-1154. http://dx.doi.org/10.1007/s11274-010-0561-x
13. Wang, W., Du, Y., Qiu, Y., Wang, X., Hu, Y., Yang, J., Cai, J. and Kennedy, J.F. (2008) A New Green Technology for Direct Production of Low Molecular Weight Chitosan. Carbohydrate Polymers, 74, 127-132. http://dx.doi.org/ 10.1016/j.carbpol.2008.01.025
14. Nwe, N. and Stevens, W.F. (2002) Production of Fungal Chitosan by Solid Substrate Fermentation Followed by Enzymatic Extraction. Biotechnology Letters, 24, 131-134. http://dx.doi.org/10.1023/A:1013850621734
15. Suntornsuk, W., Pochanavanich, P. and Suntornsuk, L. (2002) Fungal Chitosan Production on Food Processing By- Products. Process Biochemistry, 37, 727-729. http://dx.doi.org/10.1016/S0032-9592(01)00265-5
16. Bartnicki-Garcia, S. and Nickerson, W.J. (1962) Isolation, Composition, and Structure of Cell Walls of Filamentous and Yeast-Like Forms of Mucor rouxii. Biochimica et Biophysica Acta, 58, 102-119. http://dx.doi.org/10.1016/0006-3002(62)90822-3
17. Chatterjee, S., Adhya, M., Guha, A.K. and Chatterjee, B.P. (2005) Chitosan from Mucor rouxii: Production and Physico-Chemical Characterization. Process Biochemistry, 40, 395-400. http://dx.doi.org/10.1016/j.procbio.2004.01.025
18. American Soybean Association (2014) SoyStats 2014. Soybean Facts. http://soystats.com/facts/
19. American Soybean Association (2014) SoyStats 2014. US Yield and Production: Production History. http://soystats.com/u-s-yield-production-production-distory/
20. Zhang, L.Y., Li, D.F. and Qiao, S.Y. (2003) Effects of Stachyose on Performance, Diarrhea Incidence and Intestinal Bacteria in Weanling Pigs. Archives of Animal Nutrition, 57, 1-10. http://dx.doi.org/10.1080/0003942031000086662
21. Roopesh, K., Ramachandran, S., Nampoothiri, K.M., Szakacs, G. and Pandey, A. (2006) Comparison of Phytase Production on Wheat Bran and Oilcakes in Solid-State Fermentation by Mucor racemosus. Bioresource Technology, 97, 506-511. http://dx.doi.org/10.1016/j.biortech.2005.02.046
22. Crestini, C., Kovac, B. and Giovannozzi-Sermanni, G. (1996) Production and Isolation of Chitosan by Submerged and Solid-State Fermentation from Lentinus edodes. Biotechnology and Bioengineering, 50, 207-210. http://dx.doi.org/ 10.1002/bit.260500202
23. Rane, K.D. and Hoover, D.G. (1993) Production of Chitosan by Fungi. Food Biotechnology, 7, 11-33. http://dx.doi.org/10.1080/08905439309549843
24. Shigemasa, Y., Matsuura, H., Sashiwa, H. and Saimoto, H. (1996) Evaluation of Different Absorbance Ratios from Infrared Spectroscopy for Analyzing the Degree of Deacetylation in Chitin. International Journal of Biological Macromolecules, 18, 237-242. http://dx.doi.org/10.1016/0141-8130(95)0179-3
25. Nwe, N., Furuike, T. and Tamura, H. (2010) Production of Fungal Chitosan by Enzymatic Method and Applications in Plant Tissue Culture and Tissue Engineering : 11 Years of Our Progress, Present Situation and Future Prospects. In: Elnashar, M., Ed., Biopolymers, ISBN: 978-953-307-109-1, InTech. http://dx.doi. org/10.5772 /10261 http://www.intechopen.com/books/biopolymers/production-of-fungal-chitosan-by-enzymatic-method-and-applications-in-plant-tissue-culture-and-tissu
26. Khalaf, S.A. (2004) Production and Characterization of Fungal Chitosan under Solid-State Fermentation Conditions. International Journal of Agriculture & Biology, 6, 1033-1036. http://www.fspublishers.org/published_papers/4088_..pdf

CHAPTER 9

Iron Phosphate Glasses: Structure Determination and Displacement Energy Thresholds, Using a Fixed Charge Potential Model

Kitheri Joseph 1, Kenny Jolley, Roger Smith *

Mathematical Sciences, Loughborough University, Loughborough LE11 3TU, UK

ABSTRACT

Iron phosphate glass is a versatile matrix for the immobilisation of various radioactive elements found in high-level nuclear waste (HLW). Quenched glass structures of iron phosphate glasses with Fe/P ratios of 0.33, 0.67 and 0.75 and with a composition of 40 mol% Fe_2O_3 and 60 mol% P_2O_5, with 4% and 17% Fe^{2+} ion concentrations were generated using molecular dynamics and the threshold displacement energies calculated. In the minimum energy structures, we found that in nearly all cases the P atoms were 4-fold coordinated. The potential energy per atom increased with increasing concentration of Fe^{2+} ions with similar Fe/P ratio, suggesting that decreasing the Fe^{2+} content is a stabilising factor. The average bond distances between Fe^{2+}–O, Fe^{3+}–O, P–O and O–O were calculated as 2.12, 1.88, 1.5 and 2.5 Å respectively. The threshold displacement energy (E_d) was found to be dependent upon the ion specie, less for Fe^{2+} ions compared to Fe^{3+} ions, and was overall slightly lower than that determined for borosilicate glass.

Keywords: Iron phosphate glass Computer simulation Structure Displacement energy thresholds

1. INTRODUCTION

Phosphate glasses, due to their favourable properties such as: reasonably low liquid and glass transition temperatures, low viscosity, high thermal expansion coefficient, high electrical conductivity and high ultraviolet transmission, find application in a wide range of fields. For example, phosphate glasses are used in lasers [1], solid electrolytes [2], bio-medical devices [3] and nuclear waste immobilisation [4]. A good review up to the year 2000 is given in [5]. According to the literature [6], in spite of the good glass forming characteristics of phosphate glasses, their relatively poor chemical durability limits their application, especially in the field of nuclear waste immobilisation. However, a new group of phosphate glasses, iron phosphate glass, is being considered as a promising matrix for the immobilisation of high-level waste rich in alkali oxides, sulphates and chrome oxides [7–10]. Iron phosphate glass and its waste form containing simulated fast reactor waste were synthesised, characterised and reported by us earlier [11,12]. Higher waste loading, better chemical durability and better corrosion resistance [13,14] are certain promising features of iron phosphate glass compared to other phosphate glasses. Among the various compositions of iron phosphate glass, the one with 40 mol% Fe_2O_3–60 mol% P_2O_5 (referred as IPG in the present paper) is found to be more chemically durable [15–17]. It also has the ability to accommodate large amounts of certain nuclear wastes, especially those that are not well suited for borosilicate glasses [16]. The better chemical durability of iron phosphate glass is attributed to the presence of more hydration resistant Fe–O–P bonds compared to P–O–P bonds available in other phosphate glasses [16]. The synthesis, characterisation and experimental determination to elucidate the structure of a wide variety of iron phosphate glasses are available in the literature [7,17–20]. However, the available literature on the structural modelling of iron phosphate glasses is limited [21,22]. This is because, the structure of iron phosphate glasses, not only depends on composition, quenching temperature of the melt and quenching technique [23], it depends also on the concentration of Fe^{2+} and Fe^{3+} ions in the glass.

The iron phosphate glass with the composition 40 mol% Fe_2O_3–60 mol% P_2O_5, has been reported by various researchers to have different ratios of Fe^{2+}/Fe [7,12,24,25]. The concentration of Fe^{2+}/Fe in glass with the same atomic composition varies from 4 to 20%, while the density varies between 2.9 and 3.0 g cm^{-3} with the uncertainty in density ranging from ± 0.005 to 0.02 g cm^{-3}[7,24]. The concentrations of Fe^{2+}/Fe and density of IPG reported by us [12] were 4% and 2.9 g cm^{-3} respectively. The promising composition of 40 mol% Fe_2O_3–60 mol% P_2O_5 (IPG) is found with the varying density depending on the concentration of Fe^{2+} in the glass. It is evident from the literature [7,24,25] that as the concentration of Fe^{2+} in IPG increases, the density also increases for the same atomic composition. It has been shown experimentally by Mössbauer spectroscopy [24], that as the melting temperature of IPG increased from 1423 to 1673 K, the density of the glass increased along with the increase in concentration of Fe^{2+}. Thus, it becomes essential to model the structure of iron phosphate glasses specific to the composition and Fe^{2+}/Fe ratio.

It is well known that amorphous structures are more difficult to handle computationally than crystalline lattices since many different atomic configurations are possible. Therefore averaging over many different structures is very important. It is also essential to have good inter-atomic potentials that describe the amorphous systems. A recent paper [21] reported one of the first studies of iron phosphate glasses using fixed charge potentials.

Other authors have also developed potentials of a form that could be used to model phosphate glasses. However potentials that involve shell models such as that by Ainsworth et al. [26] are not really suitable for radiation cascade studies and a previous potential formulation by Pedone et al. [27], which also included the capability to model Fe–P–O systems had not been explicitly tested on these glasses. Since a key aim of our research programme is to investigate radiation effects in phosphate glasses, the potential developed in [21] was chosen as the underlying model for the work.

Thus we modify the potential in [21] so that it is suitable for radiation studies, compare the results with the previous work, and then test other compositional structures with various Fe/P ratio and Fe^{2+}/Fe ratios that we have produced experimentally. Furthermore, we determine the threshold displacement energies as the first step towards the investigation of collision cascades in these systems.

2. METHODOLOGY

Molecular dynamics simulation studies were carried out to model the structure of iron phosphate glasses. The code (LBOMD) has been used over many years for radiation damage studies, see e.g. [28] for work involving radiation damage in spinels. The two body interactions, i.e. Fe–O, P–O and O–O were modelled using a Buckingham rigid ion potential (Eq. (1)), together with a Coulomb term to model the long-range interactions between ionic charges.

$$V(r_{ij}) = \frac{q_i q_j}{4\pi\varepsilon_0 r_{ij}} + A_{ij}\exp\left[\frac{-r_{ij}}{\rho_{ij}}\right] - \frac{C_{ij}}{r_{ij}^6} \quad (1)$$

The subscripts, i and j, refer to each ion; q_i and q_j are the ion charges, r_{ij} is the inter-ionic distance and ε_0 is the permittivity of free space. The parameters: A_{ij}, ρ_{ij} and C_{ij} for each ionic bond are given in reference [21].

In addition to the two-body terms, three body terms were also used to control the local bond angles. This three-body potential is especially very important for P–O bonds due to their ionic–covalent nature similar to that of Si–O bonds found in silicate glasses [29].

The original authors [21] used a harmonic three-body potential, $V(\theta_{ijk}) = 1/2\ k(\theta_{ijk} - \theta_0)^2$, where k = 3.5 eV and $\theta_0 = 109.47°$ for O–P–O, and k = 3.0 eV and $\theta_0 = 135.5°$ for P–O–P. This function presents some computational difficulties for radiation damage studies, as it requires the computation of the derivative of arccos(θ) to obtain the forces. This derivative is infinite when θ = 180°. Furthermore, there is no smooth cut-off to zero as the atomic separation increases.

Among the various analytical forms of three-body potentials listed in the literature [30–32], the three-body Stillinger–Weber potential [31] (Eq. (2)) was chosen instead to model the three body terms. This is more suitable to use in radiation damage simulations and the construction of a potential that can be used in such simulations is also a key aim of the work.

$$\phi_3(r_{ij}, r_{ik}, \theta_{jik}) = \lambda\ \exp\left[\left(\frac{\gamma}{r_{ij}-r_c} + \frac{\gamma}{r_{ik}-r_c}\right)\left(\cos\theta_{jik} - \cos\theta_0\right)^2\right] \quad (2)$$

For an atom triplet (j–i–k), r_{ij} and r_{ik}, are the two internal atomic separations, and θ_{jik}, is the bond angle at the central atom 'i'. θ_0 represents the angle towards which the angle θ_{jik} is constrained and r_c is the cut-off radius beyond which the three-body terms do not apply. λ, γ and θ_0 are adjustable parameters.

The Stillinger–Weber potential was chosen because the cos(θ) term is obtainable directly from the atomic separations via the dot product (also, the derivatives are a function of cos(θ)). The Stillinger–Weber potential has continuous derivatives and a built-in smooth cut-off to zero at $r = r_c$. We fit the Stillinger–Weber potential to the original authors' function by computing the parameter λ. We do this by computing the Taylor series expansion of Eq. (2) at $\theta = \theta_0$ to the second order. For this calculation we set γ = 0.5 eV and $r_{ij} = r_{ik}$ = 1.5 Å (the equilibrium bond length).

For both triplets (P–O–P and O–P–O), we set this equal to the original authors' function and rearrange to find λ. These parameter values (λ, γ and θ_0) for both triplets are summarised in Table 1.

Table 1. Three-body potential terms used in the present study.

Parameter	**O–P–O**	**P–O–P**
λ_i (eV)	5.3516	8.2997
$\gamma_{ij} = \gamma_{ik}$ (Å)	0.5	0.5
θ_0	109.47°	135.5°
r_c (Å)	2.5	2.5

For radiation damage studies, the two-body potential cannot model the repulsion between nuclei when the inter-particle separation is small. A screened Coulomb potential is normally used to model such interactions and the ZBL model [33] is the commonly used model when the separation is small. As a result we have joined the two-body potential given in Eq. (1) to the ZBL potential using a splining function. The details and parameters for the splining function are given in Table 2.

Table 2. The splining parameters for joining the Buckingham + Coulomb potential to the ZBL potential. The splining function is of the form $\exp(f_0 + f_1x + f_2x^2 + f_3x^3 + f_4x^4 + f_5x^5)$ joined to the ZBL potential at x = a and to the Buckingham + Coulomb at x = b so that the function and its first 2 derivatives are continuous. The units of f_i are Å^{-i}. In the case of Fe^{3+}–O and O–P, there were offsets of 13 and 50 eV respectively added to the potential (and later subtracted) to make the potential positive at x = b.

Interaction	**a (Å)**	**b (Å)**	f_0	f_1	f_2	f_3	f_4	f_5
Fe^{2+}–Fe^{2+}	1.1	1.9	38.456891	- 115.889467	166.591541	- 122.352450	44.173231	- 6.208431
Fe^{2+}–Fe^{3+}	0.9	1.9	11.672255	- 14.893823	17.600151	- 15.321189	6.858104	- 1.147068
Fe^{2+}–O	0.2	0.85	11.274280	- 21.233242	44.802056	- 69.570862	60.519843	- 21.948215
Fe^{2+}–P	0.7	1.45	13.486801	- 28.411131	49.982100	- 53.364231	28.581882	- 5.844285
Fe^{3+}–Fe^{3+}	0.9	1.9	3.289491	20.674793	- 40.640856	30.334182	- 10.171431	1.285723

Fe^{3+}–O	0.5	1.25	13.468756	– 36.254018	82.390627	– 109.951757	74.401925	– 20.216820
Fe^{3+}–P	0.4	1.45	11.230029	– 14.585045	17.056385	– 15.965608	8.848866	– 1.935188
O_–O	0.35	1.8	9.273627	– 11.588100	8.398709	– 2.456703	– 0.030643	0.093801
O_–P	0.25	1.03	11.166080	– 26.912807	75.616106	– 136.453869	116.256770	– 36.456113
P_–P	0.24	1.26	11.028636	– 16.858035	23.118958	– 21.054756	11.153372	– 2.472515

Each of the glass structures was prepared by distributing the required number of atoms of each species randomly within a cubic box. We then use a simple temperature-rescaling algorithm [34] to quench the system. This works by measuring the temperature at each time-step. If the temperature exceeds the desired value by 7%, then, the velocities of all the atoms are rescaled, such that, the temperature is the desired value. We quench the system from 6000 K to 10 K at a rate of 5×10^{12} Ks^{-1} using this temperature-rescaling algorithm with constant volume molecular dynamics. The total simulation time required was 1.2 ns. Finally, the structures were minimised to 0 K using the conjugate gradient method. This cooling rate is slower than that given in [21] by a factor of 2. This was found to be necessary to generate the lowest energy structures.

Molecular dynamics computation was carried out on these iron phosphate glasses contained in a cubic box with a small (~ 150) number of atoms with a varying Fe/P atomic ratio. The total number of atoms varied depending on the Fe/P atomic ratio. The potential energy of iron phosphate glasses with a varying Fe/P atomic ratio was analysed as a function of density to obtain the structure with the lowest potential energy, and hence the optimal density for each case. We then computed quenched structures for each system with a larger number of atoms (1100–1550) at the optimal density. Having determined the optimised structure of IPG, the mechanical properties (bulk modulus, shear modulus and Young's modulus) were calculated using GULP [30] and compared with that of experimental values available in the literature for IPG with a composition of 40 mol% Fe_2O_3–60 mol% P_2O_5.

The threshold displacement energies of each species: iron (Fe^{2+} and Fe^{3+}), phosphorous and oxygen were computed. Since only iron ions exist in ferrous and ferric states, the oxidation state is specified as Fe^{2+} and Fe^{3+} and the ionic nature of ions (P^{5+} and O^{2-}) with single oxidation state are not specified throughout the paper. We consider each atomic species, oxidation state and local coordination separately for the calculation of threshold energies.

To find the threshold energies for each case, firstly, using the quenched lattices, a list of candidate atoms to test is generated. We then randomly select a target atom from this list and pick a random direction. To find the threshold energy, we begin by performing a trial cascade by injecting 100 eV of kinetic energy into the target atom. After 2 ps of MD simulation time we compare the initial and final lattices. If the target atom has been displaced, we know that the threshold energy is lower than 100 eV, otherwise, it must be higher. We then perform additional cascades to zero-in on the threshold energy using the binary search algorithm. The cut-off tolerance used was 0.5 eV, meaning that the results obtained are within 0.5 eV of the actual threshold energy. We then select a new random target atom from our list, and then pick a new random direction; we then find the threshold energy using the method above. We repeat this process at least 900 times to obtain good statistics.

3. RESULTS AND DISCUSSION

The details of the composition of the smaller systems of iron phosphate glass, Fe/P atomic ratio and the number of atoms quenched are given in Table 3. Fig. 1 shows the average potential energy obtained as a function of glass density for systems $Fe_3(P_2O_7)_2$, $Fe(PO_3)_3$ and $Fe_4(P_2O_7)_3$. The composition of 40 mol% Fe_2O_3–60 mol% P_2O_5 is the same as that of the $Fe_4(P_2O_7)_3$ glass and can also be referred as IPG. However, $Fe_4(P_2O_7)_3$ does not have any Fe^{2+}ions in the glass. For the small systems, the lowest energy was obtained with densities of 3.04 and 3.2 g cm^{-3} for $Fe_4(P_2O_7)_3$ and $Fe_3(P_2O_7)_2$ glasses respectively. The lowest energy structure of $Fe_3(P_2O_7)_2$ glass (containing both Fe^{2+} and Fe^{3+} ions) had a higher density compared to the other systems containing only Fe^{3+} ions. Larger systems of $Fe_3(P_2O_7)_2$ and $Fe_4(P_2O_7)_3$ were generated with the density that showed the lowest energy in the smaller atom systems. The computation was done a minimum of three times for each large system at the fixed density. Whenever the difference in energy between the structures was small (< 1%), the structure with the lowest potential energy was chosen for further analysis based on certain logical guidelines.

Table 3. The composition (in atomic %), of each atomic species, in the three small system glass structures studied.

Glass system	Composition (at.%)				Fe/P atomic ratio	Number of atoms
	Fe^{2+}	Fe^{3+}	P	O		
$Fe_3(P_2O_7)_2$	4.762	9.524	19.048	66.667	0.75	126
$Fe(PO_3)_3$	–	7.692	23.077	69.231	0.33	130
$Fe_4(P_2O_7)_3$	–	12.903	19.355	67.742	0.67	155

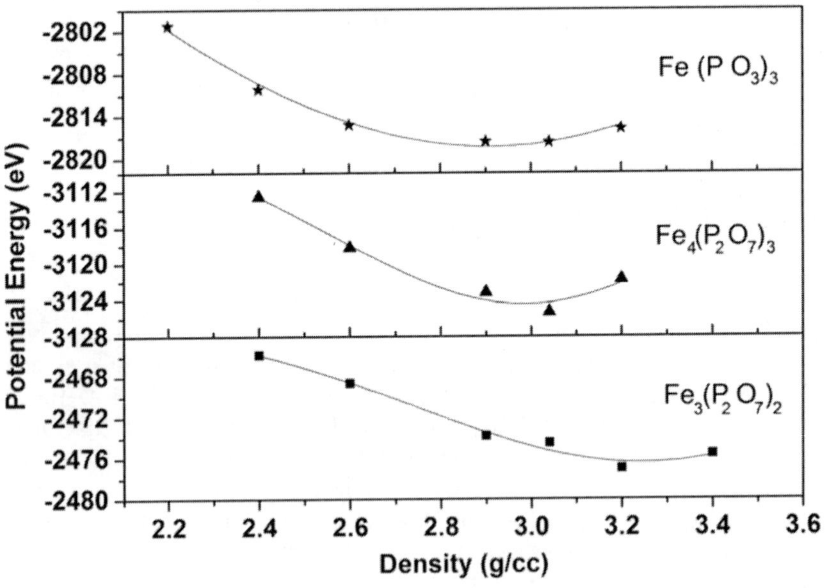

Figure. 1. Plot of potential energy as a function of density for various glass compositions; the solid line is to guide the eye.

The logical guidelines used to choose the best structure when the difference in potential energy was small are given below:

1. The coordination of P should be 4 for a maximum number of P ions in the glass.
2. The coordination of Fe^{2+} and Fe^{3+} varies between 4 and 6. The structures that contain iron ions with a coordination number of 3 are less stable.

3. The coordination number of oxygen varies between 1 and 3. The more stable structures have the maximum number of oxygen ions with coordination number 2.

For $Fe(PO_3)_3$ glass (130 atoms), the difference in potential energy between the structures with densities of 2.9 and 3.04 g cm^{-3} was very small, i.e., 3.5×10^{-3}%. Thus larger systems were modelled using both the densities (2.9 and 3.04 g cm^{-3}) for $Fe(PO_3)_3$ glass (1300 atoms). For $Fe(PO_3)_3$ glass (1300 atoms), based on the guidelines above, the structure with a density of 3.04 g cm^{-3} was chosen for further analysis. The coordination numbers of each ion (Fe^{2+}, Fe^{3+}, P and O) for the large equilibrated structures of $Fe_3(P_2O_7)_2$, $Fe_4(P_2O_7)_3$ and $Fe(PO_3)_3$, are presented in Table 4. Table 4 shows that all the phosphorous ions have 4-fold coordination in $Fe_3(P_2O_7)_2$ and $Fe_4(P_2O_7)_3$glass. However, two five-fold coordinated phosphorous ions were observed in the $Fe(PO_3)_3$ glass. This is significantly lower than the 1.9% that has been reported previously [21].

Table 4. The composition of each atom (in atomic %) of iron phosphate glasses at the optimum density for the systems containing the larger number of atoms.

Glass system	ρ (g cm^{-3})	CN	Fe^{2+} at.%	Fe^{3+} at.%	P at.%	O at.%
$Fe_3(P_2O_7)_2$	3.2	1	–	–	–	3.17
		2	–	–	–	53.49
		3	–	–	–	9.68
		4	2.78	4.84	19.05	0.32
		5	1.83	4.21	–	–
		6	0.16	0.48	–	–
$Fe_4(P_2O_7)_3$	3.04	1	–	–	–	4.06
		2	–	–	–	60.06
		3	–	–	–	3.61
		4	–	7.35	19.35	–
		5	–	5.10	–	–
		6	–	0.45	–	–
$Fe(PO_3)_3$	3.04	1	–	–	–	11.08

		2	–	–	–	57.23
		3	–	–	–	0.92
		4	–	2.92	22.92	–
		5	–	4.46	0.15	–
		6	–	0.31	–	–

CN: coordination number.

Fe^{2+} is present only in $Fe_3(P_2O_7)_2$ glass and it is found with coordination numbers 4 to 6. Generally, the Fe^{3+} ion is found with 4 to 6 coordination in $Fe_3(P_2O_7)_2$, $Fe_4(P_2O_7)_3$ and $Fe(PO_3)_3$ glasses. The oxygen ion coordination was found to vary between 1 and 3 in $Fe_3(P_2O_7)_2$, $Fe_4(P_2O_7)_3$ and $Fe(PO_3)_3$glasses. $Fe_3(P_2O_7)_2$ glass also showed 0.5% of four coordinated oxygen ions. The snapshots, shown in Fig. 2, show the coordination of ions within the $Fe_3(P_2O_7)_2$ glass structure.

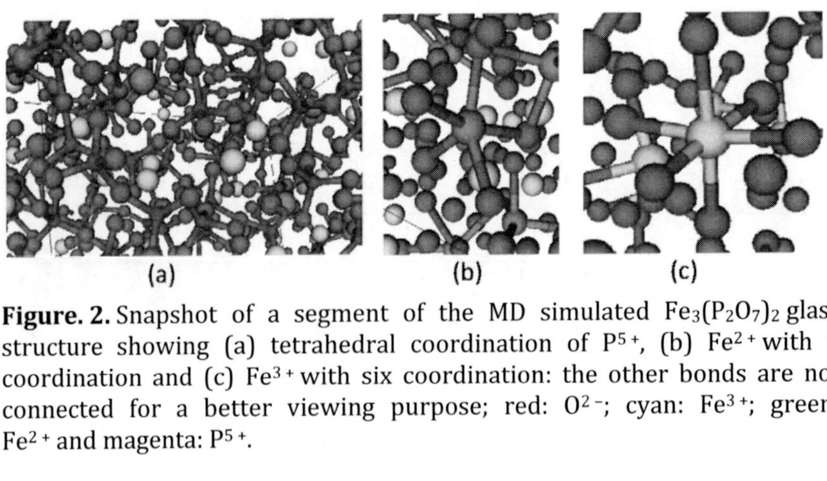

Figure. 2. Snapshot of a segment of the MD simulated $Fe_3(P_2O_7)_2$ glass structure showing (a) tetrahedral coordination of P^{5+}, (b) Fe^{2+} with 5 coordination and (c) Fe^{3+} with six coordination: the other bonds are not connected for a better viewing purpose; red: O^{2-}; cyan: Fe^{3+}; green: Fe^{2+} and magenta: P^{5+}.

In addition to these three iron phosphate glasses of varying Fe/P atomic ratio, it is useful to model the structure of the chemically durable iron phosphate glass with a composition of 40 mol% Fe_2O_3–60 mol% P_2O_5 (IPG) [15]. Before describing the results of the simulations, it is useful to understand the experimental observations on crystallisation of iron phosphate glasses. Iron phosphate glass crystallises to $FePO_4$, $Fe_3(P_2O_7)_2$, $Fe(PO_3)_3$, $Fe(PO_3)_2$, $Fe_7(PO_4)_6$, and $Fe_2P_2O_7$ depending on the composition, atmosphere of crystallisation and ratio of Fe^{2+}/Fe [24,25,35–39]. It was reported by us [40]that, on heating, IPG containing 4% Fe^{2+}, crystallises to a mixture of $Fe_3(P_2O_7)_2$, $Fe(PO_3)_3$ and $Fe_4(P_2O_7)_3$ under flowing argon atmosphere. C. S. Ray et al. [37] reported a fraction of 17% of Fe^{2+} ions within the IPG glass with each phase also of the form $Fe_3(P_2O_7)_2$, $Fe(PO_3)_3$ and $Fe_4(P_2O_7)_3$. Thus these are the data that are presented in Table 5.

Table 5. Contribution of various phases in IPG and the atomic composition of the glass structures.

System (IPG)	Contribution (%)			at.%				Fe/P atomic ratio
	$Fe_3(P_2O_7)_2$	$Fe(PO_3)_3$	$Fe_4(P_2O_7)_3$	Fe^{2+}	Fe^{3+}	P	O	
4% Fe^{2+}	14	12	74	0.51	12.24	19.53	67.71	0.65
17% Fe^{2+}	54	10	36	2.27	11.0	19.41	67.31	0.68

The structure of IPG was computed using the combination of atoms of Fe, P and O to provide concentrations of (a) 4% Fe^{2+} and (b) 17% Fe^{2+}. The atomic percentage of each ion used for constructing the structure of IPG involving 4% Fe^{2+} and 17% Fe^{2+}, is also presented in Table 5. The Fe/P atomic ratio of the final glass (Table 5), differs slightly from the starting batch composition of Fe/P atomic ratio = 0.67 in both the cases. However, it is well within the reported experimental uncertainty [37,41]. Computation was carried out to obtain structures with the lowest potential energy using densities 2.9 and 3.04 g $cm^{-3}$for IPG containing 4% Fe^{2+} ions. These densities are chosen based on the experimental result [11] and the value obtained from ref. [21]. For IPG containing 17% Fe^{2+}, the densities of 2.9, 2.95 and 3.04 g cm^{-3} were used to compute the structure of glass to obtain the glass with the lowest energy. A density of 2.95 g cm^{-3} was additionally used in IPG containing 17% Fe^{2+} ions since the experimental density reported by C. S. Ray et al. [37] was 2.95 g cm^{-3}.

As explained earlier, the lowest energy structure of IPGs containing 4% $Fe^{2+}$and 17% Fe^{2+} was computed using similar MD methods. The potential energy obtained for densities of 2.9 and 3.04 g cm^{-3} is presented in Table 6. The contribution of various ions and their coordination numbers is also given in Table 6 for both the densities (2.9 and 3.04 g cm^{-3}). As per the guidelines, the structure with the lowest number of Fe^{2+} ions with 3-fold coordination has a density of 2.9 g cm^{-3}. In the structure with a density of 3.04 g cm^{-3}, there are a greater number of 3-fold coordinated Fe^{2+} ions. This reduces the average coordination number of Fe^{2+} below 4.0 and is therefore not considered for computation of the threshold displacement energy. Under a similar analogy, the structure of IPG with 17% Fe^{2+} was the structure with a density of 2.95 g $cm^{-3}$and that the optimised structure was used for further threshold energy displacement calculations. The details of potential energy, coordination number of various ions in IPG with 17%

Fe^{2+}, are also presented in Table 6. The computed density of the structures containing 4% and 17% Fe^{2+} is in good agreement with the reported experimental values [11,37].

Table 6. The composition of each atom (in atomic %) of IPG containing 4% Fe^{2+} and 17% Fe^{2+} together with the computed potential energy of the IPG.

System (IPG)	ρ (g cm⁻³)	Potential energy (eV)	CN	Fe^{2+}at.%	Fe^{3+}at.%	P at.%	O at.%
IPG with 4% Fe^{2+}	2.9	– 27,689	1	–	–	–	4.37
			2	–	–	–	59.91
			3	0.07	0.07	–	3.42
			4	0.36	6.85	19.53	–
			5	0.07	5.25	–	–
			6	–	0.07	–	–
IPG with 4% Fe^{2+}	3.04	– 27,709	1	–	–	–	3.79
			2	–	–	–	60.79
			3	0.22	–	–	3.06
			4	0.15	6.85	19.53	0.07
			5	0.15	5.03	–	–
			6	–	0.36	–	–
IPG with 17% Fe^{2+}	2.9	– 23,777	1	–	–	–	4.37
			2	–	–	–	57.82
			3	0.08	–	–	5.13
			4	1.51	6.97	19.41	–
			5	0.67	4.03	–	–
			6	–	–	–	–
IPG with 17% Fe^{2+}	2.95	– 23,784	1	–	–	–	3.53
			2	–	–	–	58.32
			3	–	–	–	5.38
			4	1.09	6.81	19.33	0.08
			5	0.92	4.03	0.08	–
			6	0.25	0.16	–	–
IPG with 17% Fe^{2+}	3.04	– 23,799	1	–	–	–	3.70
			2	–	–	–	57.90
			3	0.25	–	–	5.63

			4	1.26	6.05	19.41	0.08
			5	0.76	4.45	–	–
			6	–	0.50	–	–

CN: coordination number.

The IPG containing 4% Fe^{2+} showed Fe^{2+} and Fe^{3+} with coordination number 3, whereas, the IPG containing 17% Fe^{2+} showed no 3 coordinated iron ions. However, the IPG containing 17% Fe^{2+} was found to contain one five-fold coordinated phosphorous ion.

In order to compare the structures, the potential energy per atom was determined and this is presented in Fig. 3. The potential energy per atom increases linearly with the Fe/P ratio. This plot also clearly shows that as the concentration of Fe^{2+} ions increases in the glass, the potential energy per atom also increases.

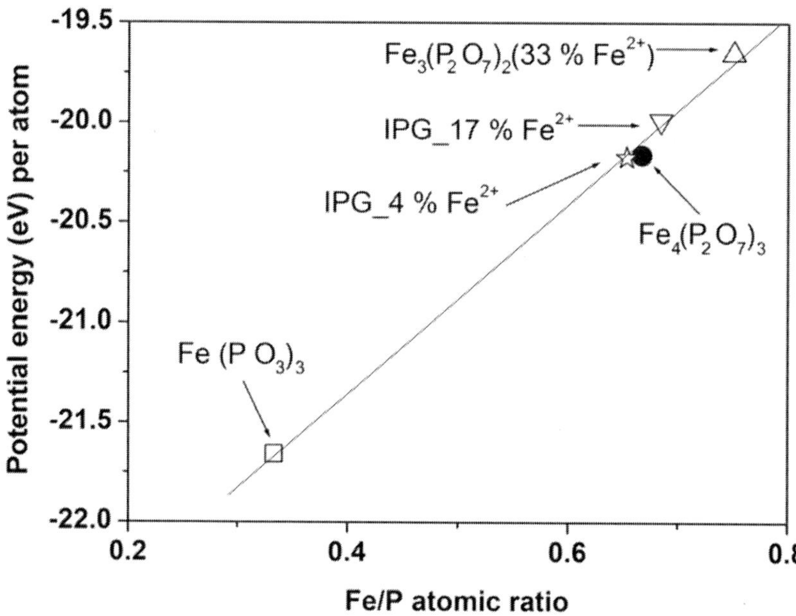

Figure. 3. Plot of potential energy per atom as a function of Fe/P atomic ratio of the glasses: the solid line is the linear fit with R^2: 0.998 and standard deviation of 0.05.

In the experiment, the starting material contains only Fe^{3+} ions but during the preparation process, some Fe^{2+} ions form. Ammonium di-hydrogen phosphate is used as the source of phosphorous during the preparation of the iron phosphate glass. Evolution of ammonia, a reducing agent, during the preparation of the iron phosphate glass (due to the decomposition of ammonium di-hydrogen phosphate), reduces some Fe^{3+} to Fe^{2+}. However, the reducing effect can be minimised by a low temperature pre-calcination, as indicated in ref. [40]. Hence the most stable structure would be expected to be the one with the lowest concentration of Fe^{2+} ions, that is the IPG containing 4% Fe^{2+}concentration.

Modelling studies on the structure of iron phosphate glasses are given in refs. [21,22]. During the structural evolution of the IPG, Stoch et al. reported [22] the formation of crystallised products in air as $FePO_4$ and $Fe_2P_2O_7$. Though the starting composition was 40 mol% Fe_2O_3–60 mol% P_2O_5, (Fe/P atomic ratio of 0.67), the final crystallised products contain Fe/P with an atomic ratio of 1, regardless of the contribution of each phase ($FePO_4$ and $Fe_2P_2O_7$) in the glass. The Fe/P ratio increases after crystallisation, indicating the loss of P and O. This can be explained as follows:

40 mol% Fe_2O_3 – 60 mol% P_2O_5 = $Fe_4P_6O_{21}$ amorphous (3)

$$Fe_4P_6O_{21\,(amorphous)} \xrightarrow{\text{crystallisation in air}} 2FePO_4 + Fe_2P_2O_7 + P_2O_5\uparrow + 1/2\ O_2\uparrow$$

Eq. (4) also justifies the formation of Fe^{2+} ions ($Fe_2P_2O_7$) from the starting composition containing only Fe^{3+} ions with the loss of oxygen. Thus some Fe^{2+} are to be expected despite the fact that the model gives a higher cohesive energy when the amount of Fe^{2+} is minimised.

The radial distribution functions of $Fe_3(P_2O_7)_2$ glass and IPG containing 4% Fe^{2+} concentration are shown in Fig. 4. The bond distance between Fe^{3+}–O in all the five iron phosphate glasses did not vary significantly. This agrees with the bond distance determined in [21]. As expected, the Fe^{2+}–O distance was longer than that of Fe^{3+}–O. These values are in good agreement with the experimental values reported in the literature [19,23,42–44] for iron phosphate glasses. Table 7 presents details of the mean bond distance between Fe^{2+}–O, Fe^{3+}–O and P–O in the iron phosphate glasses that we have modelled.

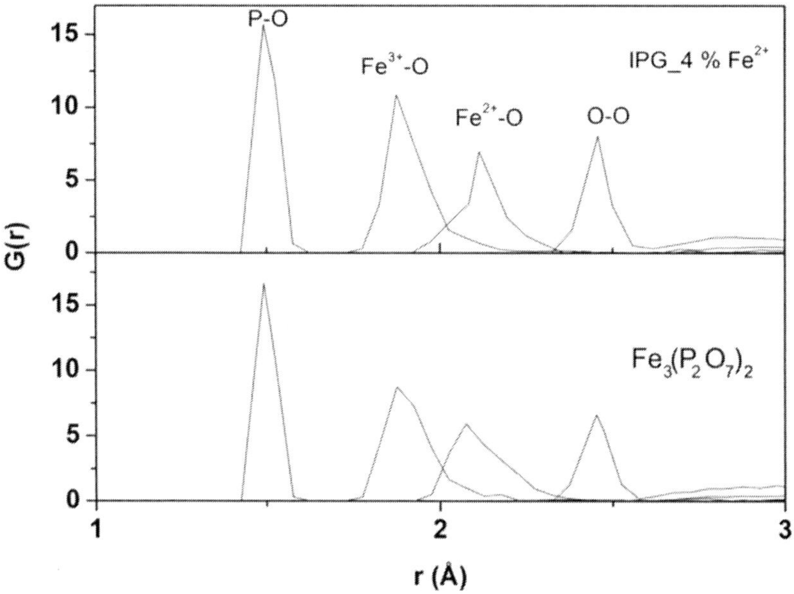

Figure. 4. Computed radial distribution functions of $Fe_3(P_2O_7)_2$ and IPG with 4% Fe^{2+}. Similar calculations for the Fe–Fe and Fe–P interactions showed no discernable structure.

Table 7. Mean bond distance found in iron phosphate glasses.

Glasses	**Mean bond distance (Å)**		
	Fe^{2+}–O (2.08–2.12)*	**Fe^{3+}–O (1.89–1.91)***	**P–O (1.48–1.55)***
$Fe_3(P_2O_7)_2$	2.07	1.87	1.5
$Fe_4(P_2O_7)_3$	–	1.87	1.5
$Fe(PO_3)_3$	–	1.89	1.5
IPG with 4% Fe^{2+}	2.12	1.88	1.5
IPG with 17% Fe^{2+}	2.12	1.87	1.5

* The values in parentheses are experimental values from refs. [19, 35–38].

The calculated elastic properties (bulk modulus and shear modulus) of IPG containing 4% and 17% concentrations of Fe^{2+} ions are presented in Table 8. The values of these properties are obtained as the average of the three methods employed in GULP [30]. Experimentally determined bulk modulus and Young's modulus of IPG with a composition of 40 mol% Fe_2O_3–60 mol% P_2O_5were 47 GPa and 70–72 GPa respectively as reported in the literature [45,46]. The calculated properties are in good agreement with the experimentally reported values.

Table 8. Mechanical properties of IPG using GULP.

Glass	Bulk modulus (GPa)	Shear modulus (GPa)	Young's modulus (GPa)
IPG_4% Fe^{2+}	48	31	81
IPG_17% Fe^{2+}	44	31	71

The threshold displacement energy (E_d) of various glasses $Fe_3(P_2O_7)_2$ (1260 atoms), $Fe_4(P_2O_7)_3$ (1550 atoms) and $Fe(PO_3)_3$ (1300 atoms) was computed. E_d was found to change with the coordination number and species of the particular ion in $Fe_3(P_2O_7)_2$ glass. The effect of the coordination of Fe^{2+} ions on E_d is shown in Fig. 5. As would be expected, the values of E_d are generally lower for the lower coordination numbers. This was found to be true for other two systems ($Fe_4(P_2O_7)_3$ and $Fe(PO_3)_3$) also. E_d also varied as a function of the coordination number and nature of the ion in IPG. Fig. 6 gives E_d for ions of fixed coordination in IPG containing both 4 and 17% concentrations of Fe^{2+} ion. It is clear from the plots that E_d varies depending on the nature of the ion with the lowest values for the Fe^{2+} ions. However, similar values of E_d were observed for IPG containing both 4% and 17% concentrations of Fe^{2+} ions. This indicates that E_d does not change with the concentration of Fe^{2+} ions in IPG but as shown in Fig. 6, Fe^{2+} has a lower E_d than Fe^{3+}.

Since phosphorous is 4-fold coordinated in all the iron phosphate glasses, the values of E_d for phosphorous ions in all these glasses are compared and shown in Fig. 7. It is apparent from the plot, that E_d remains similar for phosphorous ions in all the iron phosphate glasses. The peak values of E_d of the different ions as a function of their coordination number are presented in Table 9. Although a large number of calculations were performed to determine these values, there is a wide statistical variation, and the values for a given species are very similar for the different coordination numbers.

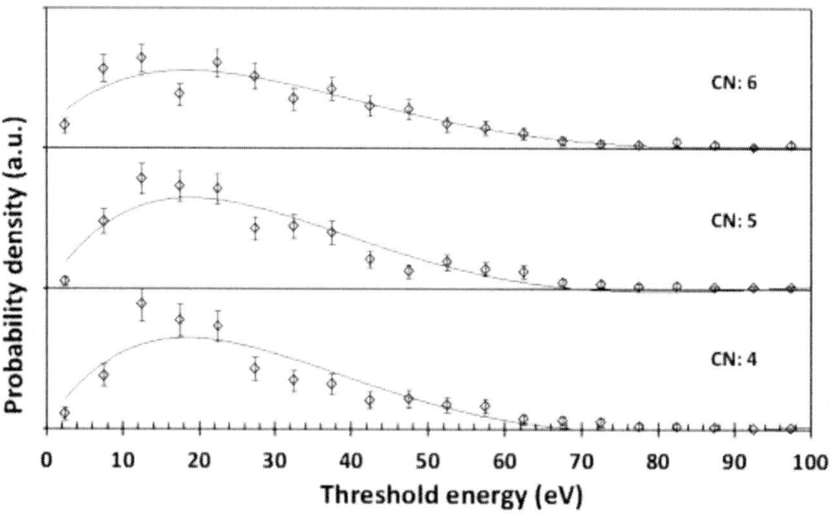

Figure. 5. Effect of coordination of the Fe^{2+} ions on the threshold displacement energy of $Fe_3(P_2O_7)_2$ glass; the solid line is a guide to the eye and the error bars refer to 2 standard deviations using the binomial distribution and a bin width of 4 eV.

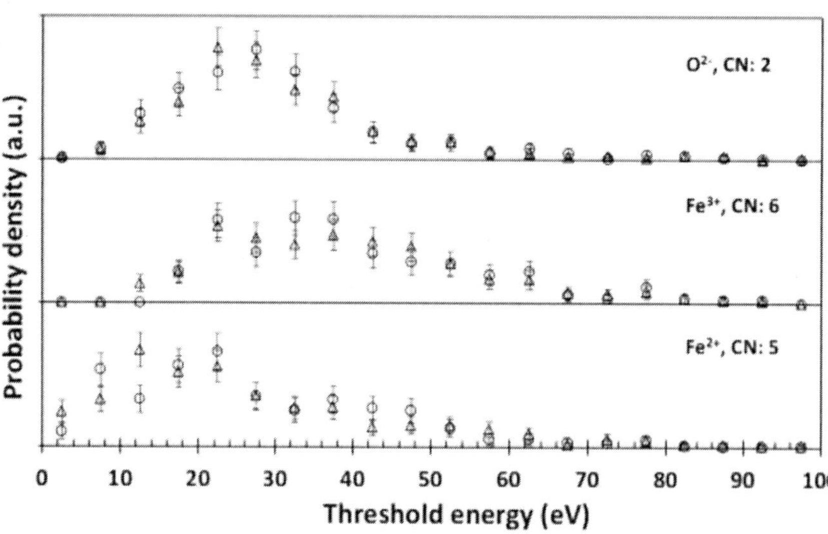

Figure. 6. The threshold displacement energy of various ions with coordination number (CN) in IPG containing 4% (circle) and 17% (triangle) concentration of Fe^{2+} ions. The error bars refer to 2 standard deviations using the binomial distribution and a bin width of 4 eV.

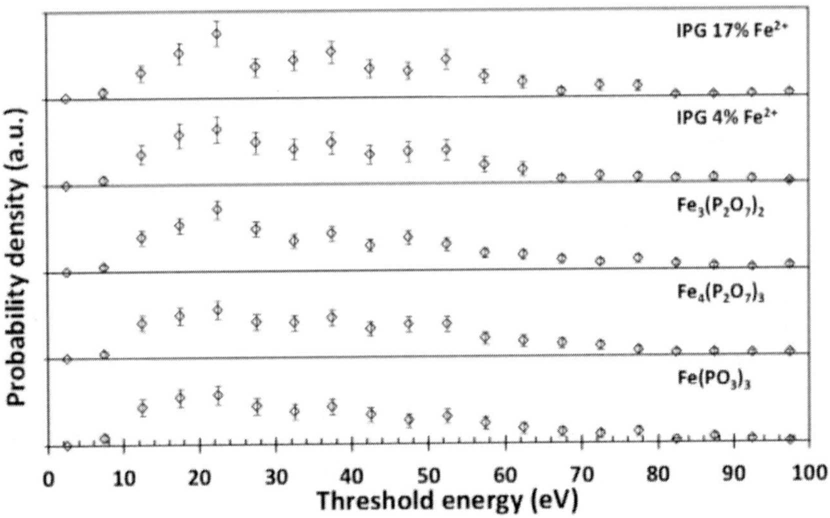

Figure. 7. Threshold displacement energies of phosphorous ions in the various iron phosphate glasses. The error bars refer to 2 standard deviations using the binomial distribution and a bin width of 4 eV.

Table 9. Peak values of the threshold displacement energies, E_d, (eV) of various iron phosphate glasses; CN is the coordination number.

Glass	CN	Threshold energy, E_d (eV)			
		Fe^{2+}	Fe^{3+}	P^{5+}	O^{2-}
$Fe_3(P_2O_7)_2$	1	–	–	–	26
	2	–	–	–	29
	3	–	–	–	28
	4	18	33	22	–
	5	20	37	–	–
	6	24	38	–	–
$Fe_4(P_2O_7)_3$	1	–	–	–	26
	2	–	–	–	27
	3	–	–	–	27
	4	–	33	22	–
	5	–	36	–	–

	6	–	34	–	–
$Fe(PO_3)_3$	1	–	–	–	26
	2	–	–	–	32
	3	–	–	–	30
	4	–	40	22	–
	5	–	40	–	–
	6	–	36	–	–
IPG with 4% Fe^{2+}	1	–	–	–	24
	2	–	–	–	26
	3	15	23	–	25
	4	22	32	22	–
	5	23	34	–	–
	6	–	36	–	–
IPG with 17% Fe^{2+}	1	–	–	–	25
	2	–	–	–	26
	3	–	–	–	24
	4	19	31	22	–
	5	19	34	–	–
	6	18	36	–	–

The values of E_d for the O ions given in Table 9 are very similar to those calculated for the borosilicate glasses [34] but the thresholds for P and $Fe^{2+}$are lower than those calculated for boron and silicon suggesting that the phosphate glasses might not be as radiation resistant as the borosilicate.

4. CONCLUSION

The structures of iron phosphate glasses with varying Fe/P atomic ratios were modelled as a function of density using the molecular dynamics simulation method. Furthermore,the structures of iron phosphate glass with a composition of 40 mol% Fe_2O_3–60 mol% P_2O_5 (IPG) containing the concentrations of 4 and 17% Fe^{2+} ions, were also computed and compared in order to understand the effect of Fe^{2+} concentration. The potential energy per atom of all these glasses indicated that the system energy increases with both increasing Fe/P atomic ratio and Fe^{2+} concentration. The computed elastic properties of 40 mol% Fe_2O_3–60 mol% P_2O_5 glass were found to be similar to that of the experimental values reported in the literature.

The threshold displacement energy varies with the nature and as well as with the coordination of ions in these iron phosphate glasses. However, similar threshold energies were obtained for IPG containing 4 and 17% Fe^{2+} ions. The threshold energy of phosphorous ions remains the same for all the iron phosphate glasses. The results show that the peak displacement energy threshold of the Fe^{2+} ions is lower than that for Fe^{3+} and O ions. The cohesive energy of the glasses is also reduced as the amount of Fe^{2+} increases. This suggests that for the purposes of nuclear waste immobilisation, it would be preferable to produce the glasses with as low a Fe^{2+} content as possible. This is not the only consideration since the glasses must also be suitable to incorporate heavy atoms such as caesium and the associated structure is also important. However, the bond lengths and densities have been shown to vary only slightly as a function of composition.

Future work will investigate the structural changes induced by collision cascades in the glasses. It was for this purpose that the modified potential was developed since it provides a better description of the collision dynamics than the simple Teter model developed only for structural studies.

ACKNOWLEDGEMENT

The authors thank Dr. Christopher Scott for his timely help to visualise the structure of glasses using the CDJsVis programme. The authors thank EPSRCgrant EP/K007882/1 for funding the project. The authors also thank the Department of Mathematical Sciences, Loughborough University, UK for providing excellent computational facilities.

REFERENCES

1. T.T. Fernandez, P. Haro-Gonzales, B. Sotello, M. Hernandez, D. Jaque, P. Fernandez, C. Domengo, J. Siegel, J. SolisOpt. Lett., 38 (2013), pp. 5248-5251
2. K. TakadaActa Mater., 61 (2013), pp. 759-770

3. J.K. Christie, R.I. Ainsworth, D. Tomaso, N.H. de LeeuwJ. Phys. Chem. B, 117 (2013), pp. 10652-10657
4. P. SenguptaJ. Hazard. Mater., 235 (2012), pp. 17-28
5. R.K. BrowJ. Non-Cryst. Solids, 263/264 (2000), pp. 1-28
6. I.W. Donald, B.L. Metcalfe, R.N.J. TaylorJ. Mater. Sci., 32 (1997), pp. 5851-5887
7. P.A. Bingham, R.J. hand, O.M. Hannant, S.D. Forder, S.H. KilcoyneJ. Non-Cryst. Solids, 355 (2009), pp. 1526-1538
8. D.E. Day, Z. Wu, C.S. Ray, P. HrmaJ. Non-Cryst. Solids, 241 (1998), pp. 1-12
9. C.W. Kim, C.S. Ray, D. Zhu, D.E. Day, D. Gombert, A. Aloy, A. Mogus-Milankovic, M. KarabulutJ. Nucl. Mater., 322 (2003), pp. 152-164
10. J.M. Perez Jr., D.F. Bickford, D.E. Day, D.S. Kim, S.L. Lambert, S.L. Marra, D.K. Peeler, D.M. Strachan, M.B. Triplet, J.D. Vienna, R.S. Wittman, **Pacific Northwest Laboratory Report, PNNL — 13582**(2001)
11. Kitheri Joseph, R. Asuvathraman, R. Raja Madhavan, H. Jena, K.V. Govindan Kutty, P.R. Vasudeva RaoEnergy Procedia, 7 (2011), pp. 518-524
12. Joseph Kitheri, R. Asuvathraman, R. Venkata Krishnan, T.R. Ravindran, R. Govindaraj, K.V. Govindan Kutty, P.R. Vasudeva RaoJ. Nucl. Mater., 452 (2014), pp. 273-280
13. D.S. Kim, W.C. Buchmiller, M.J. Schweiger, J.D. Vienna, D.E. Day, C.W. Kim, D. Zhu, T.E. Day, T. Neidt, D.K. Peeler, T.B. Edwards, I.A. Reamer, R.J. Workman, **Pacific Northwest Laboratory Report, PNNL — 14251**(2003)
14. Jen-Hsien Hsu, Joseph W. Newkirk, C.W. Kim, C.S. Ray, R.K. Brow, M.E. Schlesinger, D.E. DayCorros. Sci., 75 (2013), pp. 148-157
15. M. Karabulut, G.K. Marasinghe, C.S. Ray, D.E. Day, O. Ozturk, G.D. WaddillJ. Non-Cryst. Solids, 249 (1999), pp. 106-116
16. M. Karabulut, G.K. Marasinghe, C.S. Ray, D.E. Day, G.D. Waddill, C.H. Booth, P.G. Allen, J.J. Bucher, D.L. Caulder, D.K. ShuhJ. Non-Cryst. Solids, 306 (2002), pp. 182-192
17. D.E. Day, C.S. Ray, K. Marasinghe, M. Karabulut, X. Fang, **Final Report, Project No. 55110, 200**Available from:http://www.osti.gov/em52/final_Reports/55110.pdf
18. F. Wedgwood, A. WrightJ. Non-Cryst. Solids, 21 (1976), pp. 95-105
19. G.K. Marasinghe, M. Karabulut, C.S. Ray, D.E. Day, M.G. Shumsky, W.B. Yelon, C.H. Booth, P.G. Allen, D.K. ShuhJ. Non-Cryst. Solids, 222 (1997), pp. 144-152
20. A.C. Wright, R.N. Sinclair, J.L. Shaw, R. Haworth, G.K. Marasinghe, D.E. Day, P.A. Bingham, S.D. Forder, G.J. Cuello, H.E. FischerPhys. Chem. Glasses Eur. J. Glass Sci. Technol. B, 53 (6) (2012), pp. 227-244
21. Bushra Al-Hasni, Gavin MountjoyJ. Non-Cryst. Solids, 357 (2011), pp. 2775-2779
22. P. Stoch, W. Szczerba, W. Bodnar, M. Ciecinska, A. Stoch, E. BurkelPhys. Chem. Chem. Phys., 16 (2014), pp. 19917-19927
23. Y. Lin, Y. Zhang, W. Huang, K. Lu, Y. ZhaoJ. Non-Cryst. Solids, 112 (1989), pp. 136-141
24. X. Fang, C.S. Ray, A. Mogus-Milankovic, D.E. DayJ. Non-Cryst. Solids, 283 (2001), pp. 162-172
25. M. Karabulut, B. Yuce, O. Bozdogan, H. Ertap, G.M. MammadovJ. Non-Cryst. Solids, 357 (2001), pp. 1455-1462
26. R.J. Ainsworth, D.D. Tommaso, J.K. Christie, N.H. de LeeuwJ. Chem. Phys., 137 (2012), p. 234502
27. Pedone, G. Malavasi, M.C. Menzian, A.N. Cormack, U. SegreJ. Phys. Chem. B, 110 (2006), pp. 11780-11795
28. R. Smith, D. Bacorisen, B.P. Uberuaga, K.E. Sickafus, J.A. Ball, R.W. GrimesJ. Phys. Condens. Matter, 17 (2005), pp. 875-891
29. J.M. Delaye, D. GhalebJ. Non-Cryst. Solids, 195 (1996), pp. 239-248
30. Julian D. Gale, **GULP: General Utility Lattice Program: a computer program for the symmetry — adapted simulation of solids**J. Chem. Soc. Faraday Trans., 93 (1997), pp. 629-637
31. F.H. Stillinger, T.A. WeberPhys. Rev. B, 31 (1985), pp. 5262-5271
32. B.P. Feuston, S.H. GarofaliniJ. Chem. Phys., 89 (1988), pp. 5818-5824

33. J. Ziegler, J.P. Biersack, U. Littmark, **The Stopping and Range of Ions in Matter**Pergamon Press, New York (1985)
34. Kenny Jolley, Roger Smith, Kitheri JosephNucl. Instrum. Methods B (2014), 10.1016/j.nimb.2014.12.024(in press)
35. J. Doupovec, J. Sitek, J. KalkosJ. Therm. Anal., 22 (1981), pp. 213-219
36. A.M. Milankovic, D.E. Day, K. FuricKey Eng. Mater., 132–136 (1997), pp. 2212-2215
37. C.S. Ray, X. Fang, M. Karabulut, G.K. Marasinghe, D.E. DayJ. Non-Cryst. Solids, 249 (1999), pp. 1-16
38. F.J.M. Almeida, J.R. Martinelli, C.S.M. PartitiJ. Non-Cryst. Solids, 353 (2007), pp. 4783-4791
39. P. Stoch, M. Ciecinska, A. StochJ. Therm. Anal. Calorim., 117 (2014), pp. 197-204
40. Kitheri Joseph, Sajal Ghosh, K.V. Govindan Kutty, P.R. Vasudeva RaoJ. Nucl. Mater., 426 (2012), pp. 233-239
41. Kitheri Joseph, K.V. Govindan Kutty, P. Chandramohan, P.R. Vasudeva RaoJ. Nucl. Mater., 384 (2009), pp. 262-267
42. C.H. Booth, P.G. Allen, J.J. Bucher, N.M. Eldelstein, D.K. Shuh, G.K. Marasinghe, M. Karabulut, C.S. Ray, D.E. DayJ. Mater. Res., 14 (1999), pp. 2628-2639
43. M. Karabulut, G.K. Marasinghe, C.S. Ray, D.E. Day, G.D. Waddill, P.G. Allen, C.H. Booth, J.J. Bucher, D.L. Caulder, D.K. Shuh, M. Grimsditch, M.L. SaboungiJ. Mater. Res., 15 (2000), pp. 1972-1984
44. U. Hoppe, M. Karabulut, E. Metwalli, R.K. Brow, P. JóváriJ. Phys. Condens. Matter, 15 (2003), pp. 6143-6153
45. C.R. KurkjianJ. Non-Cryst. Solids, 263–264 (2000), pp. 207-212
46. K.H. Chang, T.H. Lee, L.G. HwaChin. J. Phys., 41 (2003), pp. 414-421

CHAPTER 10

Enhancements of Thermoelectric Performance Utilizing Self-Assembled Monolayers in Semiconductors

T.H. Wang[a], H.T. Jeng[a,b,*]

[a]*Department of Physics, National Tsing Hua University, Hsinchu 30013, Taiwan*

[b]*Institute of Physics, Academia Sinica, Taipei 11529, Taiwan*

ABSTRACT

We investigate the thermoelectric effect of the self-assembled monolayers (SAMs) in semiconductors (SCs). The thermoelectric performance of the SC-SAM-SC heterostructure is largely improved due to two reasons. First, the SAMs in semiconductors act as good thermal insulators. Second, the thickness of the SAMs can be smaller than one nanometer so that electrons could easily tunnel through the SAMs. As an example, we consider the GaAs-SAM-GaAs heterostructure at room temperature. The electrical conductivity, the Seebeck coefficient, and the electronic thermal conductivity as functions of donor concentration are calculated through the Boltzmann equation and the tunneling current model. The highly improved figure of merit and Seebeck coefficient demonstrate an excellent direction for fabricating high efficiency nano-thermalelectric devices.

1. INTRODUCTION

The thermoelectric devices, which can convert heat into electricity or vice versa, attract continuing attention because they yield many applications in power generation and refrigeration and have many advantages such as solid-state operation without moving parts, no release of greenhouse gases, good stability and high reliability [1]; [2]; [3]; [4]; [5]; [6]; [7]; [8]; [9] ; [10]. The maximum energy-conversion efficiency can be characterized by a dimensionless figure of merit $ZT = \sigma S^2 T/\kappa$ [1]. Here, σ, S, κ, and T are electrical conductivity, Seebeck coefficient, thermal conductivity, and absolute temperature, respectively. Therefore, high electrical conductivity (reducing the Joule heating), low thermal conductivity (increasing the temperature difference), and large Seebeck coefficient (increasing the potential difference) are necessary for the realization of high-performance thermoelectric materials or devices.

Different strategies have been used to enhance the figure of merit. In general, they can be classified into two categories, enhancing the power factor σS^2 and reducing the thermal conductivity. The former is usually achieved through band-structure engineering. For example, the power factor can be enhanced by doping resonant impurities which introduce a large number of resonant states near the Fermi level [11], or by choosing the materials having multiple valley [12]. Another common strategy is employing the energy filtering of the electrons at the interface. In this approach, the low-energy carriers can be effectively blocked by the energy barrier (i.e., the interface) raising the average energy of the conduction carriers and hence the magnitude of the Seebeck coefficient. However, the interface could significantly reduce the electrical conductivity. Therefore, a careful design is necessary to optimize the power factor. Rowe and Min [13] first proposed that the power factor could be enhanced by the presence of small barrier potentials. Shakouri and Bowers then proposed that the selective emission of hot electrons and thus a higher power factor can be achieved by inserting tall barrier layers into degenerate semiconductor superlattices [14]. Experimental evidence for the power-factor enhancement through the energy filtering of the electrons has been seen in the $In_{0.53}Ga_{0.47}As$-based superlattices [15]. In fact, similar phenomena have been observed in a wide range of materials such as the PbTe nanocomposites [16], $Bi_2Te_{2.7}Se_{0.3}$ nanoplatelet composites [17], $(La,Sr)TiO_3$ with Nb-doped grain boundaries [18], heavily B-doped Si nanocrystals [19]. In addition to improving the power factor, the reduction of the thermal conductivity is vital for thermoelectric performance. Many strategies have been employed to reduce the thermal conductivity such as introducing point defects, alloying, creating rattling structure, using complex crystal structure [1] ; [3]. Recently, the strategy of all-length-scale hierarchical architectures has been employed to reduce the thermal conductivity with respect to phonons of different mean free paths [20]; [21]. In general, the strategies resulting in the reduction of the thermal conductivity will simultaneously deteriorate the electrical conductivity. How to reduce the thermal conductivity without significantly compromising the electrical conductivity is a challenging problem.

In this article, we propose that the thermoelectric performance of the semiconductors can be greatly enhanced by the presence of the self-assembled monolayers (SAMs). In this approach, the SAMs in semiconductors act as good thermal insulators and thermal filters because only the phonons which are resonant with the discrete modes of the SAMs could easily pass through the SAMs. Because the thickness of the SAMs can be smaller than one nanometer, the conduction electrons could easily tunnel through the SAMs. As a result, we can greatly reduce the thermal conductivity without significantly compromising the electrical conductivity. Furthermore, the magnitude of the Seebeck coefficient is enhanced due to the energy-filtering effect. Some experiments indicate that the presence of the SAMs is promising for improving the thermoelectric performance. Wang et al. [22] have measured the interfacial thermal conductance of the Au-SAM-GaAs, in which, the SAM is composed of 1,*n*-alkanedithiol. They found the interfacial thermal conductances of *n* C long SAM interfaces (n =8,9 and 10) are about 25MW$m^{-2}K^{-1}$, being insensitive to the chain length. On the other hand, Hu et al. [23] calculate the interfacial thermal conductances of Au-SAM-Si by using molecular dynamics simulations. Their results show that, contrary to intuition, the interfacial thermal conductance of 3 C long SAM interfaces is only 6 MW m^{-2} K^{-1}, which is twice lower than that of 8 C long SAM interface, 15 MW m^{-2} K^{-1}, and even smaller than that of 32 C long SAM interface, 9 MW m^{-2} K^{-1}. These results are good news to *simultaneously* meet the high-thermoelectric-performance criteria of low thermal conductivity and high electrical conductivity. In addition, they found the interfacial thermal conductance of the Si-SAM interface is about 11 MW m^{-2} K^{-1}, and one order smaller than that of Au-SAM interface, 110 MW m^{-2} K^{-1}, for the case of 16 C long SAM interfaces. This indicates that the interfacial thermal conductance of a semiconductor-SAM-semiconductor structure could be generally smaller than that of Au-SAM-semiconductor.

As an example, we consider the case of GaAs-SAM-GaAs at room temperature with the SAM being the alkanedithiol molecule [–S–$(CH_2)n$–S–]. The interfacial thermal conductance of the GaAs-SAMs-GaAs is set at a quiet large value of 25 MW m^{-2} K^{-1}according to the measurements by Wang et al. [22] for the case of 8–10 C long Au-SAM-GaAs interfaces. Our results show that the figure of merit of the GaAs-SAM-GaAs heterostructure could be several times higher than that of the bulk GaAs. Furthermore, we find the

Seebeck coefficient can also be enhanced by the presence of the SAM due to the energy-filtering effect. While this enhancement is even more significant especially for heavily doped semiconductors, which are typically promising candidates of good thermoelectric materials. We note here that the figure of merit thus calculated could be further improved with shorter SAMs. For example the 3 C long GaAs-SAM-GaAs interfaces, the figure of merit could be several times larger because of the much smaller thermal conductance as mentioned in the previous paragraph.

2. METHOD

The electronic structure near the conduction-band edge (CBE) of GaAs is calculated through two-band Kane's model, which can be expressed as [24]

$$E_k(1+\alpha E_k)=\frac{\hbar^2k^2}{2m_c^*},\ \alpha=\frac{1}{E_g}\left(1-\frac{m_c^*}{m_0}\right)^2 \qquad (1)$$

where Eg is the direct band gap, m_c^* the effective mass of the electron at the CBE, and m_0 the free-electron mass. The Fermi level Ef is determined by the charge neutrality condition, which, for the case of n-type semiconductors, can be expressed as

$$N_d=\int_0^\infty dED(E)f(E-E_f) \qquad (2)$$

where $D(E)$ is the conduction-band density of states, f the Fermi–Dirac distribution function, and Nd the donor concentration. Here the energy zero is set at the CBE. We assume that all the donors are completely ionized. This is a good assumption for usual doping concentration at room temperature.

The thermoelectric transport properties can be derived from the Boltzmann transport equation with relaxation time approximation [25]. The electrical conductivity σb, the Seebeck coefficient Sb can be expressed as

$$\sigma_b=\frac{e^2I_0}{\hbar a_B},\ S_b=\frac{k_BI_1}{eI_0} \qquad (3)$$

where e is the elementary charge, kB the Boltzmann constant, $\hbar$ the reduced Planck constant, and aB the Bohr radius. The In is a dimensionless integral which can be written as

$$I_n = \frac{\hbar a_B}{(2\pi)^3}\int d^3k\left(\frac{E_k - E_f}{k_B T}\right)^n\left[-\frac{\partial f(E_k - E_f)}{\partial E_k}\right]\left(\frac{\partial E_k}{\hbar \partial k}\right)^2 \tau_k,\quad n = 0,\ 1,\ 2 \tag{4}$$

where τ_k is the relaxation time. Here we consider the electron–phonon scattering mechanisms including the acoustic deformation potential scattering and the polar optical phonon scattering. The optical deformation potential scattering is negligible. This is due to the fact that, in the case of zincblende-type semiconductors, the optical phonon displacement has symmetry Γ_4, so the matrix element of electron-optical-phonon interaction between two *s*-like Γ_1 conduction-band state is zero [26]. In addition, we also consider the mechanism of the ionized impurity scattering. The explicit form of τ_k can be found in Ref. [27]. The bulk thermal conductivity κ_b is the sum of contributions from the lattice κ_l and from the electronic carriers κ_e. The latter can be expressed as

$$\kappa_e = \frac{k_B^2 T}{\hbar a_B}\left(I_2 - \frac{I_1^2}{I_0}\right) \tag{5}$$

The parameters used in calculating the electronic band and the electronic relaxation time can be found in Refs. [28]; [29] ; [30]. Based on these formulae and parameters, our calculated electrical conductivities and the Seebeck coefficients as functions of donor concentration are in agreement with previous experiments [31] ; [32].

The thermoelectric transport properties of the GaAs-SAM-GaAs heterostructures can be characterized by the effective electrical conductivity σ_{eff}, the effective thermal conductivity κ_{eff}, the effective Seebeck coefficient S_{eff}, and the effective figure of merit $Z_{\mathrm{eff}} T = \sigma_{\mathrm{eff}} S_{\mathrm{eff}}^2 T / \kappa_{\mathrm{eff}}$.These effective thermoelectric parameters are defined by requiring that the two terminals of the GaAs-SAM-GaAs heterostructure have the same response as those of the homogeneous material whose thermoelectric transport parameters and size are the same as the effective ones of the GaAs-SAM-GaAs heterostructures when the temperature or the voltage bias is applied. According to this definition, the effective thermal conductivity of the GaAs-SAM-GaAs heterostructure can be expressed as

$$\kappa_{\mathrm{eff}} = L\left(\frac{L - d_{\mathrm{SAM}}}{\kappa_b} + \frac{1}{G_{\mathrm{th}}}\right)^{-1} \tag{6}$$

where L is the total thickness, d_{SAM} the thickness of the SAM, and G_{th} the interfacial thermal conductance of the SAM. The effective electrical conductivity can be written in a similar form,

$$\sigma_{\text{eff}} = L\left(\frac{L - d_{\text{SAM}}}{\sigma_b} + \frac{1}{G_e}\right)^{-1} \tag{7}$$

where Ge is the interfacial electrical conductance of the SAM. The electronic transport through the SAM is characterized as that through a rectangular barrier. By using the Tsu–Esaki formula [33], the tunneling current induced by the bias V and the temperature difference ΔT can be written as

$$J_t = -\frac{ek_B m_c^*}{2\pi^2\hbar^3}\int_0^{\infty} dE\left\{\Xi(E)T\log\left[1+\exp\left(-\frac{E-E_f}{k_BT}\right)\right]\right. \\ \left. -(T+\Delta T)\log\left[1+\exp\left(-\frac{E-E_f-eV}{k_B(T+\Delta T)}\right)\right]\right\} \tag{8}$$

where m_c^* is the GaAs conduction-band effective mass, e the elementary charge, and k_B the Boltzmann constant. The $\Xi(E)$ is the tunneling probability of electrons through the barrier with barrier height Eb, thickness d_{SAM}, and electron effective mass m_{SAM}^*:

$$\Xi(E) = \frac{1}{1+\frac{E_b^2\sinh^2(\gamma_{\text{SAM}}d_{\text{SAM}})}{4E(E_b-E)}},\ \gamma_{\text{SAM}} = \frac{\sqrt{2m_{\text{SAM}}^*(E_b-E)}}{\hbar} \tag{9}$$

The interfacial electrical conductance is $G_e \equiv \partial J_t/\partial V$, and can be expressed as

$$G_e = \frac{e^2 m_c^*}{2\pi^2\hbar^3}\int_0^{\infty} dE\,\Xi(E)f(E-E_f) \tag{10}$$

The effective Seebeck coefficient can be expressed as

$$S_{\text{eff}} = (1-r)S_b + rS_{\text{SAM}} \tag{11}$$

where $r \equiv \kappa_{\text{eff}}/LG_{th}$ equals the ratio of the temperature drop across the SAMs to that across of the entire structure. The S_{SAM}, the Seebeck coefficient of the SAM, is the ratio of the voltage drop to the temperature bias across the SAM in the absence of the electric current. It is proportional to the difference between the average energy of the conduction electrons and the Fermi level under the assumption that the relaxation time is a constant in the energy region several $k_B T$ around the average energy of the conduction electrons [34]. Therefore, the Seebeck coefficient of the SAM should be nearly equal to that of the bulk GaAs if the tunneling probability of the conduction electrons through the SAM does not correlate noticeably with the energies of electrons. In this case, the average energy of conduction electrons in the SAM is nearly the same as that in the bulk GaAs. However, the tunneling probability could correlate significantly with the energy of electrons. In general, the electrons of higher kinetic energy can more easily tunnel through the barrier. Therefore, the average energy of conduction electrons (and hence the Seebeck coefficient) in the SAM should be larger than that in the bulk GaAs. Similar phenomena have been seen by Kim and Lundstrom [35]. The Seebeck coefficient of the SAM is nearly the sum of the *Sb* and a correction term caused by the energy dependence of the tunneling probability δS, which can be expressed as

$$\delta S = \lim_{\Delta T \to 0} \left[\left(\frac{V}{\Delta T} \right)_{J_t = 0} - \left(\frac{V}{\Delta T} \right)_{J_t^{(0)} = 0} \right] \tag{12}$$

where $J_t^{(0)}$ is the same as *Jt* except that the tunneling probability Ξ is set as a constant.

3. RESULTS AND DISCUSSION

In the present calculation, the magnitude of the lattice thermal conductivity is set as 45 W m^{-1} K^{-1} given from the experimental data of undoped GaAs [30]. In the case of doped GaAs, this magnitude could be slightly overestimated because of the neglected thermal resistance from the phonon scattering by impurities and electrons. In comparison with the lattice thermal conductivity, the electronic thermal conductivity is almost negligible. When the donor concentration is as high as 10^{20}cm^{-3}, the calculated electronic thermal conductivity is 1.3 W m^{-1} K^{-1}. The tunneling barrier *Eb* and the effective mass m^*_{SAM} are set as 1.25eV and $0.15 m_0$, respectively. [36] Here, the m_0 is the free-electron mass. The thickness of the whole structure is set as L=200nm and that of the SAM is set as d_{SAM}=0.7nm. The SAM interfacial thermal conductance, as mentioned earlier, is set as a quiet large value of 25 MW m^{-2} K^{-1}. Substituting these parameters into Eq. (6), we obtain the effective thermal conductivity $\kappa_{\mathrm{eff}} = 4.5\ \mathrm{W\ m^{-1}\ K^{-1}}$, which is insensitive to the donor concentration and one order smaller than the thermal conductivity of the bulk GaAs. This is an evidence that the presence of a SAM can greatly reduce the thermal conductance.

Fig. 1 shows the effective electrical conductivity σ_{eff} of the GaAs-SAM-GaAs heterostructure and the electrical conductivity σb of the bulk GaAs as functions of donor concentration. As can be seen, the electrical conductivity is reduced due to the presence of the SAM. Both the σ_{eff} and the σb increase with increasing donor concentration. We find the former increases more rapidly than the latter so that the ratio of the former to the latter, σ_{eff}/σ_b, increases with donor concentration as shown in Fig. 2. In order to gain further insight into the electronic transport properties, the Fermi level as a function of donor concentration is also plotted in Fig. 2, in which the energy zero is set at the CBE of the GaAs. As shown, the ratio σ_{eff}/σ_b is insensitive to the donor concentration when the Fermi level is lower than the CBE, and increases remarkably when the Fermi level is higher than the CBE. When the Fermi level is lower than the CBE, almost all the conduction electrons are near the CBE. Their energy distribution and hence the tunneling probability through the SAM barrier are insensitive to the donor concentration. However, when the Fermi level is higher than the CBE, the energies of almost all the conduction electrons are near the Fermi level, and increase remarkably with the donor concentration due to the Pauli exclusion principle. Because the electrons of higher energy can more easily tunnel through the SAM barrier, the ratio σ_{eff}/σ_b increases remarkably with the donor concentration, when the Fermi level is higher than the CBE.

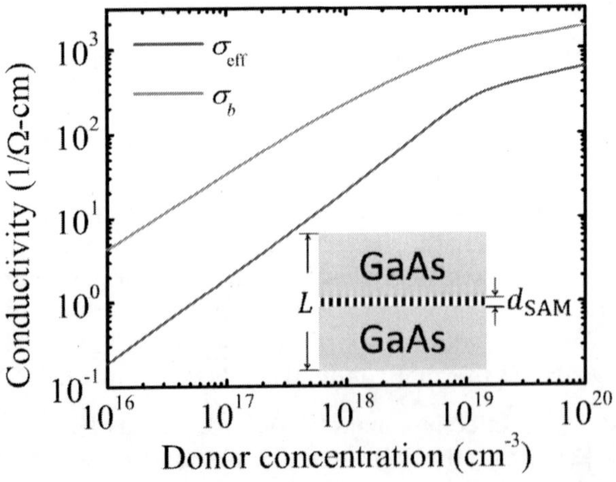

Figure. 1. The effective conductivity σ_{eff} of the GaAs-SAM-GaAs heterostructure and the electrical conductivity σ_b of the bulk GaAs as functions of donor concentration. The inset is a schematic plot of the GaAs-SAM-GaAs heterostructure.

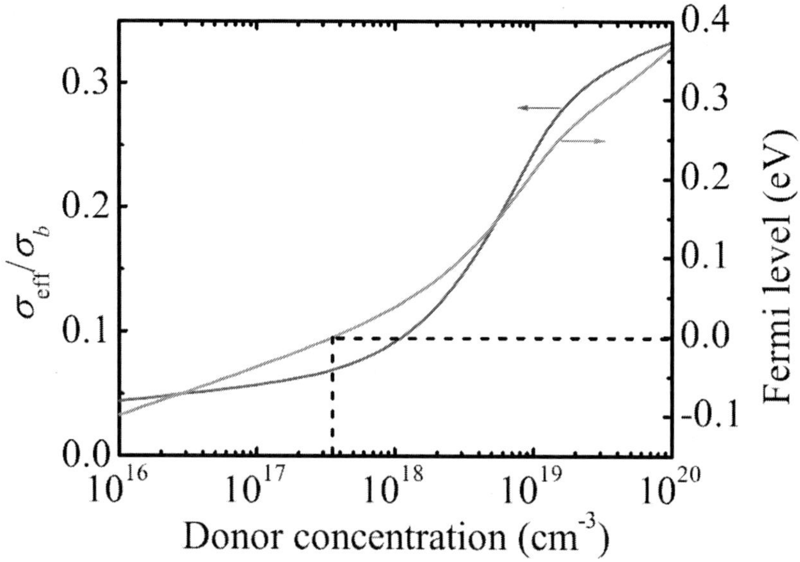

Figure. 2. The ratio of the effective electrical conductivity σ_{eff} of the GaAs-SAM-GaAs structure to the electrical conductivity σ_b of the bulk GaAs as a function of donor concentration. The Fermi level is also plotted. The energy zero is set at the CBE.

As mentioned earlier, the presence of the SAM can enhance the Seebeck coefficient. This enhancement is significant especially for the case of high donor concentration as illustrated in Fig. 3. The magnitude of the S_{eff} is about 1.5 times higher than that of the *Sb*, when the donor concentration is higher than $4\times10^{18}cm^{-3}$. When the donor concentration is $10^{16}cm^{-3}$, the S_{eff} is 1.15 times the magnitude of the *Sb*. Because for nondegenerate semiconductors, the magnitude of the Seebeck coefficient generally decreases rapidly with increasing dopant concentration, the enhancement of the Seebeck coefficient, i.e., the $S_{eff}-S_b$, which is only weakly correlated with the donor concentration, should be generally more important for the case of high dopant concentration than that of low dopant concentration. Note that good thermoelectric materials are typically heavily doped semiconductors with a carrier concentration between $10^{19}cm^{-3}$ and $10^{21}cm^{-3}$[3]. Therefore, enhancement of Seebeck coefficient caused by the presence of the SAM should be significant for most of the thermoelectric materials.

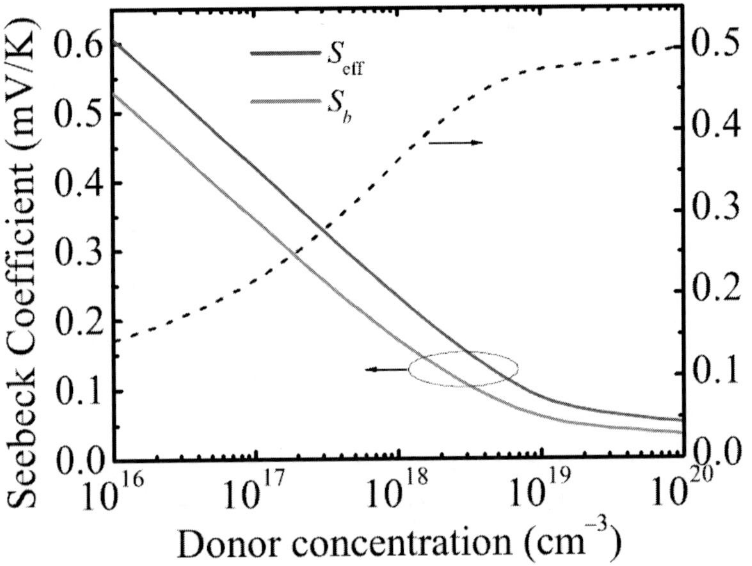

Figure. 3. The effective Seebeck coefficient S_{eff} of the GaAs-SAM-GaAs structure and the Seebeck coefficient *Sb*of the bulk GaAs as functions of donor concentration.

Fig. 4 shows the effective figure of merit $Z_{eff}T$ of the GaAs-SAM-GaAs structure and the figure of merit $Z_b T$ of the bulk GaAs as functions of donor concentration. In the case of low donor concentration ($N_d \lesssim 10^{17}$ cm^{-3}), the $Z_{eff}T$ of the GaAs-SAM-GaAs structure is slightly smaller than the $Z_b T$ of the bulk GaAs. When the donor concentration is higher than $10^{18}cm^{-3}$, the $Z_{eff}T$ of the GaAs-SAM-GaAs structure is much higher than the $Z_b T$ of the bulk GaAs. This is mainly due to the fact that in the case of $N_d \lesssim 10^{17}$ cm^{-3}, the Fermi level is lower than the CBE, and then almost all of the conduction electrons are near the CBE. The tunneling of the conduction electrons can be significantly blocked by the SAM. When the donor concentration is higher than $10^{18}cm^{-3}$, the conduction electrons gain more kinetic energy and can tunnel through the SAM more easily. In this case, although the SAM, to some extent, can reduce the electrical conductance, the effect of electrical-conductance reduction is smaller than that of the thermal-conductance reduction. When the donor concentration is higher than $10^{20}cm^{-3}$, the $Z_{eff}T$ of the GaAs-SAM-GaAs structure is nearly one order higher than the $Z_b T$ of the bulk GaAs. Furthermore, the enhancement of the Seebeck coefficient also plays an important role as mentioned earlier. This approximately doubles the figure of merit when the donor concentration is higher than $2\times10^{18}cm^{-3}$.

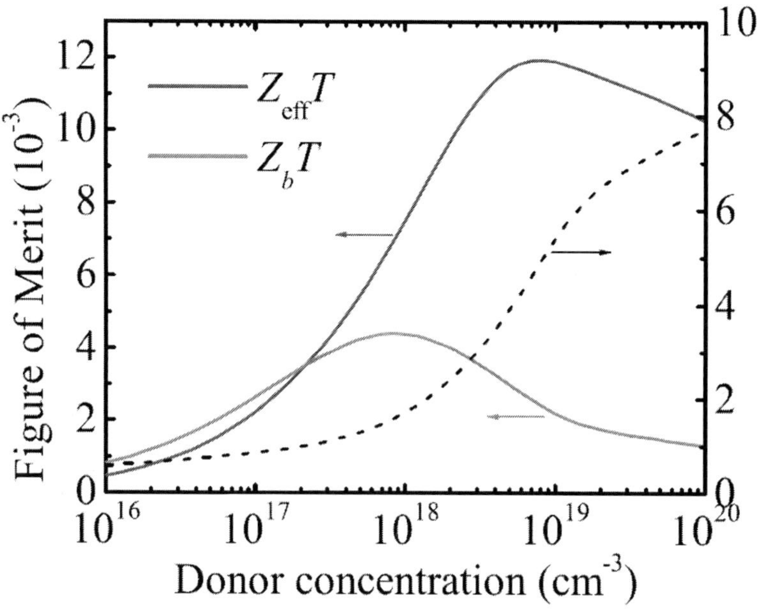

Figure. 4. The effective figure of merit $Z_{eff}T$ of the GaAs-SAM-GaAs structure and the figure of merit Z_bT of the bulk GaAs as functions of donor concentration.

To give an order parameter indicating the thermoelectric performance, we define an enhancement factor γ , which is the ratio of the peak height of the $Z_{eff}T$ to that of the Z_bT. The factor γ being greater (smaller) than unity suggests that the system is improved (deteriorated) by the presence of the SAM. In the present case, the enhancement factor $\gamma=2.71$ as can be seen in Fig. 4. Here, we set the barrier height E_b=1.25meV and the thickness d_{SAM}=0.7nm. These parameters depend on the molecule type of the SAM and the chain length. The enhancement factor γ of various magnitudes of the Eb and the d_{SAM} is plotted in Fig. 5. As shown, the magnitude of the enhancement factor γ generally decreases with increasing barrier height Eb and with increasing thickness d_{SAM}, because the SAM of higher Eb and larger d_{SAM} can block the electron tunneling more efficiently. On the other hand, the maximum of the enhancement factor γ is not observed under the Ohmic-contact condition (i.e., $E_b=0$), but under the condition of $E_b \gtrsim 0.1$ eV. This is due to the fact that the Seebeck coefficient can be enhanced by the presence of the SAM.

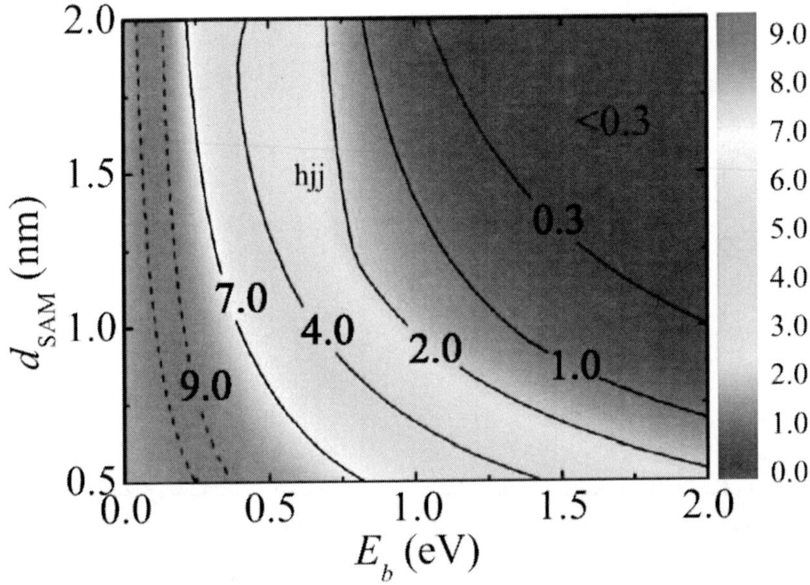

Figure. 5. The contour plot of the enhancement factor $\gamma(E_b, d_{SAM})$ in the Eb–d_{SAM} plane. The region of $\gamma \geq 9$ is that between the two dashed lines.

In this study, we only consider the most typical SAM, 1,*n*-alkanedithiol. It is worth to note that the choice of the SAM material will greatly change both the electronic transport properties and the thermal transport properties [37] ; [38]. In the present case, the Fermi level locates in the energy gap defined by the energy region between the highest occupied molecular orbital (HOMO) and the lowest unoccupied molecular orbital (LUMO). There is no molecular state of the SAM near the Fermi level. We have shown, even in this case, the TE performance can still be largely enhanced. In future studies, it would be of great interest to choose other SAM molecules whose molecular orbitals are near the Fermi level. As a result, the conduction electrons can transport more easily through the SAM by the resonant tunneling and hence result in a higher electrical conductance. In this case, the SAM should not be treated as a quantum barrier anymore, and the TE performance will be further improved by the enhanced electrical conductance.

4. CONCLUSIONS

In conclusion, we propose and demonstrate that the presence of SAMs in semiconductors can greatly enhance the thermoelectric performance. Although the presence of SAM, to some extent, reduces the electrical conductance, it also strongly decreases the thermal conductance. The electrical-conductance reduction on the thermoelectric performance is much weaker than that of the thermal-conductance reduction, particularly at high donor concentration ($N_d \gtrsim 10^{18}\ \mathrm{cm}^{-3}$). In this case, the energy of most of the conduction electron is much higher than that of the CBE, so that they can more easily tunnel through the SAM in comparison with the low donor concentration case ($N_d \lesssim 10^{17}\ \mathrm{cm}^{-3}$). Furthermore, due to the fact that the electrons of higher energy can more easily tunnel through the SAM, the Seebeck coefficient can be also enhanced due to the presence of the SAM. This effect is significant especially for heavily doped semiconductors, which are typically the cases for good thermoelectric materials. In this work we consider the case of GaAs as an example, the effect of the SAMs on the thermoelectric performance presented here should be generally applicable to other semiconductors.

ACKNOWLEDGMENTS

This work was supported by the Ministry of Science and Technology, Taiwan.

REFERENCES

1. In: D.M. Rowe (Ed.), CRC Handbook of Thermoelectrics, 1995, CRC Press, Boca Raton.
2. T.M. Tritt, M.A. Subramanian MRS Bull., 31 (2006), p. 188
3. G.J. Snyder, E.S. Toberer Nat. Mater., 7 (2008), p. 105
4. G.J. Snyder Electrochem. Soc. Interface, 17 (2008), p. 54
5. T.M. Tritt Annu. Rev. Mater. Res., 41 (2011), p. 433
6. H. Alam, S. Ramakrishna Nano Energy, 2 (2013), p. 190
7. M.H. Elsheikh, D.A. Shnawah, M.F.M. Sabri, S.B.M. Said, M.H. Hassan, M.B.A. Bashir, M. Mohamad Renew. Sustain. Energy Rev., 30 (2014), p. 337
8. W. He, G. Zhang, X.X. Zhang, J. Ji, G.Q. Li, X.D. Zhao Appl. Energy, 143 (2015), p. 1
9. D.K. Aswal, R. Basu, A. Singh Energy Convers. Manag., 114 (2016), p. 50
10. Q.H. Zhang, X.Y. Huang, S.Q. Bai, X. Shi, C. Uher, L.D. Chen Adv. Eng. Mater., 18 (2016), p. 194
11. J.P. Heremans, B. Wiendlocha, A.M. Chamoire Energy Environ. Sci., 5 (2012), p. 5510
12. Y.Z. Pei, X.Y. Shi, A. LaLonde, H. Wang, L.D. Chen, G.J. Snyder Nature, 473 (2011), p. 66

13. D.M. Rowe, G. Min, in: Thirteenth International Conference on Thermoelectrics, IEEE, Kansas City, 1995, p. 339.
14. Shakouri, J.E. Bowers, Appl. Phys. Lett. 71 (1997) 1234.
15. J.M.O. Zide, D. Vashaee, Z.X. Bian, G. Zeng, J.E. Bowers, A. Shakouri, A.C. Gossard Phys. Rev. B, 74 (2006)
16. B. Paul, V.A. Kumar, P. Banerji J. Appl. Phys., 108 (2010), p. 064322
17. Soni, *et al.* Nano Lett., 12 (2012), p. 4305
18. Y.F. Wang, X.Y. Zhang, L.M. Shen, N.Z. Bao, C.L. Wan, N.H. Park, K. Koumoto, A. Gupta J. Power Sources, 241 (2013), p. 255
19. N. Neophytou, X. Zianni, H. Kosina, S. Frabboni, B. Lorenzi, D. Narducci Nanotechnology, 24 (2013), p. 205402
20. K. Biswas, J.Q. He, I.D. Blum, C.I. Wu, T.P. Hogan, D.N. Seidman, V.P. Dravid, M.G. Kanatzidis Nature, 489 (2012), p. 414
21. L.D. Zhao, V.P. Dravid, M.G. Kanatzidis Energy Environ. Sci., 7 (2014), p. 251
22. R.Y. Wang, R.A. Segalman, A. Majumdar Appl. Phys. Lett., 89 (2006), p. 173113
23. L. Hu, L.F. Zhang, M. Hu, J.S. Wang, B.W. Li, P. Keblinski Phys. Rev. B, 81 (2010), p. 235427
24. E.O. Kane Semicond. Semimet., 1 (1966), p. 75
25. G.D. Mahan, J.O. Sofo Proc. Natl. Acad. Sci. U.S.A., 93 (1996), p. 7436
26. P.Y. Yu, M. Cardona **Fundamentals of Semiconductors** Springer, Berlin, Heidelberg (2005)
27. C. Jacoboni **Theory of Electron Transport in Semiconductors** Springer, Berlin, Heidelberg (2010)
28. O. Madelung **Semiconductors: Data Handbook** Springer-Verlag, Berlin (2004)
29. M. Lundstrom **Fundamentals of Carrier Transport** Cambridge University Press (2009)
30. S. Adachi **Handbook on Physical Properties of Semiconductors** Kluwer Academic Publishers (2004)
31. R.O. Carlson, S.J. Silverman, H. Ehrenreich J. Phys. Chem. Solids, 23 (1962), p. 422
32. S.M. Sze, K.N. Kwok **Physics of Semiconductor Devices** Wiley, New Jersey (2007)
33. R. Tsu, L. Esaki Appl. Phys. Lett., 22 (1973), p. 562
34. P. Pichanusakorn, P. Bandaru Mater. Sci. Eng. R—Rep., 67 (2010), p. 19
35. R. Kim, M.S. Lundstrom J. Appl. Phys., 110 (2011), p. 34511
36. F. Camacho-Alanis, L.L. Wu, G. Zangari, N. Swami J. Mater. Chem., 18 (2008), p. 5459
37. Z.T. Tian, A. Marconnet, G. Chen Appl. Phys. Lett., 106 (2015), p. 211602
38. S. Casalini, C.A. Bortolotti, F. Leonardi, F. Biscarini Chem. Soc. Rev. (2016) http://dx.doi.org/10.1039/c6cs00509h (Advance Article)

CHAPTER 11

Solid-State NMR Spectroscopic Approaches to Investigate Dynamics, Secondary Structure and Topology of Membrane Proteins

Shadi Abu-Baker*, Gary A. Lorigan

Department of Chemistry and Biochemistry, Miami University, Oxford, USA

ABSTRACT

Solid-state NMR spectroscopy is routinely used to determine the structural and dynamic properties of both membrane proteins and peptides in phospholipid bilayers [1-26]. From the perspective of the perpetuated lipids, ^{2}H solid-state NMR spectroscopy can be used to probe the effect of embedded proteins on the order and dynamics of the acyl chains of phospholipid bilayers [8-13]. Moreover, ^{31}P solid-state NMR spectroscopy can be used to investigate the interaction of peptides, proteins and drugs with phospholipid head groups [11-14]. The secondary structure of $^{13}C{=}O$ site-specific isotopically labeled peptides or proteins inserted into lipid bilayers can be probed utilizing ^{13}C CPMAS solid-state NMR spectroscopy [15-18]. Also, solid-state NMR spectroscopic studies can be utilized to ascertain pertinent information on the backbone and side-chain dynamics of $^{2}H$and ^{15}N-labeled proteins, respectively, in phospholipid bilayers [19-26]. Finally, specific ^{15}N-labeled amide sites on a protein embedded inside oriented bilayers can be used to probe the alignment of the helices with respect to the bilayer normal [2]. A brief summary of all these solid-state NMR approaches are provided in this minireview.

Keywords: Solid-State NMR; Structure and Dynamics; Membrane Proteins

1. INTRODUCTION

Membrane proteins make up approximately one-third of the total number of known proteins [27]. They play several significant roles in biological systems that includes transporting ions, acting as receptors, participating in membrane fusion and destabilization, and many others. Despite the abundance and clear importance of membrane-associated proteins, limited information about these systems exists. Structural studies of these membrane proteins are key to understand their biological functions. X-ray crystallography has been used to elucidate structural information of biologically significant protein systems [28-35]. However, the hydrophobic surfaces associated with membrane-bound protein systems make the crystallization process very challenging. Although investigators are making progress with X-ray techniques, still only a limited list of membrane protein structures have been obtain via X-ray crystallography [36]. Alternatively, solid-state NMR spectroscopy is a powerful technique that can be used to provide structural, orientational, and dynamic information about membrane protein systems in model membrane systems [2,37,38].

2. SOLID-STATE NMR APPROACHES

2.1. ^{2}H Solid-State NMR Spectroscopy of Lipids with ^{2}H-Lableled Acyl Chains

Multilamellar vesicles (MLVs) can be prepared from phospholipids such as 1-palmitoyl-2-oleoyl-sn-glycero-phosphocholine (POPC) and the details of sample preparation and type of lipid was reported previously [3-7]. Approximately 10% of deuterated POPC (sn-1 chain, POPC-d_{31}, see Figure 1) or any other deuterated lipid of interest can be added to each sample. Other lipids can be used such as the negatively charged lipids such as 1-palmitoyl(d_{31})-2-oleoyl-sn-glycero-3-[phospho-L-serine] phospholipids (POPS-d_{31}) used in Sap C studies (see Figure 2) and Distearoyl-d_{70}-phosphatidylglycerol (DSPGd_{70}) (see Figure 3).

The effect of embedded proteins on the order and dynamics of the acyl chains of POPC-d_{31} bilayers can be investigated utilizing ^{2}H solid-state NMR spectroscopy in the absence as well as in the presence of (X) mol% of the protein with respect to the lipids at different temperatures [40]. For example, in Figure 4(a), the central resonance doublet corresponds to the terminal CD_3 groups (Carbon # 1 in Figure 1) and the remaining overlapped doublets result from the different CD_2segments of the acyl chain of POPC-d_{31}. The addition of (X) mol% of specific protein with respect to the lipids to POPC MLVs could alter (depending on the protein of interest) the lineshape and the spectral resolution of the ^2H NMR spectra (Figure 4(b)) when compared to the control (Figure 4(a)) sample.

In Figure 4(b), the loss in spectral resolution is manifested by the disappearance of sharp edges of the ^2H NMR peaks. The changes in spectral resolution of the ^2H NMR spectra indicate that the protein interacts with the POPC-d_{31}-containing MLVs when compared to the control. If the spectral resolution and the line shapes did not change when compared to the control, this indicates that the protein does not interact significantly with the POPC MLVs as indicated in a previous study from the Lorigan Lab using the transmembrane domain segment of WTPLB [12].

All the powder-type ^2H NMR spectra of multilamellar dispersions of POPC-d_{31} can be numerically deconvoluted (dePaked) using the algorithm of McCabe and Wassal [41,42]. These spectra can be dePaked such that the bilayer normal was perpendicular with respect to the direction of the static magnetic field. Then, the quadrupolar splittings can be directly measured from the dePaked spectra and converted into order parameters as described before [11,12]. The quadrupolar splittings of the CD_3 methyl groups (see Figure 1) at the end of the acyl chains are the smallest and close to 0 kHz because they rotate at the fastest frequency. The second smallest splitting was assigned to the ^2H attached to C-14 and so forth along the acyl chain.

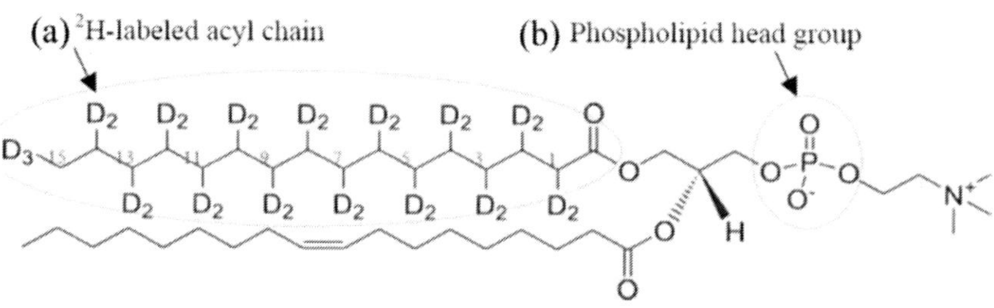

Figure 1. POPC-d_{31} structure [39]. The ^2H-labeled acyl chain and phospholipids head groups are labeled with dashed red circles. The numbering of the ^2H-labeled carbon skeleton of the acyl chain is shown is red.

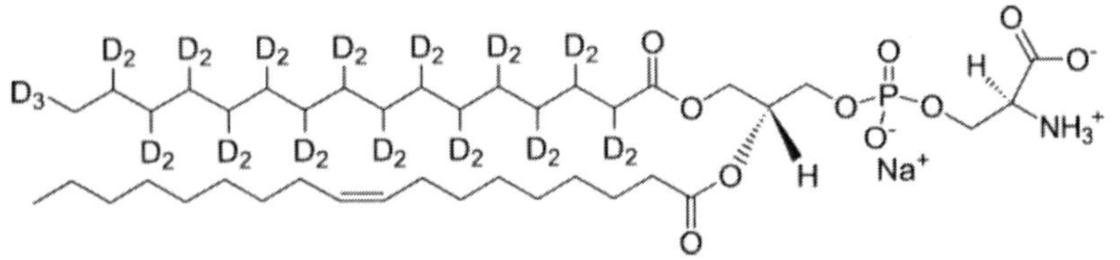

Figure 2. POPS-d_{31} structure [39]. Only one acyl chain is ^2H-labeled.

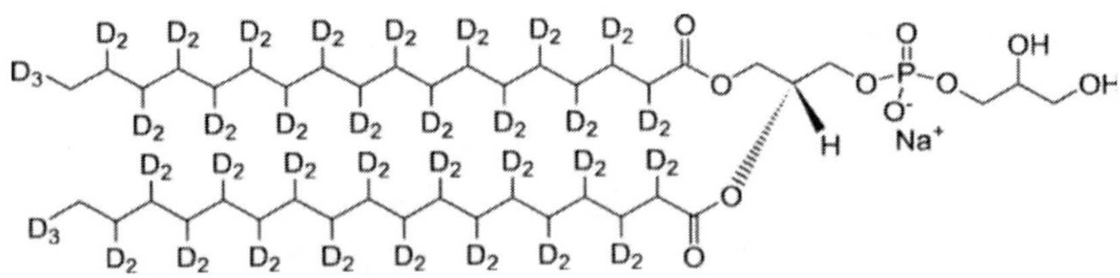

Figure 3. Distearoyl-d_{70}-phosphatidylglycerol (DSPG-d_{70}) structure [39].

The quadrupolar splittings for the deuterons in the plateau region were estimated by integration of the last broad peak.

The quadrupolar splittings can be directly measured from the dePaked spectra and converted into the S_{CD} order parameter using the following expression; $\Delta\nu^i_Q = 3/4(e^2qQ/h)\ S^i_{CD} S^i_{CD}$ [43-45]. Where $\Delta\nu^i_Q$ is the quadrupolar splitting for a deuteron attached to the ith carbon, e^2qQ/h is the quadrupolar splitting constant (168 kHz for deuterons in C-^2H bonds), and S^i_{CD} is the chain order parameter for a deuteron attached to the *i*th carbon of the acyl chain of POPC-d_{31}. The order parameters calculated for the CD_3 quadrupolar splitting should be multiplied by three according to procedures in the literature [46,47].

Figure 5 displays the smoothed segmental C-D bond order parameters (S_{CD}) of the POPC-d_{31} acyl chains calculated by dePakeing the corresponding ^2H NMR powder spectra shown in Figure 4. Figure 5 reveals a characteristic profile of decreasing order (S_{CD}) with increasing distance from the glycerol backbone of POPC-d_{31} MLVs. Any decrease (if any) in the order parameter profile of POPC-d_{31}-containing bilayers in the presence of (X) mol% of specific protein indicates more disorder and motion in the acyl chain region of the lipids when compared to the control sample (without protein). If the order parameter profile of POPC-d_{31}-containing bilayers in the presence of (X) mol% protein is similar to the control sample, this indicates that this particular protein is not altering the order and motion in the acyl chain region of the lipids.

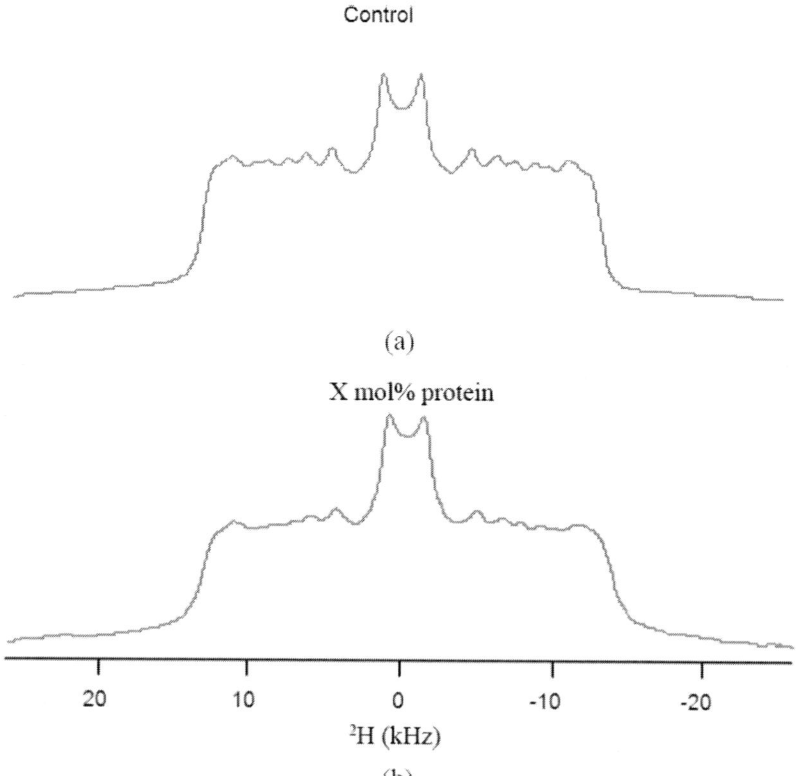

Figure 4. ^{2}H NMR powder-pattern spectra of POPC-d_{31} bilayers in the absence (a) and in the presence of (X) mol% protein with respect to the lipids.

2.2. ^{31}P Solid-State NMR Spectroscopy of Phospholipid Head Groups

The motionally averaged powder pattern spectra are characteristic of MLVs in the liquid crystalline phase (L_a) and are expected for phospholipids bilayers at a temperature well above the chain melting point transition temperature (T_m) [48]. Figure 6(a) shows the structure of the neutral phospholipid POPC ($T_m = -2$°C) [39]. In addition, other phospholipids that includes 1-palmitoyl-2- oleoyl-sn-glycero-3-[phospho-L-serine] (POPS, $T_m =$ 14 °C, Figure 6(b)), dioleoylphosphatidyl-glycerol (DOPG, $T_m = -18$°C, Figure 6(c)) and dioleoylphosphatidylserine (DOPS, $T_m = -11$°C, Figure 6(d)) structures are shown [39]. The static ^{31}P NMR spectrum of each phospholipid head group shown in **FIGURES 6**(a)-(d) have a specific ^{31}P chemical shift anisotropy (CSA; is equal s_{33}-s_{11}) width and lineshape. In several cases, the lineshapes of the static ^{31}P NMR spectra and its corresponding CSA values have been successfully used to study the perturbation effect induced by drugs and proteins on phospholipids [11,14,49,50]. The addition of (X) mol% of specific protein to specific MLVs could alter the ^{31}P CSA and line shape (Figure 7(b)) when compared to the control (Figure 7(a)) sample. This indicates that the protein of interest is interacting with the head group region of the MLVs.

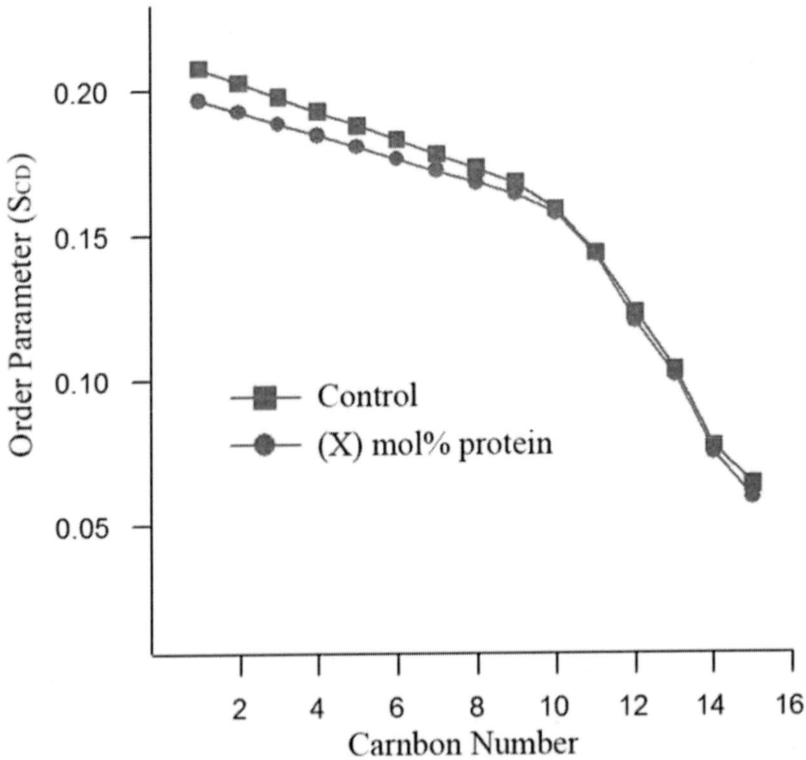

Figure 5. The smoothed acyl chains orientational order parameter (S_{CD}) profiles calculated from the dePaked ^{2}H NMR spectra of POPC-d_{31} (Figure 4). The closed squares and circles represent POPC bilayers with 0 mol% and (X) mol% protein, respectively.

2.3. ^{13}C Solid-State NMR Spectroscopy of Site-Specific ^{13}C-Labeled Peptides in MLVs

The secondary structure of $^{13}C{=}O$ site-specific isotopically labeled peptides or proteins inserted into lipid bilayers can be probed utilizing ^{13}C CPMAS solid-state NMR spectroscopy [15-18,50]. The local conformations of peptides and proteins can be characterized by examining the $^{13}C{=}O$ chemical shifts of the ^{13}C-labeled carbonyl of Ala, Leu, and Val. In general, the a-helical structure chemical shifts may vary within the data range (174 - 177 ppm) for different conformational-dependent changes (see Figure 8(a)) [17]. Also, a single ^{13}C NMR peak at approximately 172.4 ppm can be attributed to either a b-sheet or an unstructured structural conformation of the peptide (see Figure 8(b)); whereas a single peak at approximately 176 ppm can be attributed to one a-helical structure conformation (see Figure 8(c)) [15].

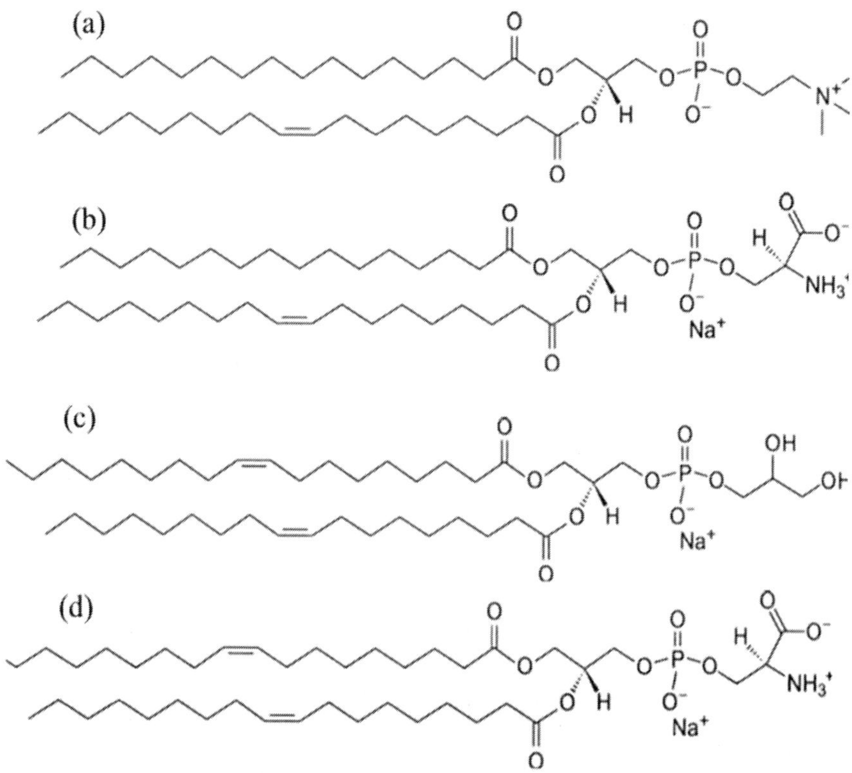

Figure 6. The structure of (a) POPC, (b) POPS, (c) DOPG and (d) DOPS phospholipids [39].

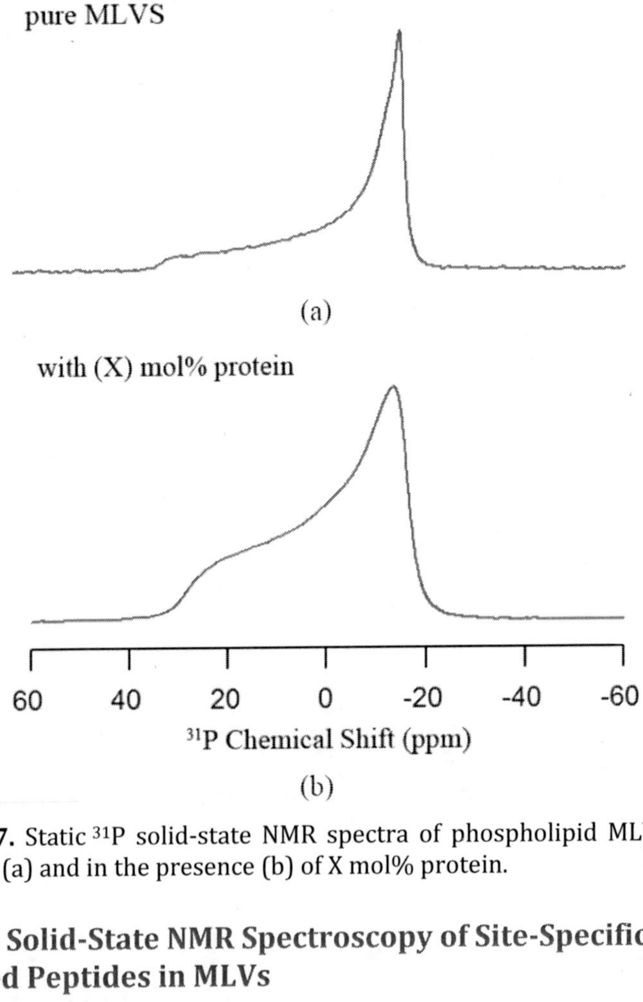

Figure 7. Static ^{31}P solid-state NMR spectra of phospholipid MLVs in the absence (a) and in the presence (b) of X mol% protein.

2.4. ^{2}H Solid-State NMR Spectroscopy of Site-Specific ^{2}H-Labeled Peptides in MLVs

As mentioned previously, probing of how both segments of phosphorylated and unphosphorylated membrane proteins are moving within the phospholipid bilayers is crucial to describe the physiological functions. ^{2}H solidstate NMR spectroscopy is a powerful well developed technique to study the structural and side-chain dynamic properties of membrane proteins in phospholipid bilayers [8,51-53]. The corresponding quadrupolar splitting and lineshapes of the ^{2}H solid-state NMR spectra can be used to probe the molecular dynamics of the sidechain of selectively labeled residues in site-specific ^{2}H-labeled integral membrane proteins [19-24].

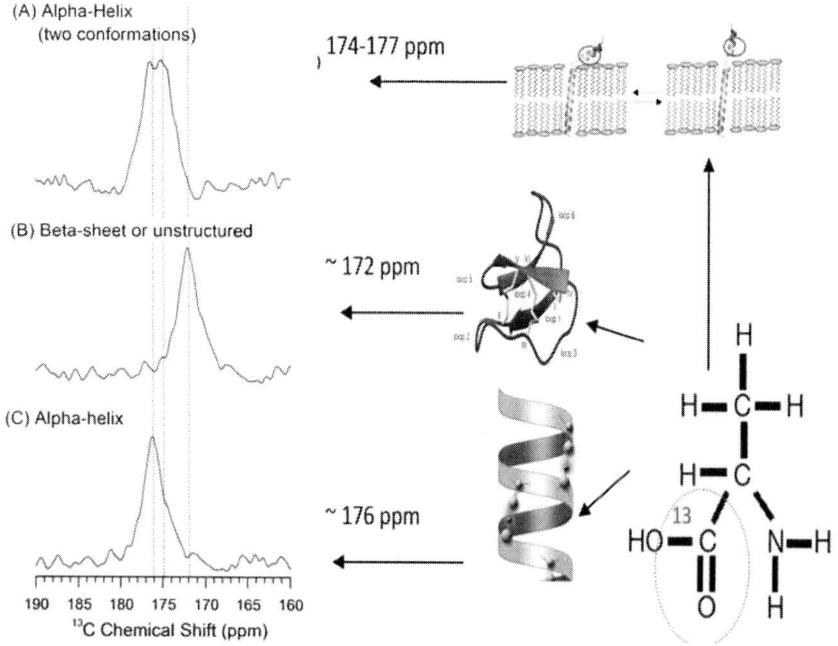

Figure 8. ^{13}C=O chemical shifts of ^{13}C-labeled carbonyl of Ala located at (A) α-helix with two structural conformations, (B) Unstructured or β-sheet structural conformation and (C) α-helical with one structure conformation.

In previous studies, methyl group motions have been well-characterized utilizing ^{2}H NMR studies of CD_3-labeled sites of alanines, valines and leucines [19,24,25,54-57]. For the isotopically labeled alanines (short aliphatic side-chains), the deuterated methyl group (CD_3) rotates along the $C\alpha$-$C_\beta$bond and allows the deuterons to make jumps between three-sites described by a tertrahedral geometry (see Figure 9(a)) [58,59]. However, for Leu, the long aliphatic side-chain can be isotopically labeled at the δ- and/or ε-CD_3 sites (see Figure 9 (b)) and the deuterium NMR powder pattern lineshapes will be strongly influenced by the motions about the C_γ-C_δ bond axis as well as by additional librational motion about the C_α-C_β and C_β-C_γ bond axes at various temperatures [24,60].

It has been reported that if the CD_3-methyl probe of a protein undergoes no motion other than those associated with the axial rotation about the C-CD_3 bond in a randomly dispersed sample, the resultant ^{2}H NMR spectra will consist of a Pake pattern with a 40 kHz quadrupolar splitting (see Figure 10(a)) [61]. However, residues located outside the membrane are expected to be more motionally averaged and yield an isotropic peak (see Figure 10(b)) [62].

2.5. ^{15}N Solid-State NMR Spectroscopy of Site Specific ^{15}N-Labeled Peptides in MLVs

^{15}N solid-state NMR spectroscopic studies are utilized to ascertain pertinent information on the backbone structure and dynamics of membrane proteins in phospholipid bilayers [25,26]. It provides important physiological and mechanistic information regarding the regulatory role of membrane proteins and it phosphorylated form in biological systems [64-66]. Generally, the immobile (without large amplitude motions) amide sites of specific ^{15}N-labeled proteins yield a broad static ^{15}N powder pattern (see Figure 11(a)); whereas, motionally averaged amide sites reveal isotropic peaks (see Figure 11(b)) [26].

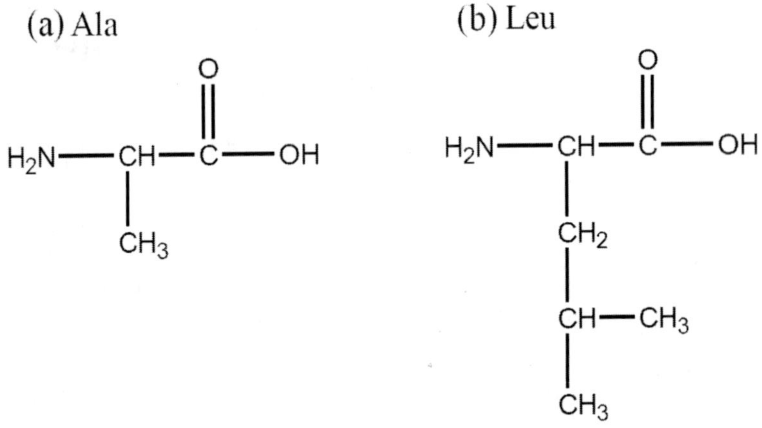

Figure 9. The structure of (a) alanine and (b) leucine residues.

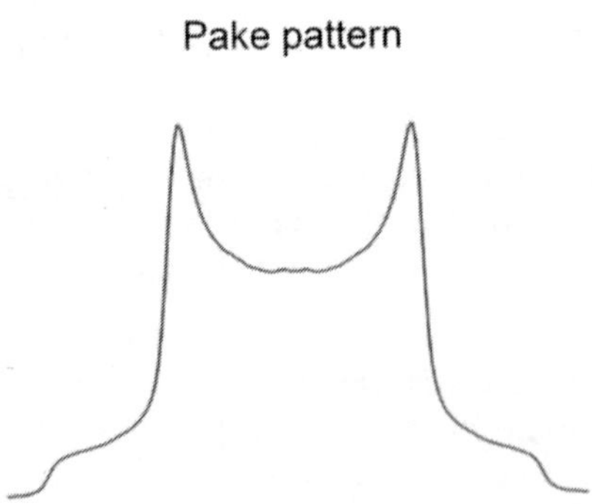

(a)

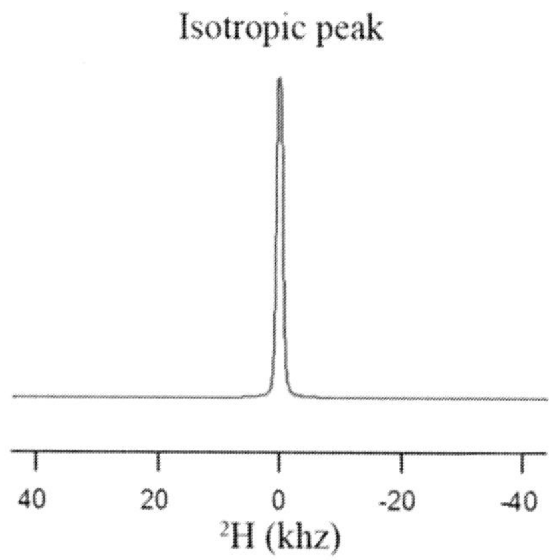

(b)

Figure 10. ^{2}H NMR simulations of CD_3-Ala using Multiple Axis Quadrupolar Echo Tailing (MXQET) program [63]. The model simulations show (a) Pake pattern and (b) isotropic peak spectra.

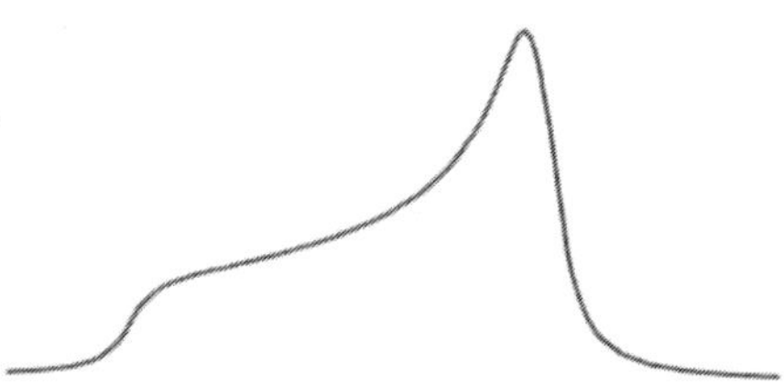

(a)

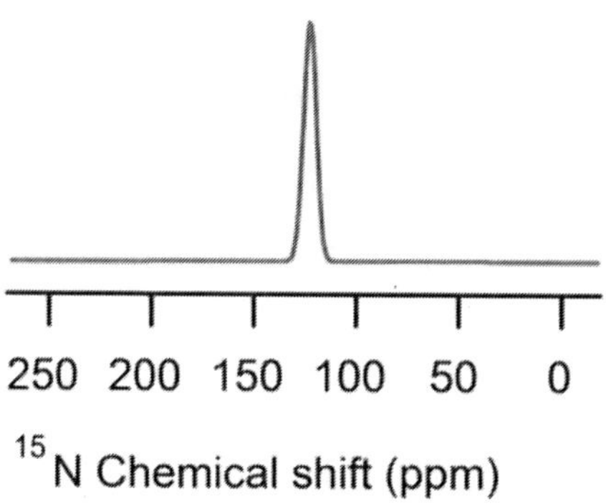

(b)

Figure 11. Simulations of the ^{15}N NMR spectra using the Dominique Massiot's Fit (DMFIT) software program [67]. The model simulations show (a) a powder pattern and (b) an isotropic peak.

2.6. ^{15}N Solid-State NMR Spectroscopy of Site-Specific ^{15}N-Labeled Peptides in Mechanically Oriented Bilayers

Determining the structural topology of membrane proteins and its interaction with the lipids is critical to understand its physiological regulatory function. ^{15}N solidstate NMR spectroscopy is a powerful tool to ascertain direct information regarding the structural topology of membrane proteins in oriented phospholipid bilayers [2].

In this approach, specific ^{15}N labeled amide sites can be used to probe the alignment of the helix with respect to the bilayer normal. For example, this technique was used to probe the orientation of both the transmembrane and cytoplasmic domains of WT-PLB embedded inside mechanically oriented phospholipids [68]. Site-specific ^{15}N-labeled WT-PLB were chosen at Ala11 (in the cytoplasmic domain) and Leu 42 and Leu 51 (both in the transmembrane helix) (see Figure 12).

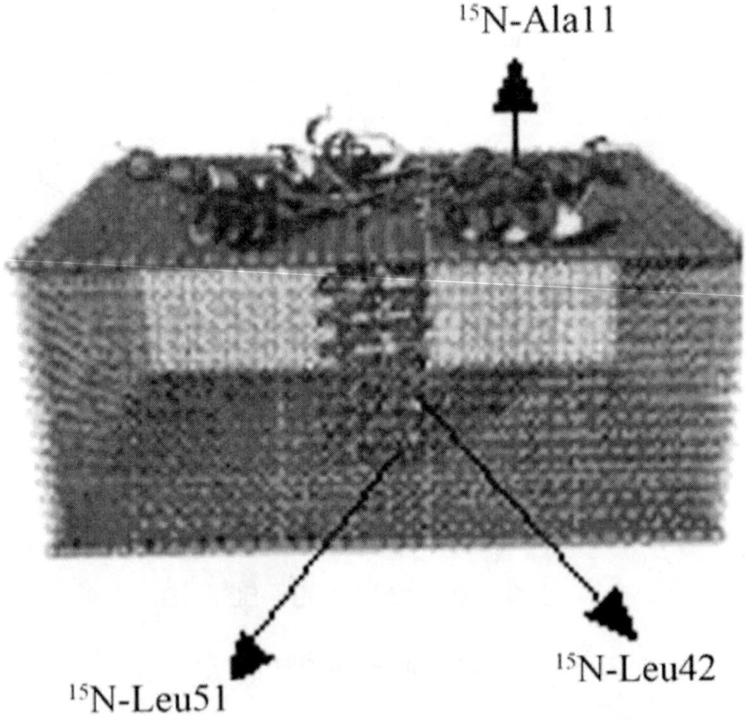

Figure 12. Side-view of a proposed structural model of site-specific ^{15}N-labeled WT-PLB at Ala11 (in the cytoplasmic domain) and Leu 42 and Leu 51 (both in the transmembrane helix) in oriented bilayers.

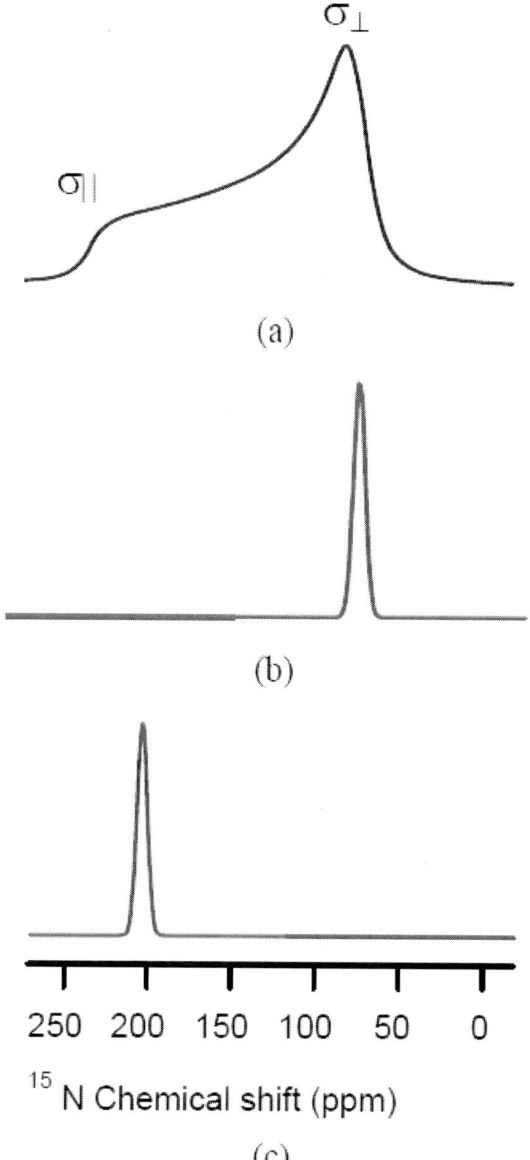

Figure 13. Simulations of the ^{15}N NMR spectra using the DMFIT software program [67]. The model simulations show (a) powder pattern and (b) ^{15}N resonance peak at approximately 70 ppm (close to the ⍰⍰ component of the chemical shift tensor of the corresponding powder spectrum) and (c) ^{15}N resonance peak at approximately 210 ppm (close to the ⍰⍰⍰component).

A resonance peak at approximately 70 ppm (close to the $\sigma_{\perp}$ component of the chemical shift tensor of the corresponding powder spectrum, see Figure 13), indicates that this residue is oriented approximately perpendicular to the bilayer normal (transmembrane helix as shown in Figure 12) [2]. Also, a ^{15}N resonance peak at approximately 210 ppm (close to the □□□ component of the chemical shift tensor of the corresponding powder spectra, see Figure 13) indicates that the amide backbone vector of this residue is nearly parallel to the bilayer normal (the helix lies on the surface of the phospholipids bilayers as shown for cytoplasmic domain of the proposed structural model of WT-PLB presented in Figure 12 [2].

3. CONCLUSION

Solid-state NMR spectroscopy is routinely used to determine the structural and dynamic properties of both membrane proteins and peptides in phospholipid bilayers. Together, ^{2}H, ^{31}P, ^{13}C and ^{15}N solid-state NMR spectroscopy can be used to probe the effect of embedded proteins on the order and dynamics of the acyl chains, phospholipid head group as well as the secondary structure of site-specific isotopically labeled amino acid and helix orientation with respect to the membrane. A summary of all these solid-state NMR approaches are provided in this mini-review.

REFERENCES

1. T. A. Cross, "Solid-State Nuclear Magnetic Resonance Characterization of Gramicidin Channel Structure," Methods in Enzymology, Vol. 289, 1997, pp. 672-696.
2. T. A. Cross and S. J. Opella, "Solid-State NMR Structural Studies of peptides and Proteins in Membranes," Current Opinion in Structural Biology, Vol. 4, No. 4, 1994, pp. 574-581.doi:10.1016/S0959-440X(94)90220-8
3. J. R. Long, W. J. Shaw, P. S. Stayton and G. P. Drobny, "Structure and Dynamics of Hydrated Staherin on Hydroxyapatite as Determined by Solid-State NMR," Biochemistry, Vol. 40, No. 51, 2001, pp. 15451-15455.
4. R. Mani, J. J. Buffy, A. J. Waring, R. I. Lehrer and M. Hong, "Solid-State NMR Investigation of the Selective Disruption of Lipid Membranes by Protegrin-1," Biochemistry, Vol. 43, No. 43, 2004, pp. 13839-13848. doi:10.1021/bi048650t
5. Marcotte, E. J. Dufourc, M. Ouellet and M. Auger, "Interaction of the Neuropeptide Met-Enkephalin with Zwitterionic and Negatively Charged Bicelles as Viewed by ^{31}P and ^{2}H Solid-State NMR," Biophysical Journal, Vol. 85, No. 1, 2003, pp. 328-339.
6. Y. Nakazawa and T. Asakura, "Structure Determination of a Peptide Model of the Repeated Helical Dmoain in Samia Cynthia Ricini Silk Fibroin before Spinning by a Combination of Advanced Solid-State NMR Methods," Journal of the American Chemical Society, Vol. 125, No. 24, 2003, pp. 7230-7237.
7. Watts, "Solid-State NMR Approaches for Studying the Interaction of Peptides and Proteins with Membranes," Biochim Biophys Acta, Vol. 37, No. 2, 1998, pp. 297-318.

8. W. Koenig, J. A. Ferritti and K. Gawrisch, "Site-Specific Deuterium Order Parameters and Membrane-Bound Behavior of a Peptide Fragment from the Intracellular Domain of HIV-1 gp-41," Biochemistry, Vol. 38, No. 19, 1999, pp. 6327-6334.
9. G. Laroche, E. J. Dufourc, M. Pezolet and J. Dufourcq, "Coupled Changes between Lipid Order and Polypeptide Conformation at the Membrane-Surface: A ^{2}H NMR and Raman-Study of Polylysine Phosphatidic-Acid Systems," Biochemistry, Vol. 29, No. 27, 1990, pp. 6460-6465. doi:10.1021/bi00479a018
10. S. Yamaguchi, D. Huster, A. Waring, R. I. Lehrer, W. Kearney, B. F. Tack and M. Hong, "Orientation and Dynamics of an antimicrobial Peptide in the Lipid Bilayer by Solid-State NMR Spectroscopy," Biophysical Journal, Vol. 81, No. 4, 2001, pp. 2203-2214.doi:10.1016/S0006-3495(01)75868-7
11. S. Abu-Baker, X. Qi, J. Newstadt and G. A. Lorigan, "Structural Changes in a Binary Mixed Phospholipid Bilayer of DOPG and DOPS upon Sap C Interaction at Acidic pH Utilizing ^{31}P and ^{2}H Solid-State NMR Spectroscopy," Biochimica et Biophysica Acta, Vol. 17, No. 1, 2005, pp. 58-66.
12. P. C. Dave, E. K. Tiburu, K. Damodaran and G. A. Lorigan, "^{31}P and ^{2}H Solid-State NMR Spectroscopic Studies of the Transmembrane Domain of the Membrane-Bound Protein Phospholamban," Biophysical Journal, Vol. 86, 2004, pp. 1564-1573.
13. J. Lu, K. Damodaran, J. Blazyk and G. A. Lorigan, "Solid-State Nuclear Magnetic Resonance Relaxation Studies of the Interaction Mechanism of Antimicrobial Peptides with Phospholipid Bilayer Membranes," Biochemistry, Vol. 44, No. 30, 2005, pp. 10208-10217.
14. J. S. Santos, D. K. Lee and A. Ramamoorthy, "Effects of Antidepressants on the Conformation of Phospholipid Headgroups Studied by Solid-State NMR," Magnetic Resonance in Chemistry, Vol. 42, No. 2, 2004, pp. 105-114.
15. D. J. Hirsh, J. Hammer, W. L. Maloy, J. Blazyk and J. Schaefer, "Secondary Structure and Location of a Magainin Analogue in Synthetic Phospholipid Bilayers," Biochemistry, Vol. 35, No. 39, 1996, pp. 12733-12741. doi:10.1021/bi961468a
16. S. Kimura, A. Naito, S. Tuzi and H. Saito, "Dynamics and Orietation of Transmembrane Peptide from Bacteriorhodopsin Incorporated into Lipid Bilayer as Revealed by Solid State ^{31}P and ^{13}C NMR Spectroscopy," Biopolymers, Vol. 63, No. 2, 2002, pp. 122-131.
17. S. Tuzi, A. Naito and H. Saito, "A High-Resolusion SolidState 13C-NMR Study on [1-13C]Ala and [3-13C] Ala and [1-13C]Leu and Val-Labelled Bacteriorhodopsin: Conformation and Dynamics of Transmembrane Helics, Loops and Termini, and Hydration-Induced Conformational Changes," European Journal of Biochemistry, Vol. 218, No. 3, 1993, pp. 837-844.
18. S. Yamaguchi, K. Shimono, Y. Sudo, S. Tuzi, A. Naito and N. Kamo, "Conformation and Dynamics of the [3- 13C]Ala, [1-13C]Val-Labeled Truncated Pharaonis Transducer, pHtrII(1-159), as Revealed by Site-Directed 13C Solid-State NMR: Changes Due to Association with Phoborhodopsin (Sensory Rhodopsin II)," Biophysical Journal, Vol. 86, No. 5, 2004, pp. 3131-3140.
19. L. S. Batchelder, C. E. Sullivan, L. W. Jelinski and D. A. Trochia, "Characterization of Leucine Side-Chain Reorientation in Collagen-Fibrils by Solid-State Deuterium NMR," Proceedings of the National Academy of Sciences of the USA, Vol. 79, No. 2, 1982, pp. 386-389.
20. K. Beshah, E. T. Olejniczak and R. G. Griffin, "Deuterium NMR Study of Methyl Dynamics in L-Alanine," Journal of Chemical Physics, Vol. 86, No. 9, 1987, pp. 4730-4736.
21. M. A. Keniry, A. Kintanar, R. L. Smith, H. S. Gutowsky and E. Oldfield, "Nuclear Magnetic Resonance Relaxation of Deuteriomethyl-Labeled Amino Acids Crystals and in Halobacterium Halobium and E. coli Cell Membranes," Biochemsitry, Vol. 23, 1984, pp. 288-298.
22. R. A. Kinsey, A. Kintanar and E. Oldfield, "Dynamics of Amino Acid Side-Chains in Membrane Proteins by High Field Solid-State Deuterium Nuclear Magnetic

Resonance Spectroscopy," The Journal of Biological Chemistry, Vol. 256, No. 17, 1981, pp. 9028-9036.

23. G. C. Leo, L. A. Colnago, K. G. Valentine and S. J. Opella, "Dynamics of fd Coat Protein in Lipid Bilayers," Biochemistry, Vol. 26, No. 3, 1987, pp. 854-862. doi:10.1021/bi00377a029

24. W. Ying, S. E. Irvine, R. A. Beekman, D. J. Siminovitch and S. O. Smith, "Deuterium NMR Reveals Helix Packing Interactions in Phospholamban," Journal of the American Chemical Society, Vol. 122, No. 45, 2000, pp. 11125-11128.

25. E. K. Tiburu, P. C. Dave, K. Damodaran and G. A. Lorigan, "Investigating Leucine Side-Chain Dynamics and Backbone Conformations of Phospholamban Incorporated in Phospholipid Bilayers Utilizing ^{2}H and ^{15}N SolidState NMR Spectroscopy," Biochemistry, Vol. 43, No. 44, 2004, pp. 13899-13909. doi:10.1021/bi0490993

26. L. A. Colnago, K. G. Valentine and S. J. Opella, "Dynamics of fd Coat Protein in Bacteriophage," Biochemistry, Vol. 26, No. 5, 1987, pp. 847-854.

27. W. L. Smith, R. M. Garavito and S. Fergusson-Miller, "Membrane Protein Structural Biology Minireview Series," The Journal of Biological Chemistry, Vol. 276, No. 35, 2001, pp. 32393-32394. doi:10.1074/jbc.R100044200

28. D. A. Doyle, J. M. Cabral, R. A. Pfuetzner, A. L. Kuo, J. M. Gulbis, S. L. Cohen, B. T. Chait and R. MacKinnon, "The Structure of the Potassium Channel: Molecular Basis of K^{+}Conduction and Selectivity," Science, Vol. 280, No. 5360, 1998, pp. 69-77.doi:10.1126/science.280.5360.69

29. G. Chang, R. H. Spencer, A. T. Lee, M. T. Barclay and D. C. Rees, "Structure of the MscL Homolog from Mycobacterium Tuberculosis: A Gated Mechanosensitive Ion Channel," Science, Vol. 282, No. 5397, 1998, pp. 2220-2226. doi:10.1126/science.282.5397.2220

30. Y. X. Jiang, A. Lee, J. Y. Chen, V. Ruta, M. Cadene, B. T. Chait and R. MacKinnon, "X-Ray Structure of a Voltage-Dependent K^{+} Channel," Nature, Vol. 423, No. 6935, 2003, pp. 33-41. doi:10.1038/nature01580

31. Y. X. Jiang, A. Lee, J. Y. Chen, M. Cadene, B. T. Chait, and R. MacKinnon, "Crystal Structure and Mechanism of a Calcium-Gated Potassium Channel," Nature, Vol. 417, No. 6888, 2002, pp. 515-522. doi:10.1038/417515a

32. Y. X. Jiang, A. Lee, J. Y. Chen, M. Cadene, B. T. Chait, and R. MacKinnon, "The Open Pore Conformation of Potassium Channels," Nature, Vol. 417, No. 6888, 2002, pp. 523-526. doi:10.1038/417523a

33. R. MacKinnon, "Nothing Automatic about Ion-Channel Structures," Nature, Vol. 416, No. 6878, 2002, pp. 261-262. doi:10.1038/416261b

34. Ben-Shem and D. Fass, "Structural Basis for Intramembrane Proteolysis by Rhomboid Serine Protease," Proceedings of the National Academy of Sciences of the USA, Vol. 102, No. 2, 2007, pp. 401-402.

35. M. J. Lemieux, S. J. Fischer, M. M. Cherney, M. M. Bateman and M. N. James, "The Crystal Structure of Rhomboid Peptidase from Haemophilus Influenzae Provides Insight into Intramembrane Proteolysis," Proceedings of the National Academy of Sciences of the USA, Vol. 104, No. 3, 2007, pp. 750-754. doi:10.1073/pnas.0609981104

36. Stephen White Laboratory Official Website, 2007. www.blanco. .uci.edu /Membrane_proteins_xtal.html

37. S. J. Opella and F. M. Marassi, "Structure Determination of Membrane Proteins by NMR Spectroscopy," Chemical Reviews, Vol. 104, No. 8, 2004, pp. 3587-3606. doi:10.1021/cr0304121

38. S. J. Opella, "NMR and Membrane Proteins," Nature Structural & Molecular Biology, Vol. 4, 1997, pp. 845-848.

39. Avanti Company Official Website, 2006. http://www.avantilipids.com /Technical Information.html

40. S. Abu-Baker and G. A. Lorigan, "Phospholamban and Its Phosphorylated form Interact Differently with Lipid Bilayers: A ^{31}P, ^{2}H, and ^{13}C Solid-State NMR Spectroscopic Study," Biochemistry, Vol. 45, No. 44, 2006, pp. 13312-13322.

41. M. A. McCabe and S. R. Wassall, "Fast-Fourier-Transform Depaking," Journal of Magnetic Resonance, Series B, Vol. 106, No. 1, 1995, pp. 80-82. doi:10.1006/jmrb.1995.1013
42. M. A. McCabe and S. R. Wassall, "Rapid Deconvolution of NMR Powder Spectra by Weighted Fast Fourier Transformation," Solid State Nuclear Magnetic Resonance, Vol 10, No. 1-2, 1997, pp. 53-61. doi:10.1016/S0926-2040(97)00024-6
43. P. C. Dave E. K. Tiburu, K. Damodaran and G. A. Lorigan, "Investigating Structural Changes in the Lipid Bilayer upon Insertion of the Transmembrane Domain of the Membrane-Bound Protein Phospholamban Utilizing ^{31}P and ^{2}H Solid-State NMR Spectroscopy," Biophysical Journal, Vol. 86, No. 3, 2004, pp. 1564-1573.
44. P. C. Dave, E. K. Tiburu, N. A. Nusair and G. A. Lorigan, "Calculating Order Parameter Profiles Utilizing Magnetically Aligned Phospholipid Bilayers for ^{2}H Solid-State NMR Studies," Solid State Nuclear Magnetic Resonance, Vol. 24, No. 2-3, 2003, pp. 137-149.doi:10.1016/S0926-2040(03)00052-3
45. D. Huster, X. Yao, K. Jakes and M. Hong, "Conformational Changes of Colicin Ia Channel-Forming Domain upon Membrane Binding: A Solid-State NMR Study," Biochimica et Biophysica Acta, Vol. 1561, No. 2, 2002, pp. 159-170.
46. D. Huster, K. Arnold and K. Garwrisch, "Influence of Docosahexaenoic Acid and Cholesterol on Lateral Lipid Organization," Biochemistry, Vol. 37, No. 49, 1998, pp. 17299-17308.
47. G. W. Stockson, C. F. Polnaszek, A. P. Tulloch, F. Hasan and I. C. P. Hasan, "Molecular Motion and Order in Single Bilayer Vesicles and Multilamellar Dispersions of Egg Lecithin and Lecithin-Cholesterol Moxtures. A Deuterium Nuclear Magnetic Resonance Study of Specifically Labeld Lipids," Biochemistry, Vol. 15, 1976, pp. 954-966.
48. J. Seelig, "^{31}P Nuclear Magnetic Resonance and the Head Group Structure of Phospholipids in Membranes," Biochimica et Biophysica Acta, Vol. 515, 1978, pp. 105-140.
49. C. D. Pointer-Keenan, D. K. Lee, K. J. Hallock, A.Tan, R. Zand and A. Ramamoorthy, "Investigation of the Interaction of Myelin Basic Protein with phospholipid Bilayers Using Solid-State NMR Spectroscopy," Chemistry and Physics of Lipids , Vol. 132, No. 1, 2004, pp. 47-54.
50. S. Abu-Baker and G. A. Lorigan, "Phospholamban and Its Phosphorylated Form Interact Differently with Lipid Bilayers: A ^{31}P, ^{2}H, and ^{13}C Solid-State NMR Spectroscopic Study," Biochemistry, Vol. 45, No. 44, 2006, pp. 13312-13322. doi:10.1021/bi0614028
51. J. W. Mack, D. A. Torchia and P. M. Steinert, "SolidState NMR Studies of the Dynamics and Structure of Mouse Keratin Intermediate Filaments," Biochemsitry, Vol. 27, No. 15, 1988, pp. 5418-5426.
52. R. S. Prosser, S. I. Daleman and J. H. Davis, "The Structure of an Integral Membrane Peptide: A Deuterium NMR-Study of Gramicidin," Biophysical Journal, Vol. 66, No. 5, 1994, pp. 1415-1428. doi:10.1016/S0006-3495(94)80932-4
53. S. Sharpe, K. R. Barber, C. W. M. Grant and M. R. Morrow, "Evidence of a Tendancy of Self-Association of the Transmembrane Domain of ErbB-2 in Fluid Phospholipid Bilayers," Biochemistry, Vol. 41, 2002, pp. 2341-2352.
54. D. H. Jones, A. C. Rigby, K. R. Barber and W. M. Grant, "Oligoerization of EGF Receptor Transmembrane Domain: A 2H NMR Study in Lipid Bilayers," Biochemsitry, Vol. 36, No. 41, 1997, pp. 12616-12624.
55. J. A. Killian, M. J. Taylor and R. E. Koeppe, "Orientation of Valine-1 Side-Chain of the Gramicidin Transmembrane Channel and Implications for Channel Functioning: A Deuterium NMR Study," Biochemsitry, Vol. 31, No. 46, 1992, pp. 11283-11290.
56. K. C. Lee, S. Huo and T. A. Cross, "Lipid-Peptide Interface: Valine Conformation and Dynamics in Gramicidin Channel in a Lipid Bilayer," Biochemsitry, Vol. 34, 1995, pp. 857-867.
57. P. C. A. Van der Wel, E. Strandberg, J. A. Killian and R. E. Koeppe, "Geometry and Intrinsic Tilt of a Trypto Phan-Anchored Transmembrane Alpha-Helix Determined

by H-2 NMR," Biophysical Journal, Vol. 83, No. 3, 2002, pp. 1479-1488. doi:10.1016/S0006-3495(02)73918-0

58. L. S. Batchelder, C. H. Niu and D. A. Torchia, "Methyl Reorientation in Polycrystaline Amino Acids in Peptides," Journal of the American Chemical Society, Vol. 105, No. 8, 1983, pp. 2228-2231.

59. S. Abu-Baker, J. X. Lu, S. Chu, C. C. Brinn, C. A. Makaroff and G. A. Lorigan, "Side Chain and Backbone Dynamics of Phospholamban in Phospholipid Bilayers Utilizing 2H and 15N Solid-State NMR Spectroscopy," Biochemistry , Vol. 46, No. 42, 2007, pp. 11695-11706. doi:10.1021/bi700749q

60. J. Janin and S. Wodak, "Conformation of Amino Acid Side-Chain in Proteins," Journal of Molecular Biology, Vol. 125, No. 3, 1978, pp. 357-386.

61. C. Rigby, K. R. Barber, G. S. Shaw and C. W. M. Grant, "Transmembrane Region of the Epidermal Growth Factor Receptor: Behavior and Interactions via ^{2}H NMR," Biochemistry, Vol. 35, 1996, pp. 12591-12601.

62. W. J. Gibbons Jr., E. S. Krap, N. A. Cellar, R. E. Minto and G. A. Lorigan, "Solid-State NMR Studies of a Diverged Microsomal Amino-Proximate Delta12 Desaturase Peptide Reveal Causes of Stability in Bilayer: Tyrosine Anchoring and Arginine Snorkeling," Biophysical Journal, Vol. 90, No. 4, 2006, pp. 1249-1259.

63. M. S. Greenfield, A. D. Ronemus, R. L. Vold, R. R. Vold, P. D. Ellis and T. E. Raidy, "Deuterium Quadrupole-Echo NMR Spectroscoppy III. Practice Aspects of line Shape Calculations for Multiaxis Rotational Processes," Journal of Magnetic Resonance, Vol. 72, No. 1, 1987, pp. 89-107.

64. K. Oxenoid and J. J. Chou, "The Structure of Phospholamban Pentamer Reveals a Channel-Like Architecture in Membrane," Proceedings of the National Academy of Sciences of the USA, Vol. 102, No. 31, 2005, pp. 10870-10875. doi:10.1073/pnas.0504920102

65. H. K. B. Simmerman and L. R. Jones, "Phospholamban: Protein Structure, Mechanism of Action, and Role in Cardiac Function," Physiological Reviews, Vol. 78, No. 4, 1998, pp. 921-947.

66. J. Zamoon, F. Nitu, C. Karim, D. D. Thomas and G. Veglia, "Mapping the Interaction Surface of a Membrane Protein: Unveiling the Conformational Switch of Phospholamban in Calcium Pump Regulation," Proceedings of the National Academy of Sciences of the USA, Vol. 102, No. 13, 2005, pp. 4747-4752. doi:10.1073/pnas.0406039102

67. D. Massiot, F. Fayon, M. Capron, I. King, S. Lecalve, B. Alonso, J. O. Durand, B. Bujoli, Z. Gan and G. Hoatson, "Modelling Oneand Two-Dimensional Solid-State NMR Spectra," Magnetic Resonance in Chemistry, Vol. 40, No. 1, 2002, pp. 70-76.

68. S. Abu-Baker, J. X. Lu, S. Chu, K. K. Shetty, P. L. Gor'kov and G. A. Lorigan, "The Structural Topology of WildType Phospholamban in Oriented Lipid Bilayers Using 15N Solid-State NMR Spectroscopy," Protein Science, Vol. 16, No. 11, 2007, pp. 2345-2349.

CHAPTER 12

Plasticity, Crack Initiation and Defect Resistance in Alkali-Borosilicate Glasses: From Normal to Anomalous Behavior

R. Limbach [a], A.Winterstein-Beckmann [a], J. Dellith[b], D. Möncke [a], L. Wondraczek [a,*]

[a] *Otto Schott Institute of Materials Research, University of Jena, Fraunhoferstrasse 6, 07743 Jena, Germany*

[b] *Leibniz Institute of Photonic Technology IPHT, Albert-Einstein-Straße 9, 07745 Jena, Germany*

ABSTRACT

We provide a comprehensive description of the defect tolerance of sodium-borosilicate glasses upon sharp contact loading. This is motivated by the key role which is taken by this particular glass system in a wide variety of applications, ranging from electronic substrates, display covers and substrates for biomedical imaging and sensing to, e.g., radioactive waste vitrification. The present report covers the mechanical properties of glasses in the Na_2O–B_2O_3–SiO_2 ternary over the broad range of compositions from pure SiO_2 to binary sodium-borates, and crossing the regions of various commercially relevant specialty borosilicate glasses, such as the multi-component Duran-, Pyrex- and BK7-type compositions and typical soda-lime silicate glasses, which are also included in this study. In terms of structure, the considered glasses may be separated into two groups, that is, one series which contains only bridging oxygen atoms, and another series which is designed with an increasing number of non-bridging oxygen ions.

Elastic moduli, Poisson ratio, hardness as well as creep and crack resistance were evaluated, as well as the contribution of densification to the overall amount of indentation deformation. Correlations between the mechanical properties and structural characteristics of near- and mid-range order are discussed, from which we obtain a mechanistic view at the molecular reactions which govern the overall deformation reaction and, ultimately, contact cracking.

Keywords: Glass Borosilicate Elasticity Plasticity Defect resistance

1. INTRODUCTION

Multi-component alkali-borosilicates represent one of the most prominent types of specialty glasses. They are an indispensable component in a wide variety of technical applications such as substrates in electronic devices or displays, household or laboratory ware, or cover glasses [1–3]. For example, they enable relatively high softening temperatures and low coefficient thermal expansion (or also matching the coefficient of thermal expansion to that of silicon), but also very specific mechanical properties [4]. This has made them a key substrate material for silicon-based electronic circuits and, e.g., photolithographic masks. Furthermore, a high transmission in the UV and low susceptibility to defect formation designate low alkali borosilicate glasses as a very suitable material for high performance optics. The comparably weak auto-luminescence has also resulted in dedicated applications as substrate for biomedical imaging and sensing. Due to their comparably high chemical and thermo-shock resistance, hardness and fracture toughness, they are also considered for waste immobilization [5,6]. At present, the understanding of the correlation between intrinsic strength, structure and mechanical properties is the key factor in the further advancement of design tools for such glasses [4,7–9]. Depending on composition, temperature or pressure, B_2O_3 containing glasses show severe structural changes, i.e., the change from three-fold (B_3) to four-fold coordinated boron (B_4), and also distinct variations of structural homogeneity [10–18]. This prevents simplistic theoretical approaches towards correlations between mechanical properties and intrinsic material characteristics [19]. Nevertheless, some models allow, up to a certain degree, the estimation of specific mechanical properties (like Young's modulus). Such a correlation was established by Phillips and later extended by Williams and Scott for alkali-free high-modulus glasses [20,21]. Their model is based on the assumption that the elastic modulus is changed in an additive manner by chemical composition. Another approach was proposed by Soga et al. [22], who correlated the packing density with the mechanical properties based on a mathematical equation representing the potential between two charged particles. Similar to this model, Bridge et al. [23] have taken the cation–anion stretching force constants to estimate elastic properties for many different network formers. Nevertheless, all these approaches regard only first neighbor influences and are therefore restricted to within certain limits. The simplistic temperature-dependent bond constraint theory as derived by Mauro and Gupta [24] on the basis of Phillips and Thorpe's topological constraint theory [25], tries to integrate the short- and mid-range architecture as defined by the presence of structural constraints, resulting in surprisingly good theoretical predictions of experimental hardness data [26–29]. More unifying efforts have been made by Rouxel et al., who tried to correlate mechanical properties such as elastic modulus and Poisson ratio to structural parameters such as the structural dimensionality, the free volume, the atomic packing density C_g, or the volume density of energy [6,30]. However, the present understanding of the mechanical response and, specifically, deformation processes in borosilicate glasses is still dominated by approximate and phenomenological expression such as, e.g., the notion of normal and anomalous behavior for describing shear and compaction-driven deformation procedures, respectively [31].

The ternary diagram of alkali-borosilicate glasses is quite complex, and offers different areas of interest. To differentiate between certain structural species that are dominant in certain compositions, two values have been established on the basis of molar composition, $R = Na_2O/B_2O_3$ and $K = Si_2O/B_2O_3$. Based on NMR data, Bray et al. [10] postulated a first model, which enabled the estimation of the fraction of fourfold coordinated boron ($N_4=B_4/(B_3+B_4)$) and the number of non-bridging oxygen ions in the silica sub-network (NBO_{Si}), from the R-value, that is, from the amount of network modifier oxides. Addition of modifier oxides to borate glasses or borosilicate glasses with $R < 0.5$ initially creates charged $[BØ_4]^-$ tetrahedra from neutral trigonal $BØ_3$ groups in pure vitreous B_2O_3 (where Ø stands for bridging oxygen atoms). That is, in contrast to vitreous SiO_2, which becomes more depolymerized with the addition of any amount of modifier oxide, the network connectivity in borate-containing glasses initially increases. However, higher modifier oxide concentrations ($R > 0.5$) cause NBO to form on trigonal borate groups, and/or on silicate tetrahedra. This coordination change makes borate containing glass systems very interesting for studying the material's response to indentation deformation, as the network can be manipulated with respect to the degree of polymerization, not only considering the number of non-bridging oxygen ions, but also considering the BO_3/BO_4 ratio [32–34].

In the present study, we provide systematic data on the elastic properties, Poisson ratio, hardness, the relation between densification and volume conservative shear flow, and correlations to damage tolerance for a range of compositions of the well investigated alkali-borosilicate system [11,12,34]. We focus on variations of the *R*- and *K*-values, i.e., BO_3/BO_4 ratios and also varying amounts of NBO, within the technologically most relevant part of the ternary (which also contains the regions of immiscibility). These data are discussed in the context of short- and mid-range structure.

2. EXPERIMENTAL

2.1. Glass preparation

Glasses from the sodium-borosilicate system were prepared from high purity (optical grade) raw materials (SiO_2, $CaCO_3$, Na_2CO_3, H_3BO_3). Batches of 100–250 g were melted in Pt or Pt-Rh20 crucibles in a conventional muffle furnace, with melting temperatures between 800 and 1750 °C, depending on the silica content. The KBS glass was prepared by melting batches of 4 l in a Pt crucible for 8 h at 1630 °C. To avoid phase separation, Al_2O_3 was added as $Al(OH)_3$ to the silica-rich glasses (NBS2, Duran, LBS and KBS). To ensure homogenous melts, the glasses with SiO_2 contents higher than 70 mol% were quenched in water and re-melted after drying. All melts were poured into preheated graphite molds (~ 20–50 K above the glass transition temperature) and cooled down to room temperature at a rate of 30 K/h, except for the Duran-type composition, which was cooled in the crucible, and the model glasses NBS A, C and D, which were quenched down to room temperature [33,34]. For comparison, commercially available SiO_2 (Suprasil, Heraeus Quarzglas GmbH & Co. KG) was analyzed as well.

Details on the glass compositions and their melting parameters are listed together with elected physical properties in Table 1. The glass transition temperature T_g was determined by differential scanning calorimetry (DSC) and the density ρ was measured by the Archimedes method, using ethanol and distilled H_2O, respectively, as immersion liquid. All experiments were conducted in laboratory air at an ambient temperature of around 25 °C.

Table 1. Nominal compositions, preparation conditions (melting temperature T_m and annealing temperature T_{cast}), and physical properties, including the fraction of $[BØ_4]^-$ units N_4, glass transition T_g, density ρ and atomic packing density C_g, of the investigated glasses. Data on T_g and ρ are taken from Refs. [33–35].

	2N8B	NBSA	NBSB	NBSC	NBSD	NBS1	NBS2	Duran	BK7	LBS	KBS[a]	NCS	SiO_2
SiO_2 (mol%)	–	25.0	48.5	42.5	60.0	74	74.0	81.5	74	81.5	72.6	74	100
B_2O_3 (mol%)	80	62.5	48.5	42.5	33.5	16	20.7	11.8	16	11.8	19.0	–	–
Na_2O (mol%)	20	12.5	3.0	15.0	6.5	10	4.3	5.4[b]	10	5.4[b]	6.4[b]	10	–
Al_2O_3 (mol%)	–	–	–	–	–	–	1.0	1.3	–	1.3	2.0	–	–
CaO (mol%)	–	–	–	–	–	–	–	–	–	–	–	16	–
K (SiO_2/B_2O_3)	–	0.40	1.00	1.00	1.79	4.63	3.57	6.91	4.63	6.91	3.82	–	–
R (Na_2O/B_2O_3)	0.25	0.20	0.06	0.35	0.19	0.63	0.21	0.46	0.63	0.46	0.24	–	–
N_4 (B)[c]	0.28	0.20	0.06	0.35	0.19	0.82	0.20	–	0.82	–	0.35	–	–
OH^- (mol%)[d]	0.400	0.190	–	0.210	0.160	0.120	0.180	0.250	–	0.120	0.140	–	0.003
T_g (°C) ± 2 K	412	415	380	490	445	550	442	540	550	489	523	530	1130
T_m (°C) ± 2 K	1200	1580	1300	1300	1580	1450	1750	1750	1450	1750	1630	1450	2000
T_{cast}(°C) ± 2 K	500	–	500	–	–	600	500	600	600	600	–	600	–
ρ(g/cm^3) ± 0.2%	2.07	2.18	2.04	2.31	2.15	2.45	2.18	2.22	2.45	2.14	2.26	2.48	2.20
C_g	0.521	0.532	0.485	0.534	0.490	0.528	0.479	0.462	0.485	0.490	0.480	0.528	0.457

[a] Glass composition was determined experimentally by Energy Dispersive X-Ray Spectroscopy (EDX).

[b] The values include 0.5 mol% K_2O in Duran, 5.4 mol% Li_2O in LBS and 0.8 mol% Li_2O as well as 4.9 mol% K_2O in KBS.

[c] NMR results for 2N8B, NBS1, NBS2, KBS and BK7 were taken from Refs. [11,36,37]. Values for NBS A, NBS B, NBS C and NBS D are from http://www.icglass.org/technical_committees/?id=1&committee=TC 03:_Glass_Structure.

[d] The OH^- content was calculated from the optical spectra, using the molar extinction coefficients ε_{2800} = 42 L mol^{-1} cm^{-1} for borosilicates and 155 L mol^{-1} cm^{-1} for SiO_2, respectively, as reported in Ref. [38].

2.2. Ultrasonic echography

The elastic properties were investigated by ultrasonic echography using an echometer (Echometer 1077, Karl Deutsch GmbH & Co KG) equipped with a piezoelectric transducer (f = 8–12 MHz), for the determination of the longitudinal t_L and transversal t_T sound wave propagation time with an accuracy of ± 1 ns. Prior to the measurements the glass specimens were polished co-planar to an optical finish and the exact thickness d of the 1–2 mm thin plates was measured with an accuracy of ± 2 µm using a micrometer screw. The longitudinal $c_L = 2d/t_L$ and transversal wave velocities $c_T = 2d/t_T$ were obtained from the specimen thickness and the time between two consecutive echoes. The shear modulus G, bulk modulus K and elastic modulus E as well as the Poisson ratio ν were calculated using the following relations, which are applicable for isotropic materials [39]:

$$G=\rho\cdot c_T^2 \quad (1)$$

$$K=\rho c_L^2-\frac{4}{3}c_T^2 \quad (2)$$

$$E=\rho\left[\frac{3c_L^2-4c_T^2}{(c_L/c_T)^2-1}\right] \quad (3)$$

$$\nu=\frac{c_L^2-2c_T^2}{2(c_L^2-c_L^2)} \quad (4)$$

2.3. Nanoindentation

Nanomechanical properties were characterized with a nanoindenter (G200, Agilent Inc.), equipped with a three-sided Berkovich diamond tip and a continuous stiffness measurement (CSM) option. Depth profiles of hardness $H$and elastic modulus E were obtained by applying a weak oscillation (Δh = 2 nm, f = 45 Hz) to the indenter tip [35]. On each glass specimen, 15 indents with a depth of 2 µm were generated at a constant strain-rate of $\dot{\varepsilon}$=0.05 s^{-1}. The resulting load–displacement curves were analyzed using the Oliver and Pharr method [40], where H is defined as the load P divided by the projected contact area of the indenter tip A_c:

$$H = \frac{P}{A_c} \tag{5}$$

The elastic modulus E was derived from the combined elastic response of the tested material and the diamond indenter (E_i = 1141 GPa, ν_i = 0.07) by means of the following equation:

$$E = \left(1-\nu^2\right)\left[\frac{1}{E_r} - \frac{1-\nu_i^{2}}{E_i}\right]^{-1}, \tag{6}$$

with the reduced elastic modulus E_r,

$$E_r = \frac{S}{2}\sqrt{\frac{\pi}{A_c}} \tag{7}$$

The symbol S represents the contact stiffness of the material and corresponds to the initial slope of the load–displacement curve upon unloading. The contact area used in Eqs. (5) and (7) was determined prior to the measurements by calibration of the Berkovich diamond tip on a fused silica reference glass sample (Corning 7980, Corning Inc.).

The indentation creep behavior was studied using a nanoindentation strain-rate jump test, as introduced by Maier et al. [41] and transferred for application on glasses by us in a previous study [35]. In this test, a three-sided Berkovich diamond tip penetrates the glass surface at a constant strain-rate of 0.05 s^{-1}down to 500 nm. During further penetration into the material, the strain-rate is changed every 250 nm, and the alteration of the hardness is determined with the CSM equipment (Δh = 5 nm, f = 45 Hz). On every glass specimen, 10 such tests with strain-rates of 0.05; 0.007 and 0.001 s^{-1} were performed. With the assumption of a depth-independent hardness ($\dot{H}/H = 0$) the applied strain-rate, defined as the loading rate divided by the actual load $\dot{P}/P$, can be translated into the indentation strain-rate $\dot{\varepsilon}_i$ by [42]:

$$\dot{\varepsilon}_i = \frac{\dot{h}}{h} = \frac{1}{2}\left(\frac{\dot{P}}{P} - \frac{\dot{H}}{H}\right) = \frac{P}{2\dot{P}} = \frac{1}{2}\dot{\varepsilon} \quad (8)$$

where the parameter $\dot{h}/h$ represents the indentation rate divided by the actual displacement of the indenter. Subsequently, the logarithm of the hardness ln*H*can be plotted against the logarithm of the indentation strain-rate lnε˙i and the strain-rate sensitivity *m* can be derived from the slope of the linear regression curve [43,44]:

$$m = \frac{\partial \ln H}{\partial \ln \dot{\varepsilon}_i} \quad (9)$$

An increase in the strain-rate sensitivity can be treated as the transition from a rigid-perfectly plastic material with $m = 0$ to an ideal Newtonian viscous solid with $m = 1$ [45,46].

2.4. Microindentation

The Vickers hardness H_V was determined using a microhardness tester (Duramin-1, Struers GmbH). On each specimen, 15 indents with a maximum load of 981 mN (= 100 g) were generated. The loading duration and dwell time were set to 15 s and 10 s, respectively. The residual imprints were subsequently analyzed with an optical microscope (AxioLab A1, Carl Zeiss Microscopy GmbH) and H_V was calculated from the following equation:

$$H_V = 1.8544\frac{P}{d^2} \quad (10)$$

where *d* stands for the length of the projected indent diagonals.

The crack resistance *CR* was approximated via Vickers indentation tests, according to Kato et al. [47]. For this purpose, the glass specimens were indented with stepwise increasing loads, and the amount of radial cracks emanating from the corners of the residual imprints were counted. From this, the probability of crack initiation *PCI* was derived as the ratio between the number of corners where a radial crack was formed and the total number of corners on all indents. The crack resistance is defined as the load at which an average of two cracks (*PCI* = 50%) occurred. The probability of crack initiation was determined for loads between 49.05 mN (= 5 g) and 19.62 N (= 2 kg). Indents with small loads up to 981 mN were generated in the nanoindenter. For higher loads, the microhardness tester was used. On every specimen, at least 25 indents with loading durations of 25s and dwell times if 10s were generated.

The topography of the residual Vickers hardness imprints was analyzed using an Atomic Force Microscope (M5, Park Scientific Instruments), equipped with a MSCT microlever tip (Bruker Nano Analytics) with a specified tip radius of approximately 10 nm. All measurements were performed in contact mode. A calibration grid was measured before and after each run to validate the tip conditions. For the evaluation of the Vickers indents an area of 10 × 10 μm was scanned with a resolution of 0.25 Å and 0.025 Å in the lateral and vertical directions, respectively. Indents with a peak load of 100 mN were created initially using the nanoindenter as described above. After this, the indentation volume V_i^-, which forms below the specimen surface, as well as the pile-up volume V_i^+, which can be observed at the periphery of the indent, were determined from the AFM profile. Subsequently, the indented glass specimen was annealed for 2 h at a temperature of $T/T_g \sim 0.9$ and both the indentation volume, now V_a^-, and the pile-up volume, now V_a^+, were determined once again. During this thermal treatment, a part of the total compaction is recovered due to structural relaxation, yielding the actual densified volume V_d[48–51]. The contribution of densification to the overall indentation deformation can finally be estimated from the temperature-induced volume recovery V_R, [49]

$$V_R = \frac{V_d}{V_i^-} = \frac{(V_i^- - V_a^-) + (V_a^+ - V_i^+)}{V_i^-} \tag{11}$$

With the progressive penetration of the indenter tip into the glass surface a part of the material deformed by volume-conservative shear flow is pushed out of the indent, leading to pile-up along the edges of the residual imprints [51,52]. The pile-up ratio V_P can therefore be regarded as an indicator for the shear-mediated deformation of the material [53]:

$$V_P = \frac{2V_i^+ - V_a^+}{V_i^-} \tag{12}$$

3. RESULTS

Experimental data on the physical and mechanical properties of the studied glasses are summarized in Table 2. The compositional dependence of some selected physical and mechanical properties in the sodium-borosilicate system is presented in Fig. 1. In a first consideration, the experimental results of T_g, ρ, E, ν and H_V are in a good agreement with literature data. Only small variations can be seen between the values of E determined by ultrasonic echography and by nanoindentation. However, for all glass compositions tested, the values of *H* are higher than H_V. The origin of this lies in the large contribution of elastic deformation to the indentation response of glasses, which results in a discrepancy between the contact area under load and the size of the residual hardness imprint [54].

Table 2. Mechanical properties of glasses in the system $Na_2O–B_2O_3–SiO_2$: Shear modulus *G*, bulk modulus *K*, elastic modulus *E*, Poisson ratio *v*, hardness *H*, strain-rate sensitivity *m*, Vickers hardness H_V and crack resistance *CR*. LBS and KBS as well as the soda-lime silicate glass, NCS, are added for comparison. Data on *v*, *E*, *H* and *m* are taken from Ref. [35], except for 2N8B, KBS and Duran.

Glass	*G* (GPa)	*K* (GPa)	*E* (GPa)[b]	*v*	*E* (GPa)[c]	*H* (GPa)	*m*	H_V (GPa)	*CR*(N)
2N8B	17.8[a]	35.2[a]	45.8[a]	0.283[a]	41.3 ± 0.2	4.09 ± 0.03	0.0283	3.53 ± 0.20	3.30
NBSA	16.8 ± 0.1	30.6 ± 0.5	42.6 ± 0.7	0.268 ± 0.006	41.6 ± 0.2	3.96 ± 0.03	0.0259	3.58 ± 0.12	6.55
NBSB	13.3 ± 0.1	23.0 ± 0.3	33.5 ± 0.5	0.257 ± 0.005	32.1 ± 0.1	3.68 ± 0.03	0.0270	3.15 ± 0.15	12.16
NBSC	23.1 ± 0.2	36.5 ± 0.7	57.3 ± 1.0	0.238 ± 0.007	57.1 ± 0.2	5.33 ± 0.03	0.0279	5.35 ± 0.33	7.26
NBSD	18.6 ± 0.1	29.1 ± 0.5	46.0 ± 0.7	0.237 ± 0.005	44.4 ± 0.1	4.98 ± 0.03	0.0240	3.75 ± 0.18	6.83
NBS1	32.7 ± 0.2	44.4 ± 0.6	78.8 ± 1.0	0.204 ± 0.004	81.4 ± 0.3	7.26 ± 0.04	0.0211	6.20 ± 0.32	0.41
NBS2	23.0 ± 0.2	32.1 ± 0.6	55.7 ± 1.0	0.210 ± 0.006	50.6 ± 0.2	5.89 ± 0.04	0.0209	4.25 ± 0.22	3.44
Duran	26.7[a]	35.6[a]	64.0[a]	0.200[a]	61.9 ± 0.2	7.18 ± 0.03	0.0180	6.47 ± 0.28	4.41
BK7	32.5 ± 0.3	44.4 ± 0.9	78.4 ± 1.5	0.206 ± 0.007	84.4 ± 0.3	7.97 ± 0.05	0.0223	6.62 ± 0.26	0.20
LBS	23.9 ± 0.2	32.0 ± 0.6	57.5 ± 1.0	0.201 ± 0.006	56.7 ± 0.3	7.01 ± 0.07	0.0149	6.42 ± 0.34	2.14
KBS	26.3 ± 0.2	36.0 ± 0.6	63.5 ± 1.1	0.206 ± 0.005	64.0 ± 0.2	6.92 ± 0.03	0.0181	6.33 ± 0.29	6.49
NCS	28.9 ± 0.2	42.3 ± 0.8	70.7 ± 1.3	0.221 ± 0.007	76.0 ± 0.4	6.67 ± 0.04	0.0164	5.36 ± 0.27	1.09
SiO_2	29.9[a]	35.4[a]	70.0[a]	0.170[a]	71.6 ± 0.3	9.30 ± 0.12	0.0068	7.53 ± 0.51	–

[a] Elastic constants of 2N8B, Duran and SiO_2 were taken from Ref. [55] and the corresponding data sheets, respectively.
b Values of *E* were determined by ultrasonic echography.
[c] Values of *E* were determined by nanoindentation.

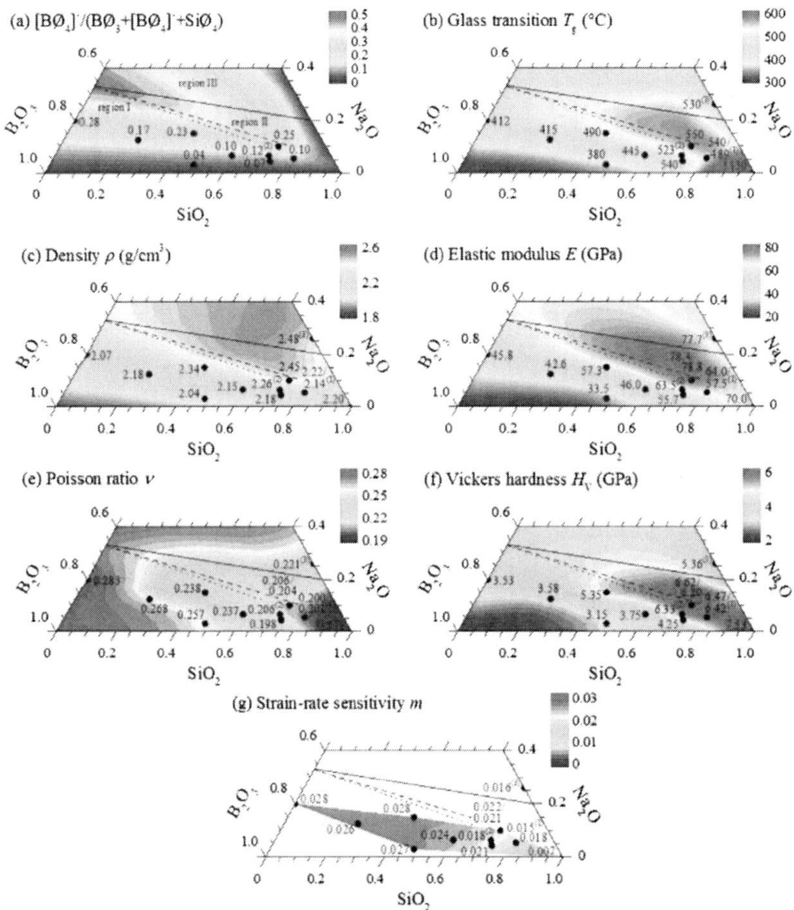

Figure. 1. Structural, physical and mechanical properties of glasses in the system Na_2O–B_2O_3–SiO_2: (a) Fraction of $[BØ_4]^-$ units per network former $[BØ_4]^-/(BØ_3 + [BØ_4]^- + SiØ_4)$, according to Ref. [56], (b) glass transition T_g, (c) density ρ, (d) elastic modulus E, (e) Poisson ratio ν, (f) Vickers hardness H_V and (g) strain-rate sensitivity m. The ternary contour diagrams of T_g[33–35,53,57–74], ρ[33–35,53,55,56,64–73,75–88], E[53,55,67–71,77–83,86–95], ν[35,53,55,67–69,79–83,86–89], H_V[31,53,68–73,84–86,95–104] and m[35] are derived from inter- and extrapolation of experimental results from the current study and literature data from previous studies. The black dots represent the glass compositions investigated in this work, where minor amounts of Al_2O_3 or other network modifier oxides (e.g. Li_2O, K_2O) are treated as B_2O_3 and Na_2O, respectively. In total, the contour diagrams result from (b) 226, (c) 184, (d) 127, (e) 77, (f) 97 and (g) 10 data points, respectively. Data of LBS (1) and KBS (2) as well as of the soda-lime

silicate glass, NCS (3) are also shown for comparison. Drawn lines correspond to the different glass-forming regions, as proposed by Yun and Bray [36], region I: $0.5 \leq R$ (dotted line), $0.5 < R < R_{max} = 0.5 + 0.0625K$ (dashed line), region II: $R_{max} < R < R_1 = 0.5 + 0.25K$ (solid line) and region III: $R_1 < R < R_2 = 1.5 + 0.75K$. $R = Na_2O/B_2O_3$ and $K = SiO_2/B_2O_3$.

In the context of deformation behavior, sodium-borosilicate glasses are commonly referred-to as so-called anomalous glasses, where the plastic deformation is controlled by the compaction of the free volume inside the glass network. This argumentation is based on the observation of ring and cone cracks around the indentation marks in vitreous SiO_2 and other commercially available glasses with high free volume and containing only small amounts of network modifiers, e.g., the Duran-type composition presented in Fig. 2a[53,71,95,105–107]. On the other hand, a pronounced contribution of volume-conservative shear flow to plastic deformation was observed for glasses with high modifier concentrations, e.g., BK7-type composition (Fig. 2b), or small values of K, e.g., NBS D (Fig. 2d) and NBS A (Fig. 2e) [34,68,71,108].

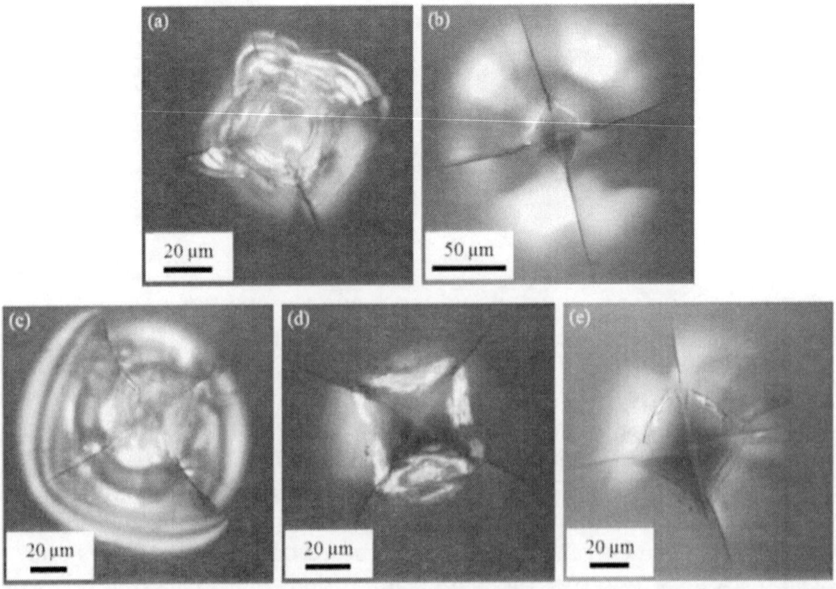

Figure. 2. Optical micrographs of the characteristic crack pattern formed by a Vickers indenter at a peak load of 9.81 N: (a) Duran, (b) BK7, (c) NBS2, (d) NBS D and (e) NBS A.

To obtain a systematic and clearer view at this classification, the influence of densification and shear-mediated plastic deformation was evaluated for the present series of sodium-borosilicate glasses. Data on the as-obtained indentation volume V_i^- and pile-up volume V_i^+, and on the post-annealing volumes V_a^- and V_a^+, the resulting volume recovery V_R and the pile-up ratio $V_P$of two technical glasses (Duran, BK7) as well as three model glasses (NBS A, C, D) are summarized in Table 3. Values for sodium borate [72] and vitreous silica [49,53] can be found in the literature.

Table 3. Indentation volume analyses of glasses in the system Na_2O–B_2O_3–SiO_2. The diagonal length *d*, indentation depth *h*, indentation volume V^- and pile-up volume V^+ of Vickers indents with a peak load of 100 mN as determined from AFM analyses are given. Subscripts "i" and "a" indicate the corresponding volumes prior to and after a thermal treatment for 2 h at $T/T_g = 0.9$, respectively. For comparison, values for 2N8B [72] and SiO_2 as taken from Ref. [49] are also reported. The parameters V_R and V_P represent the indentation volume recovery and the pile-up ratio, respectively.

Glass	**Before annealing**				**After annealing**				**Recovery**	
	d_i (μm)	**h_i (nm)**	**V_i^- (μm³)**	**V_i^+ (μm³)**	**d_a (μm)**	**h_a (nm)**	**V_a^- (μm³)**	**V_a^+ (μm³)**	**V_R (%)**	**V_P (%)**
2N8B [a]	10.89 ± 0.07	0.895 ± 0.009	8.300 ± 0.020	0.270 ± 0.020	10.13 ± 0.07	0.574 ± 0.004	2.260 ± 0.040	0.530 ± 0.040	75.9 ± 0.9	0.1
NBS A	7.51 ± 0.01	0.569 ± 0.003	2.759 ± 0.008	0.035 ± 0.002	6.66 ± 0.06	0.381 ± 0.016	1.204 ± 0.147	0.030 ± 0.009	56.2 ± 6.2	1.4 ± 0.5
NBS C	6.12 ± 0.06	0.464 ± 0.004	1.502 ± 0.018	0.025 ± 0.011	5.64 ± 0.06	0.307 ± 0.012	0.470 ± 0.012	0.039 ± 0.004	69.6 ± 3.8	0.7 ± 0.4
NBS D	7.21 ± 0.14	0.486 ± 0.017	1.904 ± 0.088	0.028 ± 0.010	6.10 ± 0.10	0.258 ± 0.023	0.416 ± 0.025	0.034 ± 0.005	78.5 ± 10.3	1.2 ± 0.7
Duran	5.12 ± 0.09	0.445 ± 0.010	1.097 ± 0.023	0.031 ± 0.007	4.53 ± 0.03	0.221 ± 0.020	0.376 ± 0.013	0.038 ± 0.003	66.4 ± 5.6	2.2 ± 0.8
BK7	5.20 ± 0.07	0.463 ± 0.005	1.176 ± 0.031	0.067 ± 0.003	4.94 ± 0.03	0.276 ± 0.006	0.594 ± 0.016	0.022 ± 0.006	45.7 ± 6.0	8.0 ± 3.5
SiO_2	5.16 ± 0.10	0.420 ± 0.025	1.030 ± 0.031	0.020 ± 0.006	4.08 ± 0.15	0.230 ± 0.014	0.200 ± 0.014	0.140 ± 0.025	92.2 ± 4.0	-

[a] Values were determined for Vickers indents with a peak load of 245 mN.

In Fig. 3 the probability for the formation of radial cracks *PCI* is shown for the glasses investigated in the current study. With increasing load, the values of *PCI*increase for all glass compositions tested, but the onset for the initiation of the first radial crack differs significantly. In order to get some information on the compositional dependence of the radial crack initiation, the experimental data was fit to a sigmoidal function of the form $y = A_2 + (A_1 - A_2)/[1 + (x/x_0)^p]$, where A_1, A_2, p as well as x_0 are fitting constants, and the crack resistance *CR*was derived from the load, at which an average of two radial cracks were formed.

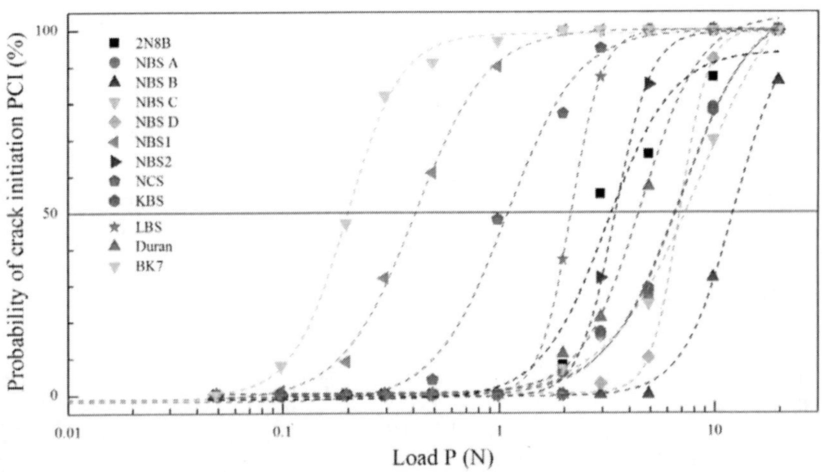

Figure. 3. Probability of radial crack formation *PCI* as a function of applied Vickers indentation load *P* for glasses investigated in the current study. The experimental data was fit to a sigmoidal function of the form $y = A_2 + (A_1 - A_2)/[1 + (x/x_0)^p]$ (dashed lines), and the crack resistance (solid line) was derived from the load at which an average of two radial cracks (*PCI* = 50%) formed at the corners of the residual Vickers hardness imprints.

4. DISCUSSION

4.1. Physical and mechanical properties

The physical and mechanical properties of glasses in the sodium-borosilicate system are well known to exhibit a strong compositional dependence. It has been shown, e.g., by Inoue et al. [109] and Barlet et al. [76], that this dependence can be attributed to distinct structural changes within the glass network. Amorphous B_2O_3 consists of corner-sharing $BØ_3$ triangles for which a major fraction (40–80%) is arranged in three-membered boroxol rings $(B_3O_6)^1$[110–113]. The incorporation of a small amount of network modifier oxides, such as Na_2O, initially converts neutral $BØ_3$ units into charged $[BØ_4]^-$tetrahedral units (see Fig. 1a) [10,114], which results in a continuous increase of ρ[55,56,72,76], but also of the network connectivity up to a maximum at $R = 0.2$. The latter represents the line of the boron oxide anomaly [32,34]. As a consequence, T_g[57,58,62,72], E[55,90,91,115] and H_V[72,90] increase as well. With the further addition of Na_2O ($R > 0.2$) the network connectivity starts to decrease due to the formation of BO_3 units containing non-bridging oxygen ions, such as $(BØ_2O)^-$, $(BØO_2)^{2-}$ and $(BO_3)^{3-}$[33,34]. However, this effect is initially counterbalanced by the progressive conversion of additional $BØ_3$ units into $[BØ_4]^-$ tetrahedra, until a sudden decrease in the mechanical stability of binary sodium borate glasses can be observed for values exceeding $R > 0.5$, where the formation of non-bridging oxygen ions becomes detectable by NMR spectroscopy [36,116].

Referring to the ternary system, sodium-borosilicate glasses with low concentrations of Na_2O ($R \leq 0.5$) can be treated as a binary sodium borate glass diluted by SiO_2[36,116,117]. SiO_2 builds-up a network of fully polymerized interconnected $SiØ_4$ tetrahedra, where the presence of Na_2O converts $BØ_3$triangles into $[BØ_4]^-$ tetrahedra [36,116,117]. Within this region T_g[61–67,109], ρ[56,64–67,85,87,109], E[67,87,109], and H_V[85,87,98,99] increase, while νdecreases [67], regardless of whether B_2O_3 is substituted by Na_2O (increasing R) or SiO_2 (increasing K). On the other hand, the indentation creep response seems to be affected primarily by the amount of SiO_2 inside the glass network. With increasing SiO_2 concentration at the expense of Na_2O and B_2O_3, respectively, the network dimensionality increases [56,76] and associated with this the values of m continuously decrease from 0.0283 for sodium borate down to 0.0067 for vitreous silica. This result is in accordance with our previous findings on the indentation creep response of glasses with small Poisson ratios up to $\nu \sim 0.33$, where a higher indentation creep resistance has been noticed for glasses with a low atomic packing density and a highly cross-linked network [35].

In contrast to binary sodium borate glasses, the maximum amount of Na_2O that can be introduced in the glass network, before NBO formation, depends not only on R but also on K. According to the NMR-based structural model for sodium-borosilicate glasses proposed by Yun and Bray [36], the $[BØ_4]^-$tetrahedra are initially present in the form of tetraborate and diborate groups. With the progressive incorporation of Na_2O, diborate groups are created at the expense of tetraborate groups. At $R = 0.5$, only diborate groups exist (Fig. 1, dotted lines). For values of $R > 0.5$ these diborate groups are converted into loose $[BØ_4]^-$ units, which allegedly bind to four $SiØ_4$ tetrahedra on the one hand to create reedmergnerite-like groups $[BSi_3Ø_8]^-$[36,116]. On the other hand danburite-like groups $[B_2Si_2Ø_8]^{2-}$ can form, which consist of two $[BØ_4]^-$ units connected to two $SiØ_4$ tetrahedra [118,119]. At $R_{max} = 0.5 + 0.0625K$, when all $SiØ_4$ tetrahedra are consumed by this process, the upper limit of the glass-forming region I (Fig. 1, dashed lines) is reached. In the glass-forming region II, additional Na_2O is shared between the borate and silicate networks, which leads to the formation of non-bridging oxygen ions at the $SiØ_4$ tetrahedra [36,116]. As a consequence of the decreasing connectivity in the silicate network ν increases and T_g[59–62,64–66,68], E[68,78,109] as well as H_V[68,85,98] exhibit a maximum in the region in-between R_{max} and $R_1 = 0.5 + 0.25K$ (Fig. 1, solid lines). Afterwards, within the subsequent glass-forming region III, the mechanical properties decrease significantly due to the formation of non-bridging oxygen ions in both the borate and silicate networks [116]. On the contrary, ρ still increases slightly with the replacement of the light boron atoms by the heavier sodium ion, and the molar volume decreases as the large voids in the loosely packed silica network are filled by the modifier cations [56,85,109].

Based on the structural models of Yun and Bray [36] as well as Budhwani and Feller [56], several authors tried to calculate the physical and mechanical properties of glasses in the system $Na_2O–B_2O_3–SiO_2$, such as T_g, ρ, E or H_V[56,76,109,120]. However, the application of this model remains still questionable. It has been shown, for example, that for glasses with $R > 0.5$ the silicate network is less depolymerized than predicted. Consequently, a substantially larger amount of non-bridging oxygen ions is attached to the borate network [34,76,116]. Another point to note is the discrepancy between the glass compositions with the minimum values of ν on the one hand and the maximum values of T_g, E and H_V on the other hand. Makishima and Mackenzie [121] and later on Rouxel [6] have demonstrated that glasses with a high degree of cross-linking and a low atomic packing density (e.g. vitreous silica) are usually associated with small values of ν, while glasses with a lower network connectivity and a higher atomic packing density (e.g. chalcogenide or metallic glasses) exhibit large values of ν. In binary sodium borate glasses ρ increases with increasing Na_2O concentration, even for $R > 0.5$ [55,56]. Nonetheless, with the addition of Na_2O, ν initially decreases, exhibits a minimum at around $R = 0.5$ and then re-increases for $R > 0.5$ [55]. This result indicates that the changes in ν can be primarily attributed to the increased connectivity in the glass network. A similar behavior is observed in the ternary sodium-borosilicate glass system. Here, ρ continuously increases with increasing R and K, while ν initially decreases until it reaches a certain minimum in a small compositional region that coincides with $0.5 \leq R \leq 0.5 + 0.0625K$. The following re-increase of ν can most likely be attributed to the incipient depolymerization of the glass network. Although contradictory to the early NMR-based structural model for sodium-borosilicate glasses, this assumption receives support from various studies in which the presence of non-bridging oxygen ions in the borate [34,76,117] and silicate networks [5,34,85], respectively, was verified.

4.2. Indentation deformation

Compared to their (hypothetic or real) crystalline analogues, glass networks exhibit a higher degree of free volume, which results in a notable degree of compressibility. In the case of vitreous silica, the application of high hydrostatic pressure can lead to an increase of the density by as much as 20% [122–124]. This behavior was confirmed for other types of glasses as well, including vitreous boron oxide [125–127], alkali–alkaline earth borates [128–130] or sodium-borosilicate glasses [13,17,128]. The ability of a glass to densify under pressure depends mainly on the atomic packing density C_g, which is defined as the ratio between the theoretical molar volume occupied by the ions and the effective molar volume of the glass [6]:

$$C_g = \rho \frac{\sum f_i V_i}{\sum f_i M_i} \tag{13}$$

In the above equation $V_i = 4/3\pi N(xr_A^3 + yr_B^3)$ represents the molar volume of an oxide A_xB_y with the molar fraction f_i and the molar mass M_i and the symbol N denotes the Avogadro number. The ionic radii r_A and r_B of the corresponding cations and anions, respectively, can be found in the literature [131].

Besides atomic packing density, a correlation between compressibility and the Poisson ratio has been established, as glasses with a more open network structure and small values of ν (such as vitreous silica) can be densified to a higher extent than close-packed structures with large values of ν, e.g., chalcogenide or metallic glasses [132]. A similar behavior was found for the indentation response of glasses [49,133,134]. Therefore, in Fig. 4 the values of V_R determined for the present series of sodium-borosilicate glasses are shown as a function of ν and C_g, together with results from earlier studies on glasses in this system [68,69,72]. In addition, literature values from other glass systems, mostly aluminosilicate and alkali–alkaline earth silicate glasses, were added for comparison [47,49,53,79,135,136]. Values from Refs. [47,68] were included using the approximation that V_R is linked to the recovery of indentation depth D_R by the simple relationship $D_R = 2.4V_R$[53].

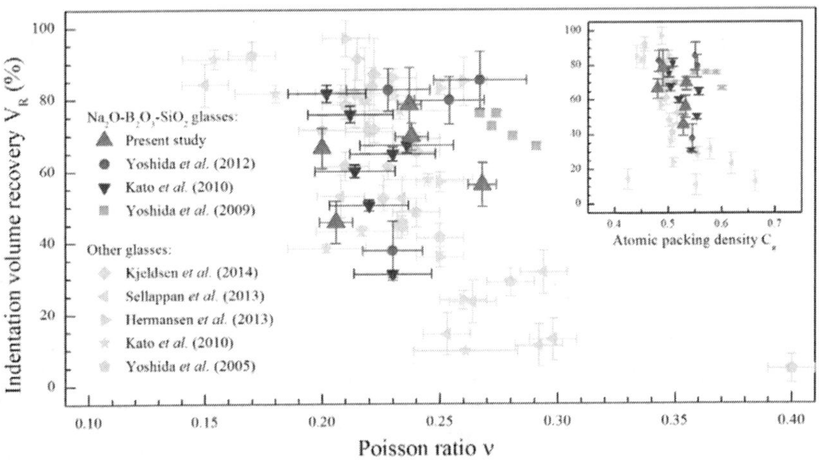

Figure. 4. Correlation between the indentation volume recovery V_R of glasses and the Poisson ratio ν. The inset shows V_R as a function of the atomic packing density C_g.

At a first glance, the contribution of densification to the indentation deformation decreases with increasing ν or increasing C_g (see inset in Fig. 4), which agrees quite well with the previously observed trend described above. However, within the range of $0.2 \le \nu \le 0.3$, where almost all sodium-borosilicate glasses are located, marked differences in V_R are observed for similar values of ν. In contrast, little or no changes of V_R are found even for large variations in ν for other glasses. Interestingly, the same observations were made for the dependency of V_R on C_g. These results imply a significant influence of additional parameters on the indentation-induced densification of glasses besides the atomic packing density and the elastic properties of the material. In earlier studies, distinct alterations in the indentation deformation behavior were reported depending on the experimental conditions, such as the maximum load or the geometry of the indenter tip [50–52,134,137–139]. Usually, blunter indenter tips (e.g. spheres) promote densification, while the high stress concentration below sharp indenter tips (e.g. Berkovich, Vickers) facilitates the activation of shear flow in the material [134,138,139]. Another important aspect, which is not considered in the above analysis, is the alteration of the resistance against elastic–plastic deformation within the densified area under the indenter tip, as a result of changes in ν, E and H_V, which might affect the further penetration of the indenter into the glass surface [17,128,130,133,140].

Despite the obvious experimental limitations of this method, a strong compositional dependence of V_R has been reported for aluminosilicate as well as alkali–alkaline earth silicate glasses [79,135,136]. Similarly, such a compositional dependence is confirmed also for the present system of Na_2O–B_2O_3–SiO_2, presented in Fig. 5. Although the available data in this system is rather limited, a local maximum of V_R is clearly visible for SiO_2-rich glasses and pure SiO_2 in particular [49,53,68,69]. These findings are quite consistent with the observations made, e.g., on alkali–alkaline earth silicate glasses [49,79]. They indicate that the SiO_2 sub-network governs the densification of SiO_2-rich sodium-borosilicate glasses. This is basically achieved by an increasing amount of smaller three- and four-membered silicate rings at the expense of larger five- and six-membered silicate rings, as well as by a decrease of the Si–O–Si bond angle [33,34,141,142]. However, increasing the atomic packing density by adding B_2O_3 or Na_2O to the glass, or creating non-bridging oxygen ions results in a continuous decrease of V_R and the transition from densification-controlled towards shear-mediated indentation deformation, where the minimum values of V_R are located in a narrow compositional range which corresponds to the region of maximum T_g, E or H_v and minimum ν, as shown in Fig. 1 in Section 3. At first view, this result is in contradiction to previous studies in which decreasing V_R with increasing ν has been reported [49,53,79]. Recalling the NMR-based structural model of Yun and Bray [36], the incorporation of Na_2O initially converts $BØ_3$ triangles into $[BØ_4]^-$ tetrahedra [36,116,117]. For $R > 0.5$, the $[BØ_4]^-$ tetrahedra were expected to be linked to the $SiØ_4$ tetrahedra and form either reedmergnerite $[BSi_3Ø_8]^-$ or danburite $[B_2Si_2Ø_8]^{2-}$-type structural units [118,119], which seem to impede the densification of the silicate sub-network. However, very recent thermodynamic considerations have demonstrated that these borosilicate ring structures do not form spontaneously when the Na_2O to B_2O_3 ratio exceeds a value of 0.5. Instead, the connectivity between the silicate and borate sub-networks increases continuously as the Na_2O and B_2O_3 concentrations increase (see Fig. 3b) [143], while at the same time, the ability of the glass to densify decreases. Furthermore, a prevalent signal due to B_3–O–Si bonds was verified in a Pyrex-type glass by 2D NMR, with no evidence for preferential B_4–O–Si bonding, despite the high N_4 ratio in this glass composition [144]. With respect to these NMR results, some of the assumptions on which the empirical structural models are based, need to be reconsidered carefully.

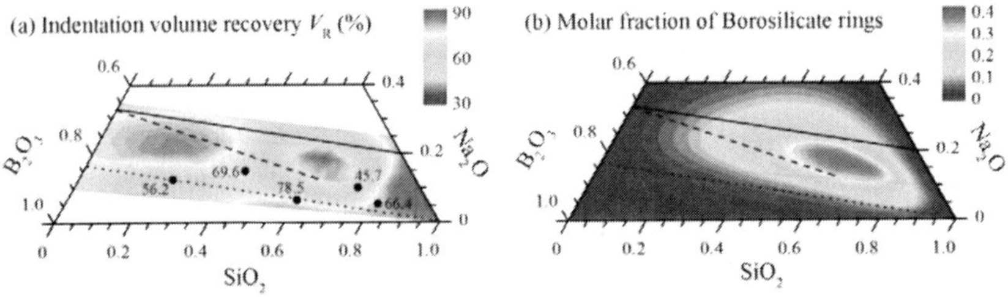

Figure. 5. (a) Compositional dependence of the indentation volume recovery V_R of glasses in the system Na_2O–B_2O_3–SiO_2. Black dots represent the glass compositions investigated in the current study. Lines drawn correspond to constant Na_2O/B_2O_3 ratios of $R = 0.2$ (dotted line), $R = 0.5$ (dashed line) and $R = 0.5 + 0.25K$ (solid line), with $K = SiO_2/B_2O_3$. The ternary contour diagram is build up of 25 data points. (b) Molar fraction of four-membered borosilicate rings, reedmergnerite $[BSi_3Ø_8]^-$ and danburite $[B_2Si_2Ø_8]^{2-}$-type structural units, according to Ref. [143].

Besides the SiO_2-rich region, a second local maximum of V_R can be seen within the B_2O_3-rich region. The large contribution of densification to the indentation deformation of B_2O_3-containing glasses can be attributed to structural alterations in the B_2O_3 sub-network. High-pressure studies on amorphous B_2O_3[145–147] as well as borate glasses [129,148] revealed that the densification of B_2O_3-containing glasses is based on the breakage of three-membered boroxol rings during loading and the conversion of $BØ_3$ units into temporarily over-coordinated $[BØ_4]^-$ groups. Although this over-coordination of boron is released after unloading, only a small fraction of the previously existing boroxol rings is restored and a permanent densification is achieved as a consequence of the reduced free volume inside the glass network [147,149]. Nonetheless, in borate glasses with minor amounts of network modifier oxides, such as Na_2O, a slightly reduced contribution of densification to the indentation deformation has been noticed in comparison to glasses with larger amounts of Na_2O [72]. It has been suggested that this effect originates from the simultaneous activation of a shear flow mechanism, which is thought to proceed over a sliding motion of the planar boroxol rings [72,150]. Since the fraction of boroxol rings decreases continuously with increasing Na_2O concentration [143], the influence of the corresponding shear and densification mechanisms diminishes. At the same time, the densification mechanism was found to shift towards a change in the statistics of borate rings containing at least one $[BØ_4]^-$-group [33,72]. For that reason an initial increase in V_R can be observed with increasing Na_2O concentration, until the creation of non-bridging oxygen ions facilitates plastic deformation through localized shear flow [72]. It is therefore not unexpected that the high contribution of densification in the B_2O_3-rich region is found for glass compositions with Na_2O/B_2O_3 ratios at exactly the cross-over where the first non-bridging oxygen ions start to form, $R = 0.2$ [33,34], and at the resolution limit for the detection of non-bridging oxygen ions by NMR, $R = 0.5$ [36,116], respectively.

4.3. Crack initiation

With the progressive penetration of an indenter tip into a glass surface the stresses are initially accommodated by elastic–plastic deformation, visible in a residual hardness imprint [151]. However, at sufficiently high loads the mismatch between the plastically deformed volume, which is assumed to be restricted to a hemispherical area below the indenter tip, and the adjacent elastically deformed material can cause the formation of ring, median, radial, or lateral cracks [107]. By analyzing the stress fields around the indented region, Yoffe [134] developed a model which has been applied successfully for the description of the differences in the crack initiation behavior for a variety of glasses [53,107,139]. In Yoffe's model, the elastic stress field of a point load Pnormal to the surface of a semi-infinite half-space can be described by the combination of a Boussinesq stress field with the spherical polar coordinates r, θ and ϕ, and a blister field. The corresponding stress components σ of this field are re-written from [134],

$$\sigma_{rr} = \frac{P}{2\pi r^2}\left[1-2\nu-2(2-\nu)\cos^2\theta\right] + \frac{4B}{r^3}\left[(5-\nu)\cos^2\theta-(2-\nu)\right] \tag{14}$$

$$\sigma_{\theta\theta} = \frac{P}{2\pi r^2}\frac{(1-2\nu)\cos\theta}{(1+\cos\theta)} - \frac{2B}{r^3}(1-2\nu)\cos^2\theta \tag{15}$$

$$\sigma_{\phi\phi} = \frac{P}{2\pi r^2}(1-2\nu)\left[\cos\theta-\frac{1}{1+\cos\theta}\right] + \frac{2B}{r^3}(1-2\nu)\left(2-3\cos^2\theta\right) \tag{16}$$

$$\sigma_{r\theta} = \frac{P}{2\pi r^2}\frac{(1-2\nu)\sin\theta\cos\theta}{(1+\cos\theta)} + \frac{4B}{r^3}(1+\nu)\sin\theta\cos\theta \tag{17}$$

$$\sigma_{r\phi} = \sigma_{\theta\phi} = 0 \tag{18}$$

where the first term describes the Boussinesq stress field around a conical indenter tip with the radius $r$and the half-included angle ψ. In particular, Yoffe's model can also be used to estimate the elastic stress fields around pyramidal indenter tips, since the contact stresses induced by both tip geometries are the same [152]. Moreover, the shape of a Vickers indenter with $\psi = 68°$ can be approximated by an equivalent conical indenter with $\psi = 70.3°$, which implies that the corresponding conical indenter has the same projected contact area ($A_c = \pi r^2$) as the Vickers indenter ($A_c = 2a^2$) [153]. As a consequence, the term πr^2 in Eqs. (14)–(17) can be replaced by $2a^2$, where a is one half of the indentation diagonal d. If the hardness is load-independent, $P/2a^2$ may be substituted by H[53].

The second term in Yoffe's stress field analysis represents the blister field, which accounts for the free surface created by densification and shear-mediated change of the shape and volume of the previously undeformed material. Here the parameter B is a measure of the strength of the blister field. For a Vickers indenter, B can be determined from the indentation topography by the following equation [53]:

$$B = \frac{Ea^3}{2\pi(1+\nu)(1-2\nu)\tan\psi}(1-V_R-V_P) \tag{19}$$

Based on these considerations the elastic stress field around a Vickers indenter can be calculated and the driving force for the initiation of ring $\sigma_{rr}(r = a, \theta = \pi/2)$, median $\sigma_{\theta\theta}(r = a, \theta = 0)$, radial $\sigma_{\phi\phi}(r = a, \theta = \pi/2)$ and lateral cracks $\sigma_{rr}(r = a, \theta = 0)$ can be estimated, according to Sellappan et al. [53], by:

$$\sigma_{rr}(r=a,\ \theta=\pi/2)/H = \frac{1-2\nu}{2} + \frac{2(\nu-2)(1-V_R-V_P)}{\pi(1+\nu)(1-2\nu)\tan\psi}\left(\frac{E}{H}\right) \tag{20}$$

$$\sigma_{\theta\theta}(r=a,\ \theta=0)/H = \frac{1-2\nu}{4} - \frac{(1-V_R-V_P)}{\pi(1+\nu)\tan\psi}\left(\frac{E}{H}\right) \tag{21}$$

$$\sigma_{\phi\phi}(r=a,\ \theta=\pi/2)/H = \frac{2\nu-1}{2} + \frac{2(1-V_R-V_P)}{\pi(1+\nu)\tan\psi}\left(\frac{E}{H}\right) \tag{22}$$

$$\sigma_{rr}(r=a,\ \theta=0)/H = -\frac{3}{2} + \frac{6(1-V_R-V_P)}{\pi(1+\nu)(1-2\nu)\tan\psi}\left(\frac{E}{H}\right) \tag{23}$$

where all stress components are normalized to H. It has to be noted that the above equations were developed for the stresses, which occur during loading at the maximum load P. Therefore, systematic differences occur in crack initiation as predicted by such a stress field analysis, and as experimentally observed on crack patterns (which typically occur after unloading): a substantial fraction of cracking occurs during unloading rather than during loading [107,154,155]. In addition, other effects, which have a significant influence on the indentation deformation and defect resistance, such as the indentation size effect [156], the load-dependence of V_R[50,137,138], or the presence of atmospheric moisture [107,157], are not considered in Yoffe's model.

The initiation of cracks is driven by the presence of tensile stresses, which continuously decrease in intensity with increasing distance from the center of pressure r. Since the stresses within the plastically deformed area are compressive, the highest tensile stresses can be expected at the region between the plastically deformed volume below the indenter tip and the surrounding elastically deformed material, at $r = a$[134]. However, the exact position of the tensile stress maximum depends significantly on B. Eq. (19)indicates that this value is mainly controlled by the term $(1 - V_R - V_P)$, whereas the magnitude of this parameter continuously decreases as the contribution of densification on the indentation deformation increases [53]. The application of νas an indicator for the densification and shear-mediated deformation of glasses was discussed in the previous section. Contrary to alkali–alkaline earth silicate glasses [79,135,136], ν is not a suitable parameter for the description of the indentation deformation behavior of sodium-borosilicate glasses, as seen from Fig. 5. Nonetheless, ν has become a useful parameter for the comparison of crack formation in different glass systems [53]. For this reason, the elastic stresses around a Vickers indenter estimated for the present series of sodium-borosilicate glasses are shown in Fig. 6 as a function of ν, together with literature values from Refs. [49,53].

To obtain insights on the influence of the glass composition on the elastic stress fields in the system Na_2O–B_2O_3–SiO_2, the further discussion will focus on the technical glasses with Duran- and BK7-type compositions, as well as one glass series with compositions along boron oxide anomaly ($R = 0.2$), at which distinct changes in the glass properties have been reported previously [5,55,158]. Starting at $B = 0$, or $(1 - V_R - V_P) = 0$ (inset in Fig. 6), and for small values of ν, the tensile stresses exhibit a maximum at the surface $\sigma_{rr}(r = a, \theta = \pi/2)$ and the initiation of ring cracks occurs during loading [134]. This result is consistent with the observation of ring and cone cracks in so-called anomalous glasses (e.g. vitreous silica), which are known to deform mainly by compaction of the free volume inside their glass network [31,107,159]. The magnitude $(1 - V_R - V_P)$ and the stresses at the surface decrease rapidly with higher contribution of volume-conservative shear flow on the indentation deformation, while the sub-surface tensile stresses along the indenter axis $\sigma_{\theta\theta}(r = a, \theta = 0)$ diminish only slightly. As a consequence, median cracks form below the indenter tip during loading in addition to the previously mentioned ring cracks [134].

During unloading these circular cracks expand towards the surface and appear as half-penny shaped cracks [107,160]. At sufficiently high loads, lateral and radial cracks may occur, but these cracks remain relatively short since their propagation is confined by the presence of ring and cone cracks, as can be seen for example in the optical micrograph of a residual Vickers hardness imprint in the Duran-type composition in Fig. 2a in Section 3.

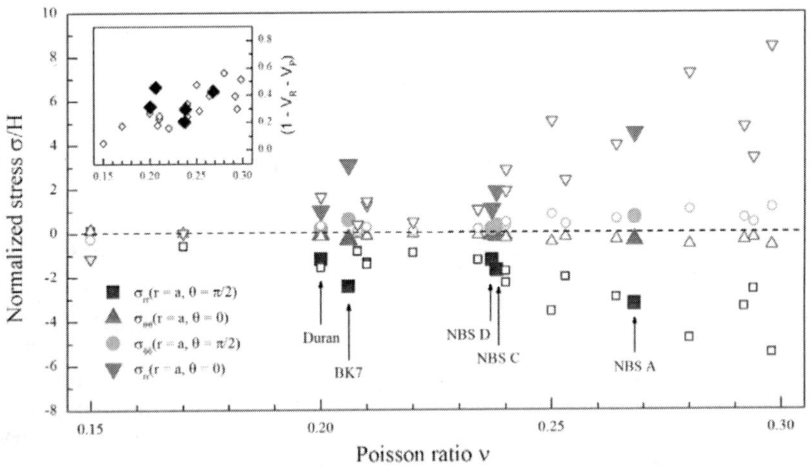

Figure. 6. Elastic stresses σ around Vickers indents in glasses as a function of the Poisson ratio ν. The stress components $\sigma_{rr}(r = a, \theta = \pi/2)$, $\sigma_{\theta\theta}(r = a, \theta = 0)$, $\sigma_{\phi\phi}(r = a, \theta = \pi/2)$ and $\sigma_{rr}(r = a, \theta = 0)$, normalized to the hardness H, represent the driving force for the initiation of ring, median, radial and lateral cracks, respectively. Filled symbols denote the glasses investigated in the current study and open symbols were taken from the literature [49,53]. In the inset, the parameter $(1 - V_R - V_P)$ is shown as a function of ν, where V_R and $V_P$denote indentation volume recovery and pile-up ratio, respectively.

Moving from the Duran- to the BK7-type composition (Fig. 2b), where a considerable amount of non-bridging oxygen ions has been verified by NMR and Raman spectroscopy [11,34], the contribution of densification to the indentation deformation becomes less pronounced and the magnitude of $(1 - V_R - V_P)$ increases. In the stress field analysis, such an effect is associated with a further decrease of the sub-surface tensile stresses along the indenter axis $\sigma_{\theta\theta}(r = a, \theta = 0)$. However, at the same time the tensile stresses at the periphery of the indenter $\sigma_{\phi\phi}(r = a, \theta = \pi/2)$ increase and radial cracks start to emanate from the corners of the indent during loading. During unloading these cracks propagate further along the specimen surface, and the parallel initiation of additional tensile stresses below the indenter tip $\sigma_{rr}(r = a, \theta = 0)$ results in the formation of sub-surface lateral cracks [134]. This crack initiation behavior is typical for so-called normal glasses (e.g. soda-lime silicates), which – *as a hypothesis* – deform through shear-flow along weak ionically bonded interfaces in the network modifier-rich regions [31,52,107].

Based on these theoretical considerations, Sellappan et al. [53] evaluated the transition from ring and cone cracks via median to radial and lateral cracks and noticed a correlation with ν. On the one hand, ring/cone (subscript "C") and median cracks (subscript "M") can be expected for glasses with small values of ν, on the other hand radial (subscript "R") and lateral cracks (subscript "L") occur predominantly in glasses with higher values of ν. Depending on the elasto-plastic ratio E/H of the glass, the transition from cone and median to median and radial cracking, ν_{CM-MR} (~ 0.18), as well as the subsequent transition from median and radial to lateral and radial cracking, ν_{MR-LR} (~ 0.21), can shift to smaller values of ν, in the case of an increasing E/H ratio, or larger values of ν if the E/H ratio decreases.

The transition from an anomalous to a normal glass can also be demonstrated in the present series of sodium-borosilicate glasses, when the SiO_2concentration decreases along the boron oxide anomaly from vitreous SiO_2 to NBS A. For high concentrations of SiO_2 and large values of $K = 3.57$ (NBS 2, Fig. 2c), the formation of ring and cone cracks is still visible, but with additional Na_2O and B_2O_3 (and the continuous decrease of K, NBS D: $K = 1.79$, Fig. 2d; NBS A: $K = 0.40$, Fig. 2e), ring and cone cracks gradually disappear. Simultaneously, the initiation of radial cracks and sub-surface lateral cracks was noticed, especially in the NBS A glass specimen. However, in contrast to the predictions from the stress field analyses, ring and cone cracks are still seen in the NBS D glass sample.

Although oxide glasses are termed as brittle materials, they can undergo distinct plastic deformation in the nano- and micrometer scales. The concept of brittleness is based on the relation between deformation and fracture [161]. Glasses with values of ν lower than ~ 0.31, such as SiO_2 or soda-lime silicates, exhibit only very small fracture energies. As a consequence, catastrophic failure occurs before a permanent plastic deformation can be achieved [162]. It is well known that the theoretical strength of brittle materials is reduced by the presence of small microscopic flaws or cracks. The resistance against the initiation and propagation of such defects, namely the toughness, is a crucial factor in technical exploitation of high-strength glasses [4,8].

In the previous section it has been shown, that the initiation of cracks in an indentation test can be attributed to elastic stresses, which originate from the mismatch between the plastically deformed volume and the surrounding elastically deformed material. Depending on the deformation mechanism and the elastic–plastic properties of the material, several different crack patterns can develop, but of practical importance are only median and radial cracks, since their propagation perpendicular to the glass surface is responsible for the fracture under bending or tensile loading conditions [95,106]. The resistance against the formation of median and radial cracks can be approximated by a pragmatic Vickers indentation test. In this test, the glass specimen is indented with stepwise increasing loads until the appearance of cracks can be observed at the corners of the residual hardness imprint [70]. In the current study, the sodium-borosilicate glasses were indented with loads ranging from ~ 50 mN up to ~ 20 N and the amounts of radial cracks emanating from the corners of the indent were counted using an optical microscope. On this basis, the crack resistance *CR* was derived from the load, at which an average of two radial cracks were formed. The results are shown in Fig. 7, together with literature values from Kato et al. [68]. Despite the limited number of data points, some distinct compositional trends are obvious. At first, *CR* seems to decrease with increasing Na_2O concentration, with the smallest values of *CR* corresponding to glass compositions with preferentially shear-mediated indentation deformation. This result is in good agreement with previous observations, in which an increasing resistance against radial crack formation with increasing contribution of densification to the indentation deformation was reported [53,68,107,139]. Additional support for these findings arises from the stress field analysis, where it has been demonstrated that the initiation of radial cracks is favored in glasses which deform through localized shear flow, while the residual stresses which are responsible for the formation of radial cracks are substantially reduced in glasses with a marked contribution of densification to the indentation deformation [134]. However, the highest values of *CR* do not coincide with the maximum values of V_R. In fact, maximum values of *CR* were obtained for sodium-borosilicate glasses with small amounts of Na_2O and less than 60 mol% SiO_2. In previous studies, large values of *CR* were attributed to the more open structure of borate networks containing a high amount of $BØ_3$ units, which facilitates the plastic deformation through densification and makes the borate network less rigid in comparison to $[BØ_4]^-$-rich glasses [68,130]. However, this argumentation is contradictory to the observations made by Yoshida et al. [72], as discussed in Subsection 4.2. Furthermore, with respect to this argument, the second local maximum of *CR* close to the binary Na_2O–SiO_2 glasses has to be highlighted (interestingly, both compositional regions fall in the compositional region corresponding to the miscibility gaps in the system Na_2O–B_2O_3–SiO_2[32,143,163], but a discussion of the effect of phase separation on the mechanical properties of sodium-borosilicate glasses is out of the scope of this study).

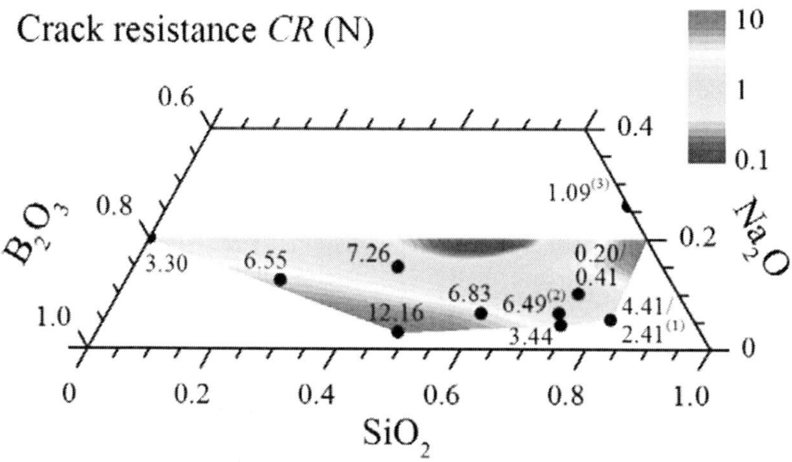

Figure. 7. Crack resistance *CR* of glasses in the system Na_2O–B_2O_3–SiO_2. Black dots represent the glass compositions investigated in the current study. LBS (1) and KBS (2), as well as the soda-lime silicate glass, NCS (3), were added for comparison. The contour diagram is build-up of 17 data points.

5. CONCLUSIONS

A detailed picture of the physical and mechanical properties of glasses from the sodium-borosilicate ternary is obtained through combining new experimental data on low-alkali glasses, including some technical multi-component alkali-borosilicates, with an extensive literature review. It is demonstrated that physical properties such as the glass transition temperature T_g and the density ρ can be correlated with mechanical properties such as the elastic modulus E and the Vickers hardness H_v predominantly through the SiO_2 content and the fraction of $[BØ_4]^-$ tetrahedral units within the glass network. That is, for low alkali-borosilicate glasses, the addition of a small amount of network modifier oxides converts neutral $BØ_3$ groups into charged $[BØ_4]^-$ units, which is accompanied by the aforementioned property increase. However, when the Na_2O-to-B_2O_3ratio, R, exceeds a value of $R = 0.2$, the boron oxide anomaly, the network connectivity starts to decrease due to the formation of non-bridging oxygen ions. This effect is initially counterbalanced by the progressive conversion of $BØ_3$groups into $[BØ_4]^-$ units, until a sudden decrease in the thermal (T_g) and mechanical stability (E, H_V) can be observed, most notably for values of $R > 0.5$, where the presence of non-bridging oxygen ions is also verified by NMR spectroscopy.

The contribution of densification to indentation-induced deformation, expressed as the indentation volume recovery V_R, reaches a maximum for silica-rich glasses, where the densification is based on congruent compaction of the SiO_2sub-network. In such glasses, permanent densification is achieved through a re-arrangement of large five- and six-membered rings into smaller three- and four-membered ring structures, as well as through a decrease of the average Si–O–Si bond angle. The incorporation of Na_2O and/or B_2O_3 reduces the silica sub-network's ability to compact of the silica sub-network. On the one hand, this is simply due to the increased atomic packing density. At high network modifier concentrations, on the other hand, the plastic deformation through localized shear flow is facilitated by the creation of non-bridging oxygen ions. Besides this, a strong correlation between densification and the interconnectivity between the silicate and borate sub-networks is noticed, where the formation of four-membered reedmergnerite $[BSi_3O_8]^-$ and danburite $[B_2Si_2O_8]^{2-}$-type ring structures and, therefore, higher connectivity between $SiØ_4$ groups and $[BØ_4]^-$units seems to result in a reduced compaction capacity.

The above-mentioned correlations qualitatively hold also for the crack initiation behavior and the phenomenological crack resistance *CR*. In SiO_2-rich glasses, where the deformation is mainly based on densification, ring and cone cracks are observed upon indentation, while the transition towards shear-mediated deformation occurs with increasing B_2O_3 concentration and/or with the formation of non-bridging oxygen species. This transition later manifests in the well-known change from ring and cone crack patterns via median cracks to a predominantly lateral and radial crack arrangement. Additional support for this is obtained from a consideration of the elastic stress fields which form around the indent, where it has previously been demonstrated that the contribution of densification to indentation-induced deformation leads to a reduction of tensile stresses in the surface plane, which are in turn responsible for the initiation of radial cracks. As a consequence, high values of *CR* are achieved in glasses where the indentation deformation is driven by compaction. This is in accordance with previous observations, which were made in other silicate glass systems, e.g., aluminosilicates or alkali–alkaline earth silicates.

In some contrast to the classical calculations of elastic stress fields, the appearance of cone cracks is still visible in glasses with relatively large amounts of Na_2O and B_2O_3. Most notably, the largest values of *CR* were obtained for glass compositions, which do not exhibit the highest values of V_R. All these findings indicate that the structural evolution, such as the conversion of $BØ_3$groups into $[BØ_4]^-$ units, the formation of non-bridging oxygen ions or the connectivity between the silicate and borate sub-networks, play an important role on the physical and mechanical properties as well as the indentation deformation and crack initiation behavior, and that for the generation of a full picture of this complex glass system, these phenomena have to be treated equally and beyond the simplified consideration of congruent compaction and shear flow alone.

ACKNOWLEDGMENTS

The financial support through the priority program SPP 1594 of the German Science Foundation under grant nos. WO1220/10-1 and MO1375/3-1 is gratefully acknowledged. The authors would further like to thank Doris Ehrt, Efstratios I. Kamitsos, Gregory Tricot, and Sindy Fuhrmann for their valuable discussions regarding the structure of the studied borosilicate glasses.

REFERENCES

1. Ellison, I.A. CornejoInt. J. Appl. Glass Sci., 1 (2010), pp. 87-103
2. M.M. Smedskjaer, R.E. Youngman, J.C. MauroAppl. Phys. A-Mater., 116 (2014), pp. 491-504
3. J.C. Mauro, **FMATS 1**(2014), 10.3389/fmats.2014.00020
4. L. Wondraczek, J.C. Mauro, J. Eckert, U. Kuhn, J. Horbach, J. Deubener, T. RouxelAdv. Mater., 23 (2011), pp. 4578-4586
5. D. EhrtGlass Technol., 41 (2000), pp. 182-185
6. T. RouxelJ. Am. Ceram. Soc., 90 (2007), pp. 3019-3039
7. L. Wondraczek, J.C. MauroJ. Eur. Ceram. Soc., 29 (2009), pp. 1227-1234
8. C.R. Kurkjian, P.K. Gupta, R.K. BrowInt. J. Appl. Glass Sci., 1 (2010), pp. 27-37
9. J.C. Mauro, C.S. Philip, D.J. Vaughn, M.S. PambianchiInt. J. Appl. Glass Sci., 5 (2014), pp. 2-15
10. P.J. Bray, S.A. Feller, G.E. Jellison, Y.H. Yun, J. Non-CrystSolids, 38–9 (1980), pp. 93-98
11. D. Möncke, D. Ehrt, H. Eckert, V. MertensPhys. Chem. Glasses, 44 (2003), pp. 113-116
12. D. Möncke, D. Ehrt, C.P.E. Varsamis, E.I. Kamitsos, A.G. KalampouniasGlass Technol.-Part A, 47 (2006), pp. 133-137
13. L. Wondraczek, S. Sen, H. Behrens, R.E. YoungmanPhys. Rev. B, 76 (2007), p. 014202(1–9)
14. L. Wondraczek, H. BehrensJ. Chem. Phys., 127 (2007), p. 154503(1–10)
15. L. Wondraczek, S. Krolikowski, H. Behrens, J. Non-CrystSolids, 356 (2010), pp. 1859-1862
16. Mantisi, S. Adichtchev, S. Sirotkin, L. Rafaelly, L. Wondraczek, H. Behrens, C. Marcenat, N.V. Surovtsev, A. Pillonnet, E. Duval, B. Champagnon, A. MermetJ. Phys.-Condens. Mat., 22 (2010), p. 025402(1–5)
17. S. Reibstein, L. Wondraczek, D. de Ligny, S. Krolikowski, S. Sirotkin, J.P. Simon, V. Martinez, B. ChampagnonJ. Chem. Phys., 134 (2011), p. 204502(1–6)
18. F. Angeli, O. Villain, S. Schuller, T. Charpentier, D. de Ligny, L. Bressel, L. WondraczekPhys. Rev. B, 85 (2012), p. 054110(1–15)
19. Makishima, J.D. Mackenzie, J. Non-CrystSolids, 12 (1973), pp. 35-45

20. C.J. PhillipsGlass Technol., 5 (1964), pp. 216-223
21. M.L. Williams, G.E. ScottGlass Technol., 11 (1970), pp. 76-79
22. N. Soga, J. Non-CrystSolids, 73 (1985), pp. 305-313
23. Bridge, N.D. Patel, D.N. WatersPhys. Status Solidi A, 77 (1983), pp. 655-668
24. P.K. Gupta, J.C. MauroJ. Chem. Phys., 130 (2009), p. 094503(1–8)
25. J.C. Phillips, M.F. ThorpeSolid State Commun., 53 (1985), pp. 699-702
26. M.M. Smedskjaer, J.C. Mauro, S. Sen, Y.Z. YueChem. Mater., 22 (2010), pp. 5358-5365
27. M.M. Smedskjaer, J.C. Mauro, Y.Z. YuePhys. Rev. Lett., 105 (2010), p. 115503(1–4)
28. M.M. Smedskjaer, J.C. Mauro, S. Sen, J. Deubener, Y.Z. YueJ. Chem. Phys., 133 (2010), p. 154509(1–6)
29. B.P. Rodrigues, L. WondraczekJ. Chem. Phys., 138 (2013), p. 244507(1–5)
30. T. RouxelC. R. Mécanique, 334 (2006), pp. 743-753
31. Arora, D.B. Marshall, B.R. Lawn, M.V. Swain, J. Non-CrystSolids, 31 (1979), pp. 415-428
32. Ehrt, **Untersuchungen über die Verteilung von CoO und Fe_2O_3 auf die Mikrophasen des Glassystems Na_2O–B_2O_3–SiO_2, Mathematisch-Naturwissenschaftlich-Technische Fakultät**, Friedrich-Schiller-Universität Jena (1973)(Phd-thesis)
33. Winterstein-Beckmann, D. Möncke, D. Palles, E.I. Kamitsos, L. Wondraczek, J. Non-CrystSolids, 401 (2014), pp. 110-114
34. Winterstein-Beckmann, D. Möncke, D. Palles, E.I. Kamitsos, L. Wondraczek, J. Non-CrystSolids, 405 (2014), pp. 196-206
35. R. Limbach, B.P. Rodrigues, L. Wondraczek, J. Non-CrystSolids, 404 (2014), pp. 124-134
36. Y.H. Yun, P.J. Bray, J. Non-CrystSolids, 27 (1978), pp. 363-380
37. S. Fuhrmann, **Druckabhängigkeit der topologischen Heterogenität von Gläsern, Technische Fakultät Fachbereich Glas und Keramik**, Friedrich-Alexander-Universität Erlangen-Nürnberg (2013)(Phd-thesis)
38. J.E. Shelby, **Water in Glasses and Melts**, The Royal Society of Chemistry, Cambridge, UK (2005)pp. 222–236
39. Thieme, D. Möncke, R. Limbach, S. Fuhrmann, E.I. Kamitsos, L. Wondraczek, J. Non-CrystSolids, 410 (2015), pp. 142-150
40. W.C. Oliver, G.M. PharrJ. Mater. Res., 7 (1992), pp. 1564-1583
41. V. Maier, K. Durst, J. Mueller, B. Backes, H.W. Höppel, M. GökenJ. Mater. Res., 26 (2011), pp. 1421-1430
42. B.N. Lucas, W.C. OliverMetall. Mater. Trans. A, 30 (1999), pp. 601-610
43. V. Maier, B. Merle, M. Göken, K. DurstJ. Mater. Res., 28 (2013), pp. 1177-1188
44. L. Shen, W.C.D. Cheong, Y.L. Foo, Z. ChenMat. Sci. Eng. A, 532 (2012), pp. 505-510
45. A.F. Bower, N.A. Fleck, A. Needleman, N. OgbonnaProc. Roy. Soc. Lond. A, 441 (1993), pp. 97-124
46. R. Goodall, T.W. ClyneActa Mater., 54 (2006), pp. 5489-5499
47. Y. Kato, H. Yamazaki, S. Yoshida, J. Matsuoka, J. Non-CrystSolids, 356 (2010), pp. 1768-1773
48. J.D. MackenzieJ. Am. Ceram. Soc., 46 (1963), pp. 470-476
49. S. Yoshida, J.C. Sangleboeuf, T. RouxelJ. Mater. Res., 20 (2005), pp. 3404-3412
50. S. Yoshida, S. Isono, J. Matsuoka, N. SogaJ. Am. Ceram. Soc., 84 (2001), pp. 2141-2143
51. J.E. Neely, J.D. MackenzieJ. Mater. Sci., 3 (1968), pp. 603-609
52. K.W. Peter, J. Non-CrystSolids, 5 (1970), pp. 103-115
53. P. Sellappan, T. Rouxel, F. Celarie, E. Becker, P. Houizot, R. ConradtActa Mater., 61 (2013), pp. 5949-5965
54. W.C. Oliver, G.M. PharrJ. Mater. Res., 19 (2004), pp. 3-20
55. M. KodamaJ. Mater. Sci., 26 (1991), pp. 4048-4053
56. K. Budhwani, S. FellerPhys. Chem. Glasses, 36 (1995), pp. 183-190
57. S. Suzuki, Y. Abe, J. Non-CrystSolids, 43 (1981), pp. 141-143
58. J.C. Mauro, P.K. Gupta, R.J. LoucksJ. Chem. Phys., 130 (2009), p. 234503(1–8)

59. Grandjean, M. Malki, C. Simonnet, D. Manara, B. PenelonPhys. Rev. B, 75 (2007), p. 054112(1–7)
60. M.J. Pascual, A. Duran, L. PascualPhys. Chem. Glasses, 43 (2002), pp. 25-31
61. I.G. Polyakova, E.V. Tokareva, **Alkali borosilicate systems: phase diagrams and the structure of glass**, A.C. Wright, S.A. Feller, A.C. Hannon (Eds.), Borate Glasses, Crystals and Melts, The Society of Glass Technology, Sheffield, UK (1997), pp. 223-230
62. S.A. Feller, J. Kottke, J. Welter, S. Nijhawan, R. Boekenhauer, H. Zhang, D. Feil, C. Parameswar, K. Budhwani, M. Affatigato, A. Bhatnagar, G. Bhasin, S. Bhowmik, J. Mackenzie, M. Royle, M. Kambeyanda, P. Pandikuthira, M. Sharma, **Physical properties of alkali borosilicate glasses**, A.C. Wright, S.A. Feller, A.C. Hannon (Eds.), Borate Glasses, Crystals and Melts, The Society of Glass Technology, Sheffield, UK (1997), pp. 246-253
63. T. Furukawa, W.B. WhiteJ. Am. Ceram. Soc., 64 (1981), pp. 443-447
64. X.W. Wu, R. Dieckmann, J. Non-CrystSolids, 357 (2011), pp. 2846-2856
65. X.W. Wu, A.K. Varshneya, R. Dieckmann, J. Non-CrystSolids, 357 (2011), pp. 3661-3669
66. X.W. Wu, R.E. Youngman, R. Dieckmann, J. Non-CrystSolids, 378 (2013), pp. 168-176
67. C.A. Maynell, G.A. Saunders, J. Non-CrystSolids, 12 (1973), pp. 271-294
68. Y. Kato, H. Yamazaki, Y. Kubo, S. Yoshida, J. Matsuoka, T. AkaiJ. Ceram. Soc. Jpn., 118 (2010), pp. 792-798
69. S. Yoshida, Y. Nishikubo, A. Konno, T. Sugawara, Y. Miura, J. Matsuokalnt. J. Appl. Glass Sci., 3 (2012), pp. 3-13
70. S. Yoshida, A. Hidaka, J. Matsuoka, J. Non-CrystSolids, 344 (2004), pp. 37-43
71. M. Bertoldi, V.M. SglavoJ. Am. Ceram. Soc., 85 (2002), pp. 2499-2506
72. S. Yoshida, Y. Hayashi, A. Konno, T. Sugawara, Y. Miura, J. MatsuokaPhys. Chem. Glasses B, 50 (2009), pp. 63-70
73. J. Sehgal, S. Ito, J. Non-CrystSolids, 253 (1999), pp. 126-132
74. S. Sakka, J.D. Mackenzi, J. Non-CrystSolids, 6 (1971), pp. 145-162
75. R. Jabra, J. Pelous, J. Phalippou, J. Non-CrystSolids, 37 (1980), pp. 349-358
76. M. Barlet, A. Kerrache, J.M. Delaye, C.L. Rountree, J. Non-CrystSolids, 382 (2013), pp. 32-44
77. S. Inaba, S. Fujino, K. MorinagaJ. Am. Ceram. Soc., 82 (1999), pp. 3501-3507
78. R.J. Eagan, J.C. SwearengenJ. Am. Ceram. Soc., 61 (1978), pp. 27-30
79. Hermansen, J. Matsuoka, S. Yoshida, H. Yamazaki, Y. Kato, Y.Z. Yue, J. Non-CrystSolids, 364 (2013), pp. 40-43
80. Abd El-MoneimPhysica B, 325 (2003), pp. 319-332
81. S.Y. Marzouk, M.S. GaafarSolid State Commun., 144 (2007), pp. 478-483
82. Y.B. Saddeek, L. Abd El LatifPhysica B, 348 (2004), pp. 475-484
83. Pedone, G. Malavasi, A.N. Cormack, U. Segre, M.C. MenzianiChem. Mater., 19 (2007), pp. 3144-3154
84. P. Paufler, S.K. Filatov, I.P. Shakhverdova, R.S. Bubnova, M. Reibold, B. Muller, A.A. Levin, D.C. MeyerGlass Phys. Chem., 33 (2007), pp. 187-198
85. H. Doweidar, J. PhysChem. Solids, 53 (1992), pp. 807-814
86. Q.J. Zheng, M. Potuzak, J.C. Mauro, M.M. Smedskjaer, R.E. Youngman, Y.Z. Yue, J. Non-CrystSolids, 358 (2012), pp. 993-1002
87. L.H. Kieu, J.M. Delaye, L. Cormier, C. Stolz, J. Non-CrystSolids, 357 (2011), pp. 3313-3321
88. Q. Zhao, M. Guerette, L.P. Huang, J. Non-CrystSolids, 358 (2012), pp. 652-657
89. ArticlePDF (1MB)
90. M.S. Gaafar, S.Y. MarzoukPhysica B, 388 (2007), pp. 294-302
91. G.H. Frischat, W. Krause, H. HubenthalJ. Am. Ceram. Soc., 67 (1984), pp. C10-C12
92. Vernaz, F. Larche, J. Zarzycki, J. Non-CrystSolids, 37 (1980), pp. 359-365
93. S.M. WiederhornJ. Am. Ceram. Soc., 52 (1969), pp. 99-105
94. E.I. KoslovskayaProc. Third All-Union Conference on the Glassy State, 2 (1960), pp. 299-301

95. N.P. Lower, R.K. Brow, C.R. Kurkjian, J. Non-CrystSolids, 349 (2004), pp. 168-172
96. T.E. Wilantewicz, J.R. VarnerJ. Mater. Sci., 43 (2008), pp. 281-298
97. T.E. Wilantewicz, J.R. Varner, J. Non-CrystSolids, 354 (2008), pp. 1553-1558
98. V.A. Mukhanov, O.O. Kurakevich, V.L. Solozhenko, J. SuperhardMater., 30 (2008), pp. 71-72
99. J. Ramkumar, V. Sudarsan, S. Chandramouleeswaran, V.K. Shrikhande, G.P. Kothiyal, P.V. Ravindran, S.K. Kulshreshtha, T. Mukherjee, J. Non-CrystSolids, 354 (2008), pp. 1591-1597
100. A.K. Seal, P. Chakraborti, N.R. Roy, S. Mukherjee, M.K. Mitra, G.C. PasB. Mater. Sci., 28 (2005), pp. 457-460
101. M. Yamane, J.D. Mackenzie, J. Non-CrystSolids, 15 (1974), pp. 153-164
102. X.C. Zhang, O.L. He, C.Z. Xu, Y.H. ZhengJ Non-Cryst Solids, 80 (1986), pp. 313-318
103. J. de Bonfils, S. Peuget, G. Panczer, D. de Ligny, S. Henry, P.Y. Noel, A. Chenet, B. Champagnon, J. Non-CrystSolids, 356 (2010), pp. 388-393
104. L. Ainsworth, J. SocGlass Technol., 38 (1954), pp. 501-535
105. L. Ainsworth, J. SocGlass Technol., 38 (1954), pp. 536-547
106. Z. Burghard, A. Zimmermann, J. Rodel, F. Aldinger, B.R. LawnActa Mater., 52 (2004), pp. 293-297
107. H. Ishikawa, N. ShinkaiJ. Am. Ceram. Soc., 65 (1982), pp. C124-C127
108. R.F. Cook, G.M. PharrJ. Am. Ceram. Soc., 73 (1990), pp. 787-817
109. M. Bertoldi, V.M. Sglavo, J. Non-CrystSolids, 344 (2004), pp. 51-59
110. H. Inoue, A. Masuno, Y. Watanabe, K. Suzuki, T. IsedaJ. Am. Ceram. Soc., 95 (2012), pp. 211-216
111. G.E. Jellison, L.W. Panek, P.J. Bray, G.B. RouseJ. Chem. Phys., 66 (1977), pp. 802-812
112. P.A.V. Johnson, A.C. Wright, R.N. Sinclair, J. Non-CrystSolids, 50 (1982), pp. 281-311
113. R.E. Youngman, S.T. Haubrich, J.W. Zwanziger, M.T. Janicke, B.F. ChmelkaScience, 269 (1995), pp. 1416-1420
114. A.C. Hannon, D.I. Grimley, R.A. Hulme, A.C. Wright, R.N. Sinclair, J. Non-CrystSolids, 177 (1994), pp. 299-316
115. J. Krogh-MoePhys. Chem. Glasses, 6 (1965), pp. 46-54
116. J. Lorösch, M. Couzi, J. Pelous, R. Vacher, A. Levasseur, J. Non-CrystSolids, 69 (1984), pp. 1-25
117. W.J. Dell, P.J. Bray, S.Z. Xiao, J. Non-CrystSolids, 58 (1983), pp. 1-16
118. M.E. Milberg, R.A. Verhelst, J.G. Okeefe, H.O. HooperPhys. Chem. Glasses, 13 (1972), pp. 79-84
119. Manara, A. Grandjean, D.R. Neuville, J. Non-CrystSolids, 355 (2009), pp. 2528-2531
120. A.A. Osipov, L.M. Osipova, V.E. EremyashevGlass Phys. Chem., 39 (2013), pp. 105-112
121. M.M. Smedskjaer, J.C. Mauro, R.E. Youngman, C.L. Hogue, M. Potuzak, Y.Z. Yue, J. PhysChem. B, 115 (2011), pp. 12930-12946
122. Makishima, J.D. Mackenzie, J. Non-CrystSolids, 17 (1975), pp. 147-157
123. H.M. Cohen, R. RoyPhys. Chem. Glasses, 6 (1965), pp. 149-161
124. J.D. MackenzieJ. Am. Ceram. Soc., 46 (1963), pp. 461-470
125. S. Susman, K.J. Volin, D.L. Price, M. Grimsditch, J.P. Rino, R.K. Kalia, P. Vashishta, G. Gwanmesia, Y. Wang, R.C. LiebermannPhys. Rev. B, 43 (1991), pp. 1194-1197
126. L. Huang, J. Nicholas, J. Kieffer, J. BassJ. Phys. Condens. Mat., 20 (2008), p. 075107(1–8)
127. TakadaPhys. Chem. Glasses, 45 (2004), pp. 156-159
128. K. Trachenko, V.V. Brazhkin, G. Ferlat, M.T. Dove, E. ArtachoPhys. Rev. B, 78 (2008), p. 172012(1–4)
129. K. Hirao, Z. Zhang, H. Morita, N. Soga, J. SocMater. Sci. Jpn, 40 (1991), pp. 400-404
130. S.K. Lee, P.J. Eng, H.K. Mao, J.F. ShuPhys. Rev. B, 78 (2008), p. 214203(1–6)
131. S. Striepe, M.M. Smedskjaer, J. Deubener, U. Bauer, H. Behrens, M. Potuzak, R.E. Youngman, J.C. Mauro, Y.Z. Yue, J. Non-CrystSolids, 364 (2013), pp. 44-52
132. R.D. ShannonActa Crystallogr. A, 32 (1976), pp. 751-767
133. T. Rouxel, H. Ji, T. Hammouda, A. MoreacPhys. Rev. Lett., 100 (2008), p. 225501(1–4)

134. T. Rouxel, H. Ji, J.P. Guin, F. Augereau, B. RuffleJ. Appl. Phys., 107 (2010), p. 094903(1–5)
135. E.H. YoffePhilos. Mag. A, 46 (1982), pp. 617-628
136. J. Kjeldsen, M.M. Smedskjaer, J.C. Mauro, Y.Z. YueAppl. Phys. Lett., 104 (2014), p. 051913(1–5)
137. J. Kjeldsen, M.M. Smedskjaer, J.C. Mauro, Y.Z. Yue, J. Non-CrystSolids, 406 (2014), pp. 22-26
138. S. Yoshida, J.C. Sangleboef, T. RouxelInt. J. Mater. Res., 98 (2007), pp. 360-364
139. S. Yoshida, H. Sawasato, T. Sugawara, Y. Miura, J. MatsuokaJ. Mater. Res., 25 (2010), pp. 2203-2211
140. T.M. Gross, J. Non-CrystSolids, 358 (2012), pp. 3445-3452
141. Z.Y. Zhang, N. Soga, K. HiraoJ. Mater. Sci., 30 (1995), pp. 6359-6362
142. S. Fuhrmann, T. Deschamps, B. Champagnon, L. WondraczekJ. Chem. Phys., 140 (2014), p. 054501(1–7)
143. D.A. Kilymis, J.M. DelayeJ. Chem. Phys., 141 (2014), p. 014504(1–10)
144. N.M. VedishchevaI. G. Polyakova, 55 (2014), pp. 225-336
145. Tricot, J. Trébosc, F. Pourpoint, R. Gauvin, L. DelevoyeChapter Four — The D-HMQC MAS-NMR Technique: An Efficient Tool for the Editing of Through-space Correlation Spectra Between Quadrupolar and Spin-1/2 (^{31}P, ^{29}Si, ^{1}H, ^{13}C) NucleiAcademic Press (2014), pp. 145-184
146. S.K. Lee, K. Mibe, Y.W. Fei, G.D. Cody, B.O. MysenPhys. Rev. Lett., 94 (2005), p. 165507(1–4)
147. A.C. Wright, C.E. Stone, R.N. Sinclair, N. Umesaki, N. Kitamura, K. Ura, N. Ohtori, A.C. HannonPhys. Chem. Glasses, 41 (2000), pp. 296-299
148. M. Grimsditch, A. Polian, A.C. WrightPhys. Rev. B, 54 (1996), pp. 152-155
149. M.M. Smedskjaer, R.E. Youngman, S. Striepe, M. Potuzak, U. Bauer, J. Deubener, H. Behrens, J.C. Mauro, Y.Z. YueSci. Rep. UK, 4 (2014), p. 3770(1–6)
150. S.K. Lee, P.J. Eng, H.K. Mao, Y. Meng, M. Newville, M.Y. Hu, J.F. ShuNat. Mater., 4 (2005), pp. 851-854
151. K. Hirao, J. Matsuoka, N. Soga, J. Non-CrystSolids, 112 (1989), pp. 336-340
152. TaborRev. Phys. Tech., 1 (1970), pp. 145-179
153. W.P. Yu, J.P. BlanchardJ. Mater. Res., 11 (1996), pp. 2358-2367
154. N.A. Sakharova, J. Fernandes, J.M. Antunes, M.C. OliveiraInt. J. Solids Struct., 46 (2009), pp. 1095-1104
155. B.R. Lawn, M.V. SwainJ. Mater. Sci., 10 (1975), pp. 113-122
156. D.B. Marshall, B.R. Lawn, A.G. EvansJ Am Ceram Soc, 65 (1982), pp. 561-566
157. W.D. Nix, H.J. Gao, J. MechPhys. Solids, 46 (1998), pp. 411-425
158. T.M. Gross, M. Tomozawa, J. Non-CrystSolids, 354 (2008), pp. 5567-5569
159. L. Shartsis, W. Capps, S. SpinnerJ. Am. Ceram. Soc., 36 (1953), pp. 35-43
160. J.T. HaganJ. Mater. Sci., 14 (1979), pp. 462-466
161. J.T. Hagan, M.V. SwainJ. Phys. D, 11 (1978), pp. 2091-2102
162. B.R. Lawn, D.B. MarshallJ. Am. Ceram. Soc., 62 (1979), pp. 347-350
163. J.J. Lewandowski, W.H. Wang, A.L. GreerPhil. Mag. Lett., 85 (2005), pp. 77-87
164. W. Haller, D.H. Blackburn, F.E. Wagstaff, R.J. CharlesJ. Am. Ceram. Soc., 53 (1970), pp. 34-39

NOTE

[1]For ring member classification the nomenclature of silicate glasses is employed, which counts the number of SiO_4-tetrahedra or network forming polyhedra per ring, not the nomenclature commonly used in borate chemistry, in which the number of participating atoms per ring unit is counted, e.g., six-membered ring for boroxol.

CHAPTER 13

A Study of the Factors Affecting Transporting Solid— Liquid Suspension through Pipelines

A. I. Abd Al Aziz, H. I. Mohamed

Public Authority of Applied Education & Training, Kuwait, Kuwait

ABSTRACT

An experimental investigation was carried out on the transport of solid liquid mixture through pipelines. The principal aim of this was to study how to transport slurries through pipeline systems. The experimental tests include measurements of main parameters affecting transport of solid liquid mixture, like sand slurry and mud slurry. These parameters are deduced by applying non-dimensional approach, which includes Reynolds number, Froude number, concentration, specific gravity, and ratio of particle to pipe diameter. Preliminary results include the following general trends: 1) Increasing input concentration increases the pressure gradient, whereas decreases the efficiency of solid transport; 2) As specific gravity of solid material increases, the pressure gradient increases and the efficiency of transport decreases; 3) As mixture velocity increases, the efficiency of transport increases; 4) Solids with fine grain size are preferred than with coarse grain size from the view points of pressure gradient and efficiency of transport. Also, the present experimental data has been compared with the correlations developed before by different authors. Such correlations relate the pressure gradient to flow velocity, specific gravity, and efficiency of transport to grain size of solid material, and input solid concentration.

Keywords: Sand Slurry; Solid Concentration; Pressure Gradient; Efficiency of Transport; Solid Diameter

1. INTRODUCTION

The present study is originated in order to investigate the problem of blocking the pipelines that transport water and sold mixture. The slurry is the product of many industrial process like oxygen converter in iron and steel industries, and drilling industries.

These lines have a considerably low flow velocity, which leads to enhanced rate of deposition. As a result of this operation, pipeline systems suffer from fast blocking.

The present study is also of interest because it is applicable to problems related to the handling and transporting of solid raw materials and products, e.g. coal.

The experimental study for determining the pressure gradient and efficiency of transport is burdened by some different constraints arising mainly from the following three characteristics of the problem.

1) The abundance of variables which govern the flow.

2) The vast range over which most of the variables may vary.

3) The inherent limitation on accuracy and reproducibility of the data that can be obtained with heterogeneous settling slurries.

The basic problem in hydraulic transport of solids is the determination of the fluid forces acting upon the solid particles and the effects of these forces upon the behavior of such particles and upon the resultant energy losses.

The experimental results were compared with the correlations (13-15) developed later by other authors.

2. LITERATURE SURVEY

2.1. Introduction

The transport of solid particles by a liquid flowing in a pipeline is governed by a number of parameters that can be classified as follows: pipeline parameters, liquid parameters, solid particle parameters and system parameters e.g. mixture flow velocity.

It would be useful to study the effects of these parameters upon the performance characteristics such as: the pressure gradient ($\Delta P/L$) and efficiency of transport (η).

During the last few years, there have been several investigations concerning the connections between these parameters with the pressure drops in addition to the effect of these parameters upon the characteristics of centrifugal pumps. The findings of the previous investigations for the flow of slurries are summarized in the following sections:

2.2. Flow Regimes and Critical Velocity

2.2.1. Flow Regimes

There are four distinct regimes of particle conveyance, (see Newit et al. [1]) who carried out experiments with solids of sizes from 0.062 to 4.67 mm. and specific gravity ranging between 1.18 and 4.6. The experiments were performed in a 2.54 cm. pipe and resulted in the following regimes.

1) Conveyance as a homogenous suspension in which mean mixture velocity is high to prevent sedimentation during transport process.

2) Conveyance as a heterogeneous suspension, which is maintained by turbulence heterogeneous flow.

3) Conveyance by saltation, in which the particles are alternatively picked up by the liquid and deposited further along the pipe.

4) Conveyance as a sliding bed, that is a regime in which the principal mechanism is sliding motion of solids along the bottom of the pipe.

Another classification of these regimes is derived by Durand [2]. This classification depends on particle size ranges as follows:

1) Particles of a size less than 40 μm, are transported as a homogenous suspension.

2) Particles of a size between 40 μm and 0.15 mm are transported as suspension that is maintained by turbulence.

3) Particles of a size among 0.15 and 1.5 mm are transported by a suspension and saltation.

4) Particles of a size greater than 1.5 mm are transported by saltation.

The flow regimes given in this classification are related having a specific gravity of 2.65.

2.2.2. Critical Velocity

The critical velocity conditions have been the study of extensive research effort and yet can still be confusing, as many definitions have been used for this term.

The most important critical velocities are those which another phase of flow is produced as laminar/turbulent transport velocity in case of homogenous slurry flow or the deposit velocity in case of settling slurry flow.

2.2.3. Critical Deposit Velocity

A further result of Durand's work is an empirical equation for predicting the critical deposit velocity of slurry, i.e. the velocity below which a stationary deposit of solids forms in the pipe. The critical deposit velocity correlation is given by:

$$v_c = F_L\sqrt{2gD(s-1)} \tag{1}$$

where F_L is a dimensionless function of particle diameter given by Durand and can be obtained using **Figure 1**, which requires only the mean particle size and concentration distribution of solids to find the value of F_L. The result in equation is of considerable practical value for the critical velocity, if sufficiently high in a given case, may be the limiting parameter.

2.3. Pressure Drop through Slurry Pipeline

The need to predict pressure drop for flowing slurries is an eminently practical one.

2.3.1. Durand's Equation for Pressure Losses in Horizontal Pipes

The empirical correlation published by DURAND [2] is related to the pressure drops associated with the flow of sand-water and gravel-water mixtures with particles of sizes ranging from 0.2 to 25 mm. and pipe diameters from 3.8 to 58 cm. with solids concentrations up to 60% by volume.

A principal result of the work by DURAND [2] is the correlations:

$$\frac{i-i_w}{i_w C_o} = k\left[\frac{v^2}{gD}\sqrt{C_D}\right]^{-1.5} \tag{2}$$

where i and i_w are the pressure drop of slurry and of water respectively, k is a constant and C_D is the drag coefficient for the free failing particle at its terminal velocity in the stagnant unbounded liquid, assumed spherical and of diameter d_s given by:

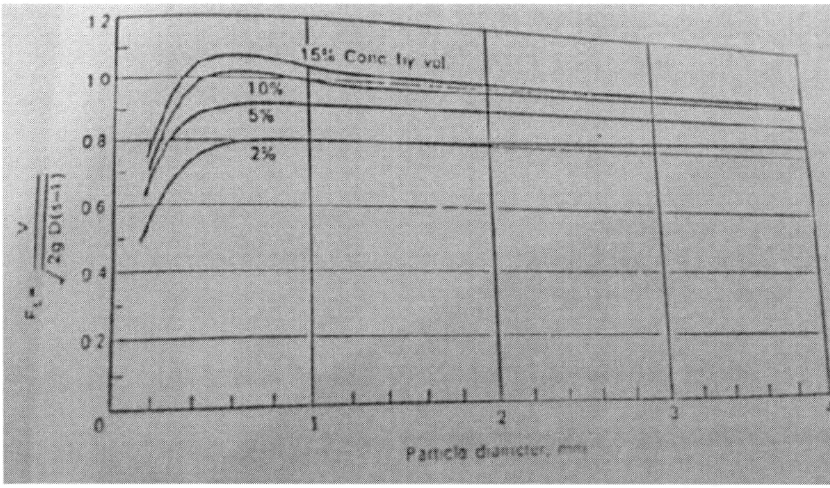

Figure 1. Values of F_L (Durand et al. (1953)).

$$C_D = \frac{4gd_s\left[s-1\right]}{3v_\infty^2} \tag{3}$$

The constant *k* is found to be 3.18 for sand slurry according to WORSTER [3]. To account for the effect of particle density, moreover, a modification of Equation (1) is given by

$$\frac{i-i_w}{i_w C_o} = k\left[\frac{v^2}{gD(S-1)}\sqrt{C_D}\right]^{-1.5} \tag{4}$$

where S is the specific gravity of solid material.

In this equation Durand's group used (S – 1) = 1.65, and, by comparing Equation (2) and Equation (4) *k* = 150.

Durand's correlation, Equation (2) gave 18.0% absolute average deviation when compared with data by Turain and Yuan [6].

2.3.2. Worster's Equation for Pressure Losses in Horizontal Pipes

Worster [3] studied flow phenomena associated with large particles in particular coal particles, and produced a similar correlation to that of Durand and his coworkers except for the inclusion of a term in take account of specific gravity of the solid material and the absence of any parameter of particle size.

$$\frac{i-i_w}{i_w C_o}=120\left[\frac{\sqrt{gD}}{v}\sqrt{\frac{\rho_{s-p}}{\rho}}\right]^3 \quad (5)$$

2.3.3. Improved Durand—Equation for Multiple Applications

The applicability of Durand's equation could be improved for general use by applying suitable parameters representing the grain size distribution. Thus, the Durand's equation can not only describe polydisperse (pseudo)-homogenous or heterogeneous transportation, but also solidliquid mixtures containing a certain amount of fine particles. Even non Newtonian influences can be taken into account.

The applicability of the extended Durand's equation for polydisperse mixtures was demonstrated by measurement data. With respect to this, the transition between pseudohomogeneous and heterogeneous transport has been considered on the basis of measured concentration profiles.

WAGNER [4] empirically could show with special polydisperse mixtures the grain size distribution can be taken into account more efficiently by using d_{s90} and d_{s10} in the following correlation factor, m; viz.

$$m=2-\left(\frac{d_{s90}}{d_{s10}}\right)^{-0.4} \quad (6)$$

where d_{s90} and d_{s10} are particle diameter at 90% and 10% passing sieve, respectively.

Thus the following extended version of Durand's equation can be written for polydisperse-heterogenous Newtonian mixture:

$$\Delta P=\left[83^{\frac{1}{m}}\left[\frac{gD}{\upsilon^2}\left(\rho_s-\rho_f\right)\rho_f\sqrt{C_D}\right]^{\frac{1.5}{m^3}}C_o+1\right] \cdot\lambda_f\rho_f 2\upsilon^2\frac{L}{D} \quad (7)$$

where ρ_s, ρ_f are solid and fluid densities, respectively.

C_o is the delivered total concentration, λ_f is the pipe friction factor. Adapting this equation also to homogeneous mixtures by taking account of grain size distribution, the heterogeneous part of the equation can be partly or totally eliminated. This can be done by defining a homogeneous part '*F*' of the grain size distribution, which is namely fine grain and a heterogeneous part (1 – *F*).

The delivered concentration of the fine material could be represented as follows.

$$C_F = F - C_o \quad (8)$$

The delivered concentration of the coarse material could be represented as follows.

$$C_c = (1-F)C_o \quad (9)$$

The density of the enriched carrier fluid is given by:

$$\rho_{fs} = FC_o\rho_s + (1-FC_o)\,\rho_f \quad (10)$$

Substituting ρ_f in Equation (7) by ρ_{fs} and adding (1 – F) as a factor of C_o.

The generally valid equation for polydisperse heterogeneous and homogeneous Newtonian mixtures that was improved by M. WEBER, [5] can be stated as follows:

$$[C\_o+1]\lambda_f \rho_{fs} 2v_m^2 \frac{L}{D} \quad (11)$$

$$\Delta P = \left[83^{\frac{1}{m}} \left[\frac{gD}{v_m^2} (\rho_s - \rho_{fs}) \rho_f \sqrt{C_D} \right]^{\frac{1.5}{m^3}} (1-F)C_o + 1 \right] \cdot \lambda_f \rho_{fs} 2v_m^2 \frac{L}{D} \quad (12)$$

2.3.4. Turain's Equation for Pressure Drop in Slurry Pipeline

The slurry flow correlation established in Turain and Yuan [6] is represented by relations:

$$\lambda = \lambda_w + 0.3202\xi \quad \text{for} \quad \xi \geq 1 \quad (13)$$

$$\lambda = \lambda_w + 0.3202\xi^{1.338} \quad \text{for} \quad \xi \leq 1 \quad (14)$$

where

$$\xi = C_o^{0.4925} \lambda_w^{0.5586} C_D^{-0.1095} \left[\frac{gD(S-1)}{v^2} \right]^{0.9982} \quad (15)$$

C_o is the discharge concentration of solids, volume %.

$C_D = \dfrac{4gd(\rho_s - \rho)}{3\rho v_\infty^2}$, Drag coefficient for free falling sphere, dimensionless, where S is the specific gravity of solids, D is the inside diameter of pipe, g is the gravitational constant, $\lambda=(PD/2\rho v^2L)$, friction factor for pipe flow of slurry, dimensionless, and $\lambda_w=(P_wD/2\rho v^2L)$, friction factor for pipe flow in absence of solids, dimensionless.

This empirical correlation was based upon a collection of 1511 data points pertained to flow in 5.04, 2.54, and 1.27 on pipelines of water suspensions of twelve different sizes of solid particles ranging from 31 μm. to 4380 μm. and covering, in round figures, suspended solid densities of 2300, 3000, 4400, 7500, and 11300 kg/m^3.

The correlation represented by Equation (8) and (9) predicts the slurry friction factor with absolute average deviations of 9.43% for all 1511 data points, with 76 points having deviations exceeding 30%, and only 9 points having deviations exceeding 50%.

2.4. Optimal Grain Distribution for Transport with High Concentration

Hydraulic conveying at high solid's concentration is very attractive and a large amount of experimental data has accumulated to date, but many fundamental problems still need to be answered. H.C. HOU [7] carried out theoretical analysis and answered some questions dealing with concentration transport.

2.4.1. Two Forms of Flow

There are two forms of transport. The first case is the case of dilute concentration with fine and/or coarse materials, in which the solid particles are suspended individually in the flowing water. In this case the energy expenditure for suspension of particle is compensated by the turbulence energy, which is taken from the kinetic flow field with the result that the von Karman constant for flow with sediment suspension should be less than that for pure water. For a given particle—composition it was found that the higher the value of concentration, the lower the value of von Karman constant. The reduction of von Karman's constant with the presence of sediment means that the increase of dissipated energy that is directly withdrawn from the mean flow field needs a corresponding increase in the energy-supply from outside the system. Therefore, the hydraulic conveying of solids by means of Newtonian fluids is not economic.

Another form of hydrotransport is the transport with hyper concentration including some of fine particles in the range($d_s \leq 0.02$-0.04). This forms a non-Newtonian system which is complex and requires thorough analysis.

2.4.2. Critical Volume Concentration

There are critical volume concentrations$(C_i)_{CR}$=6%, by which the system—transfers from Newtonian fluid (dilute concentration) to hyper concentration i.e. nonNewtonian fluid. This limit was determined by H.C. HOU [7] after carrying some current experiments concerning solid-liquid mixture transport.

2.4.3. Specific Energy Consumption in the Slurry Pipelines

An experimental approach had been carried out by M.S. Lee, V. Matousek, C.K. Chung and Y.N. Lee [8] the results indicate that the specific energy consumption at velocities near the deposition limit is not very sensitive to the pipe size.

It also indicate that the low concentrated slurries of about 7% exhibit higher specific energy consumption values for all three pipes of diameters (155, 204, and 305 mm).

2.4.4. Closure

The above review has shown a lot of experimental data that have been accumulated during the last period. However most of these data concentrates on general flowtypes rather than on specific types. The present work, on the other hand concentrates on the special problem of transport slurry of the by-product solids of oxygen converter shop.

3. THE PRESENT WORK

The present work divided into two branches; the first one was:

Theoretical and dimensional analysis, the second one was the experimental tests.

3.1. Theoretical and Dimensional Analysis

In this section the dimensional analysis is used to plan experiments and present data compactly. It is noteworthy to mention that many workers also used it in theoretical studies, as well.

Basically, dimensional analysis is a method for reducing the number and complexity of experimental variables that affect a given physical phenomenon, using a sort compacting technique. By applying this techniques for the problem of transporting solid-liquid mixture we proceed as follows:

Under isothermal conditions, at least 12 variables are needed to describe suspension flow behavior, provided that pipeline roughness and solids density distribution are excluded from consideration.

The input variables can be listed as follows:

Pipeline parameters:

1. Pipe diameter, assuming section to be circular D

2. Inclination to horizontal g

3. Pipe length. L Liquid parameters:

1. Density. ρ

2. Viscosity. μ

Solid particles parameters:

1. Density. ρ_s

2. Size distribution (dimensionless) ψ

3. Shape factor $\boldsymbol{\varepsilon}$

System parameters:

1. Velocity of flow. v

2. Solid-liquid ratio of the flowing mixture at inlet. C_i

The output variables or the performance characteristic variables are:

1. The pressure drop for pipeline flow of solid – liquid suspension (ΔP).

2. The efficiency of transport η defined as the ratio between output to input concentrations per unit volumeand can be represented as follows: $\eta = \dfrac{C_o}{C_i} \times 100\%$.

The total number of 12 variables can be reduced to 9 independent dimensionless groups by applying the conventional Π – theorem. One way of expressing this, is through a relation of the type.

$$\left(\eta, \frac{\Delta P}{2\rho \upsilon^2} \frac{D}{L}\right) = \phi\left(\frac{D\rho\upsilon}{\mu}, \frac{d_s}{D}, \frac{L}{D}, \frac{\rho_s}{\rho}, \frac{v^2}{gD}, C_o, \psi, \varepsilon\right) \quad (16)$$

Under some conditions, certain simplifications can be made in Equation (16).

For example, if the pipe length is large the dependence on the $\left(\frac{L}{D}\right)$ ratio becomes insignificant, and this term can be dropped, based on the measurement of the axial variations in pressure drop. Few other simplifications drop both ψ and $\boldsymbol{\varepsilon}$.

Accordingly, for this case Equation (16) takes the form:

$$\left(\eta, \frac{\Delta P}{2\rho \upsilon^2} \frac{D}{L}\right) = \phi\left(\frac{D\rho\upsilon}{\mu}, \frac{d_s}{D}, \frac{\rho_s}{\rho}, \frac{v^2}{gD}, C_o\right) \quad (17)$$

Despite the substantial reduction in the number of variables in going from Equation (16) to Equation (17), the functional dependence is still unknown.

3.2. Experimental Test Rig

To establish the relationship stated in Equation (17) a purely experimental approach designed. The model used for test is shown in **Figure 2**.

The model was chosen in such a way to permit the observation of effects of dimensionless groups that control the rate of deposition of solids and pressure gradient through pipeline system. These dimensionless groups are, Reynolds number, Froude number, relative density, particle grain size, and input concentration. The model consists of two cast steel reservoirs 6 cubic meter capacity each of conical-shaped bottom, and connected to a slurrytype centrifugal pump with diameters of 5.08 cm suction and 2.54 cm discharge pipes. The pumps are connected to 7.5 Kw. A/C induced motor with variable speeds. the pipeline is 285 meter long, and the solid concentration by volume was varied between 0.027% and 10%. Average grain size range between 0.25 mm and 1 mm and specific gravities ranging from 2.65 to 3.540 the Reynolds number was varied from 10907 to 63699.

The measured velocity was varied from 0.5 m/s to 2.5 m/s.

In the tests five variables were determined namely:

1. Pressure gradient along the pipeline system.

2. Flow rate measurements.

3. Input concentration to pipeline (C_i).

4. Output concentration from pipeline (C_o).

5. Input grain size.

3.2.1. Pressure Measurement

Pressure measurements were determined by means of bourdon tube manometers that were calibrated by means of dead weight testers. The bourdon gauges are fitted to the pipeline at an equal distance of 16 meter between any two successive manometers.

To avoid wrong readings of manometers as a result of existing solid material through the flow field, special connections were made by mounting a helical-shaped pipe full of oil that is lighter than water.

3.2.2. Flow Measurements

Flow measurements were determined by measuring the flow rate during one hour operation from suction tank to discharge tank.

During the experimental work, the flow rate controlled by means of adjustment of delivery valves only while keeping the suction valves fully opened.

3.2.3. Concentration Measurement

In this experimental work, the following steps have been carried out to determine the input and output concentrations

1. Prepare a clean, empty bottle and weigh it.

2. Fill the empty bottle with 50 cm^3 sample from the homogenous mixture of the suction tank.

3. Put the bottle on a heater until water is completely evaporated and hence the solids are completely dried.

4. Put the solid sample into cold dryer to cool it.

5. Weigh the solid sample.

6. Subtract the two weights in steps 5 and 1 respectively.

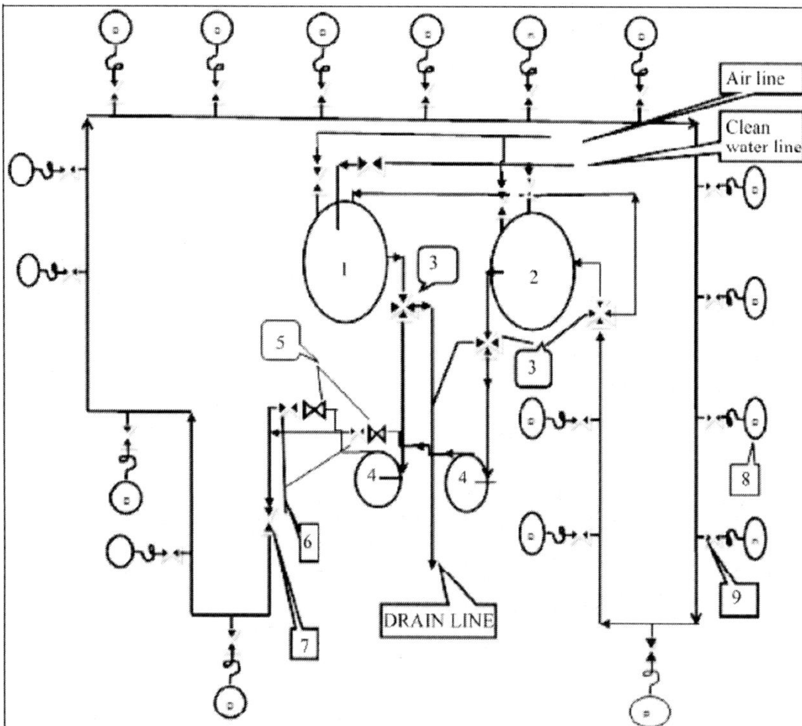

No	Description	No	Description
1	Tank 1	6	Delivery valves
2	Tank 2	7	Control valve
3	Three way suction valves	8	Pressure gauges
4	Centrifugal pumps	9	Pressure gauge valve
5	Non return valves		

Figure 2. Schematic outline of experimental test rig.

The difference is the total solid existed in the sample (suspension solid + soluble solid).

7. Take another 50 cm^3 sample from the same mixture of the suction tank and filter it, so as the water constitutes the soluble solids only.

8. Repeat the steps from (3 to 4) listed before to determine the weight of soluble solids.

9. Subtract the result of step number 8 from that of number 6 to obtain the T.S.M (Total Suspended Material) which represents the transported material through the pipeline system.

3.2.4. Input Grain Size Measurements

The input grain size was determined using a set of standard screen meshes that are available in the laboratories of the factory.

3.3. Experimental Procedure

Preliminary measurements were made to ensure that the system produces a stable (no pulsation), with no leakage from test devices. The Reynolds number (Re_D) for the tests is varied from (10907 to 63699).

The model was provided with 18 pressure gauges, fitted at locations of equal distance of 16 meter apart to ensure precise plotting of the pressure drop variation with length and velocity.

4. RESULTS AND DISCUSSIONS

In the present section the results obtained from the experimental model are presented and discussed.

4.1. Performance Characteristics of Water in Absence of Solids

From section (3) it was noted that the governing equation of slurry flow system may be written in the form:

$$\left(\eta, \frac{\Delta P}{2\rho\upsilon^2}\frac{D}{L}\right) = \phi\left(\frac{D\rho\upsilon}{\mu}, \frac{d_s}{D}, \frac{\rho_s}{\rho}, \frac{v^2}{gD}, C_o\right) \tag{17}$$

This equation can be reduced to the following equation for the clean water:

$$\left(\frac{\Delta P}{2\rho\upsilon^2}\frac{D}{L}\right) = \phi\left(\frac{D\rho\upsilon}{\mu}\right) \tag{18}$$

From this equation, the parameter describing the flow performance is represented by the pressure gradient, whereas the Reynolds number of clean water Re_w is the only affecting variable.

4.2. Effect of Varying Reynolds Number Re and Froude Number upon the Performance Characteristics

In the present experimental work, it was found that the simplest way to vary the value of Reynolds number would be through adjustment of the flow velocity. Therefore, Reynolds number as well as Froude number vary together.

1) Effect of varying Re, Fr, upon pressure gradients

The Reynolds number has been varied from (10907 – 63699) and Froude number from (0.5 - 17).

It has been noted from Figures 3 to 8 that as Reynolds $\left(\frac{\Delta P}{L}\right)$, which is a waste of power available increases.

Also these experimental results concerning pressure gradient have close results in comparison to Turian and Yuan [6] correlations. this comparison is shown in Figures 9 and 10.

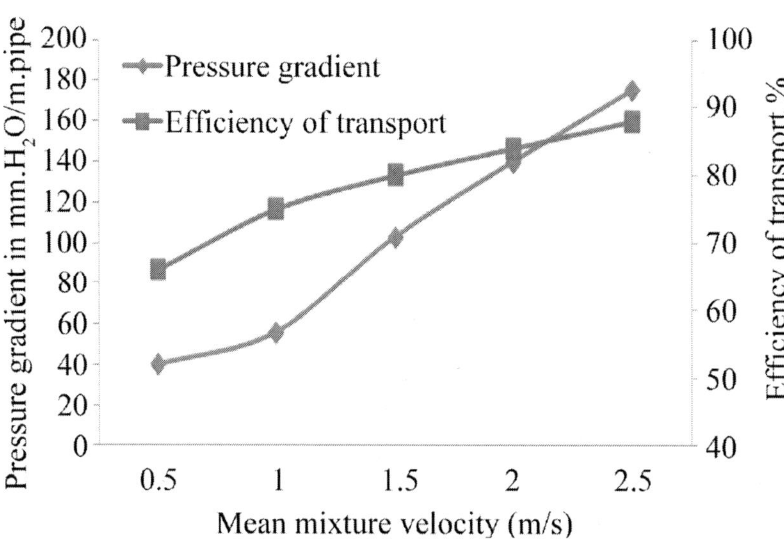

Figure 3. Variation of slurry performance characteristics for sand slurry with solid concentration 1% by volume and particle diameter d_s = 0.25 mm. versus mean mixture velocity.

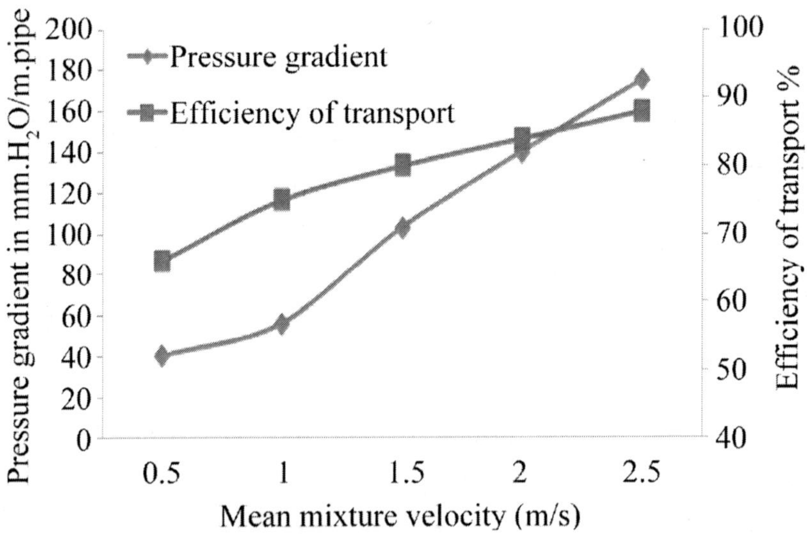

Figure 4. Variation of slurry performance characteristics for sand slurry with solid concentration 1% by volume and particle diameter d_s = 0.5 mm. versus mean mixture velocity.

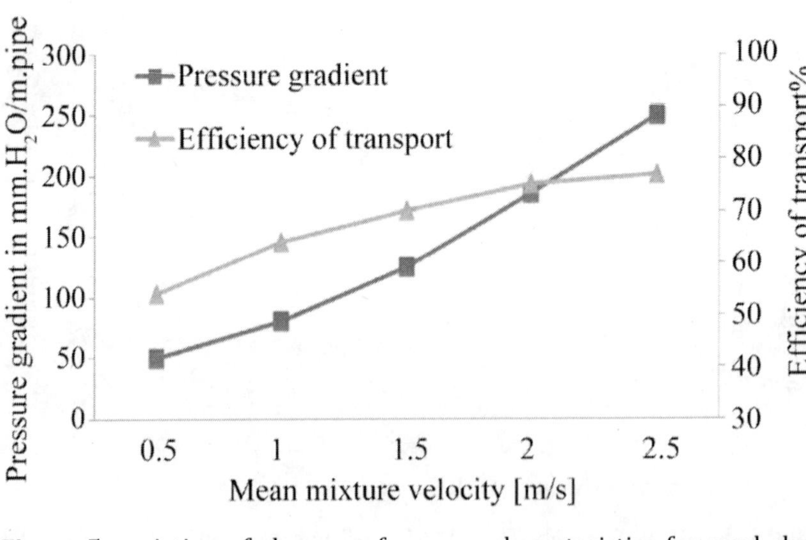

Figure 5. variation of slurry performance characteristics for sand slurry with solid concentration 1% by volume and particle diameter d_s = 1 mm. versus mean mixture velocity.

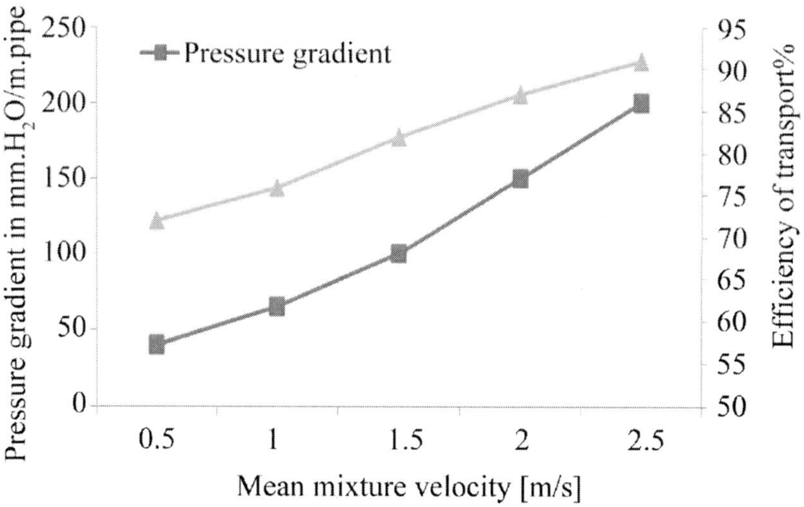

Figure 6. Variation of slurry performance characteristics for slurry of iron oxide with solid concentration 1% by volume and particle diameter d_s = 0.25 mm. versus mean mixture velocity.

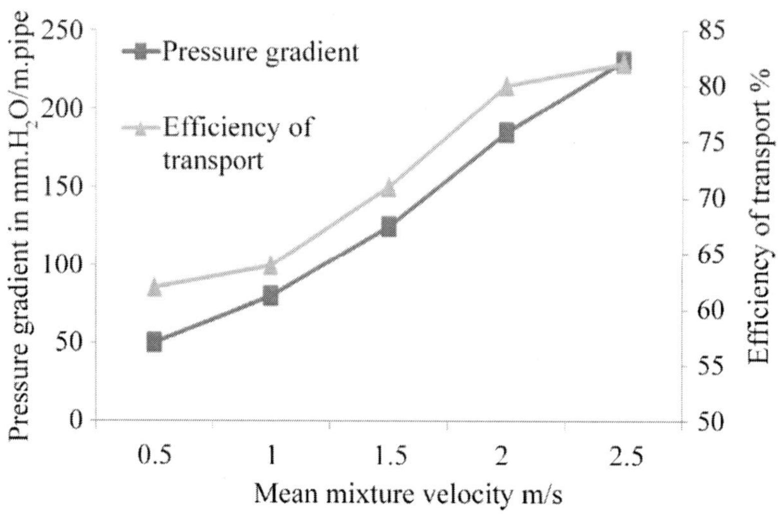

Figure 7. Variation of slurry performance characteristics for slurry of iron oxide with solid concentration 1% by volume and particle diameter d_s = 0.5. mm. versus mean mixture velocity and Froude numbers increase the pressure gradient

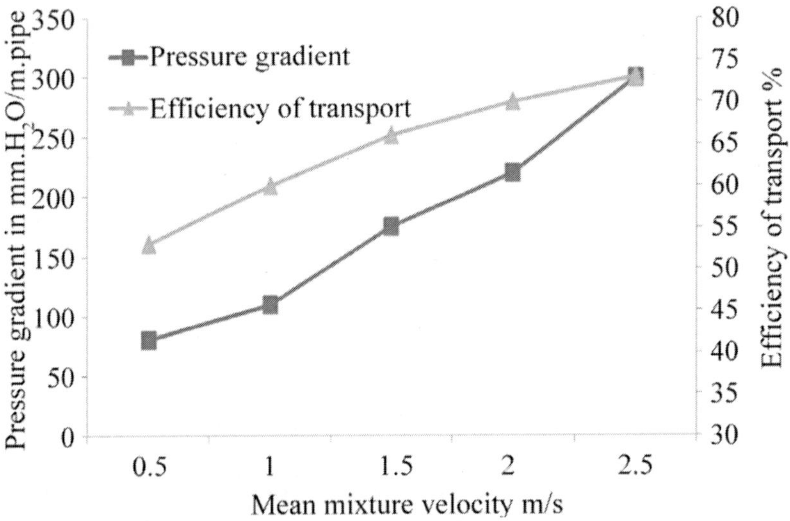

Figure 8. Variation of slurry performance characteristics for slurry of iron oxide with solid concentration 1% by volume and particle diameter d_s = 1 mm. versus mean mixture velocity.

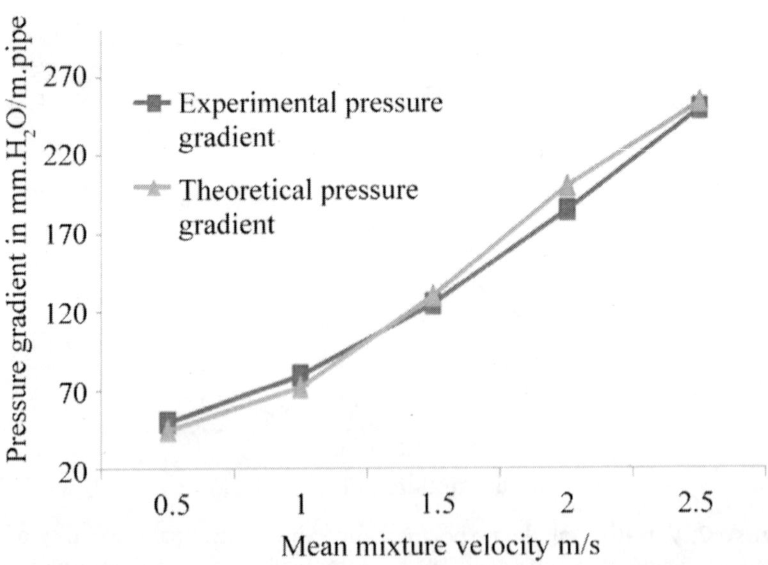

Figure 9. Comparison between experimental and theoretical pressure gradient for sand slurry with solid concentration 1% by volume with grain size diameter d_s = 1 mm.

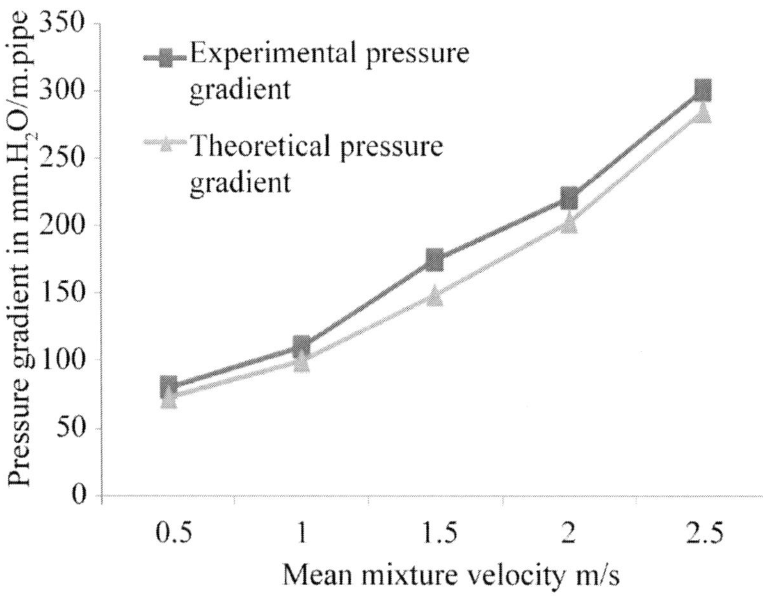

Figure 10. Comparison between experimental and theoretical pressure gradient for slurry of iron oxide with solid concentration 1% by volume with grain size diameter d_s = 1 mm.

2) Effects of varying Re, Fr, upon efficiency of transport

The effect of varying Reynolds and Froude numbers upon the efficiency of slurry transport (η) is shown in Figures from 3 to 8. It can be noticed that as Reynolds and Froude numbers increases the efficiency of transport (η) is increased. This remark also encourages the use of high velocities to avoid blocking slurry pipelines.

4.3. Effect of Varying Particle Diameter (d_s) upon Performance Characterstics of Slurry Pipeline

1) Effect of varying particle diameter (d_s) upon pressure gradient $\left(\frac{\Delta P}{L}\right)$:

Experimental tests have been carried out using samples of solid materials of arithmetic mean diameter from 0.25 mm to 1 mms. at the same input concentration by volume. The effects of changing particle diameter on pressure gradient are shown in Figures from 3 to 8 From these figures it was noticed that as the particle diameter increases the pressure drop increases slightly.

2) Effects of varying solid particle diameter (d_s) upon efficiency of transport (η):

Figures from 3 to 8, indicates also the effects of varying particle grain diameter on efficiency of transport (η). As particle diameter increases the efficiency of transport (η) decreases.

4.4. Effect of Varying Specific Gravity of Particles upon Performance Characteristics

1) Effects of varying specific gravity upon pressure gradient:

The specific gravity was varied from 2.65 for sand and 3.540 for iron oxide.

Figure 11 indicates the effects of varying specific gravity upon the pressure drop $\left(\frac{\Delta P}{L}\right)$. as specific gravity increases, the pressure drop increases; indicating the relative merit of transporting light materials other than heavy materials through pipeline system 2) Effects of varying specific gravity upon efficiency of transport (η):

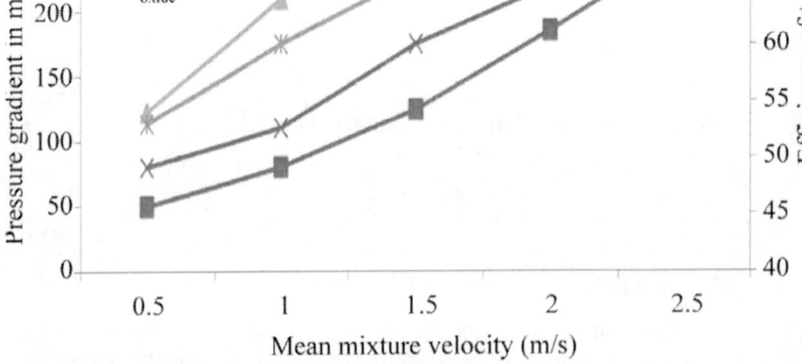

Figure 11. Effect of varying specific gravity upon characteristics of transporting solid-liquid mixtures with solid concentration 1% and particle diameter 1 mm.

From **Figure 12** as specific gravity increases the efficiency of transport decreases which encourages using light materials through hydraulic transport of solid liquid mixtures in pipelines.

4.5. Effect of Varying input Concentration upon Performance Characterstics

1) Effects of varying input concentration on pressure gradient:

The input concentration of slag of iron oxide and sand are varied from (1% - 10%) by volume. The effect of varying input concentration on pressure gradient is shown in Figures from 13 to 17. From these figures, as input concentration increases, the pressure gradient increases. This increase is directly proportional to input concentra tion by volume and this increase may be from probability of separating a part from input solid material.

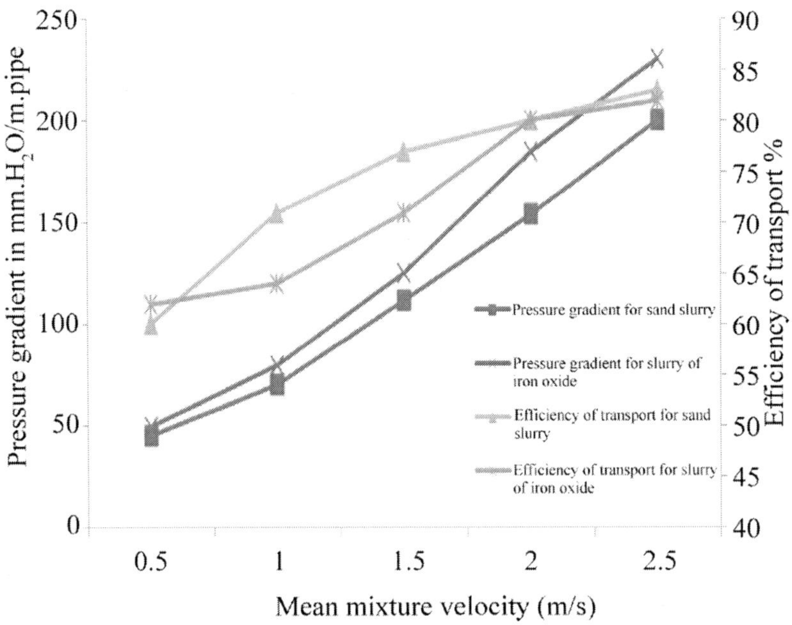

Figure 12. Effect of varying specific gravity upon characteristics of transporting solid-liquid mixtures with solid concentration 1% and particle diameter = 0.5 mm.

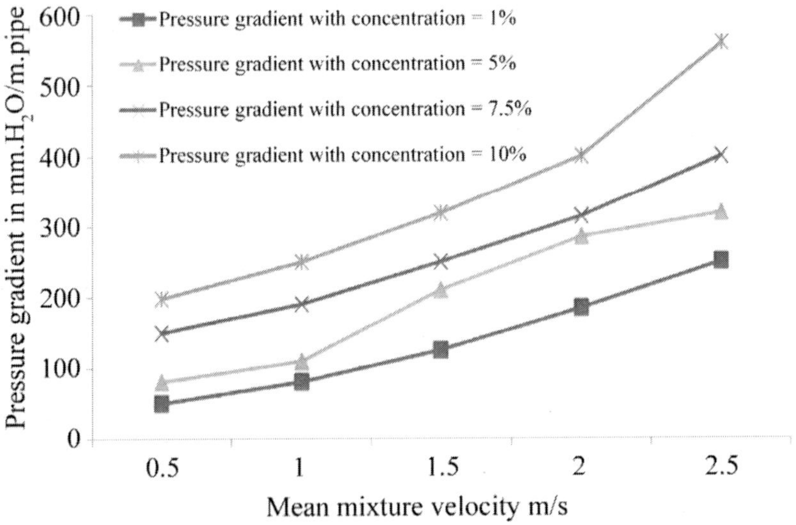

Figure 13. Effect of varying input solid concentration on pressure gradient for sand slurry of grain size diameter d_s = 1 mm. versus mean mixture velocity.

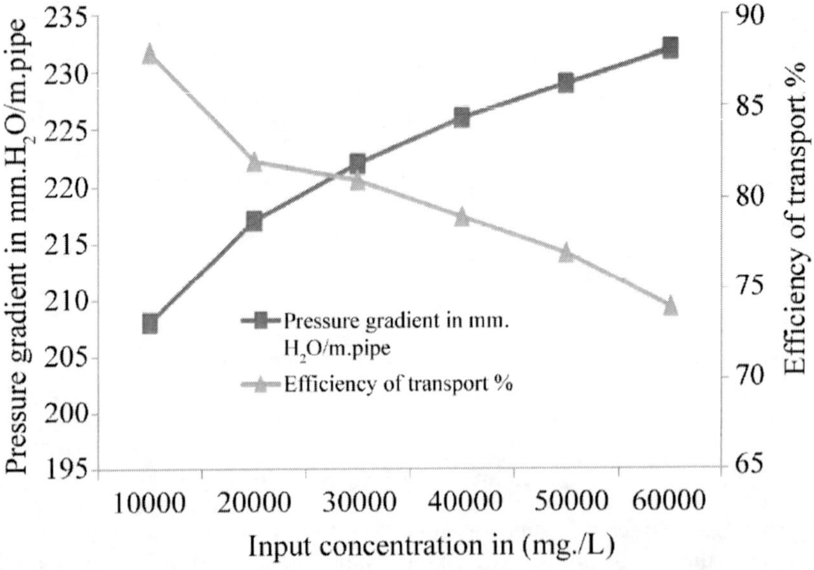

Figure 14. Variation of slurry performance characteristics for slurry of iron oxide versus input solid concentration at Reynolds number = 63699.

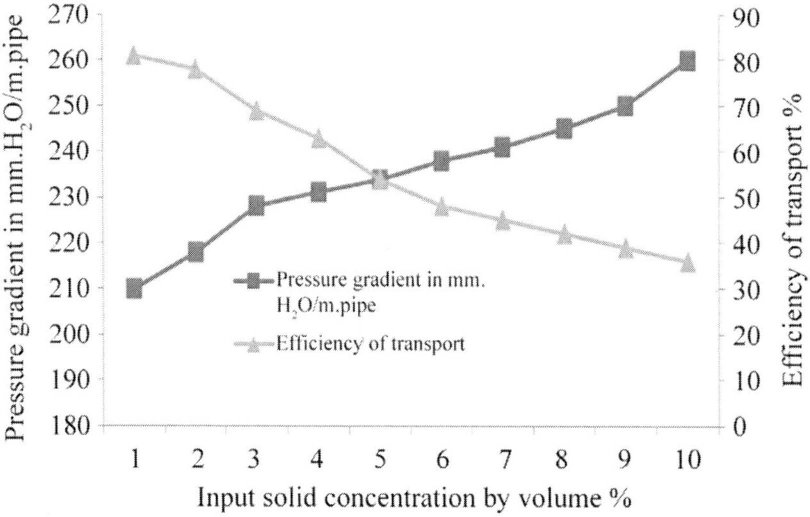

Figure 15. Variation of slurry performance characteristics for sand versus input solid concentration at Re = 63699.

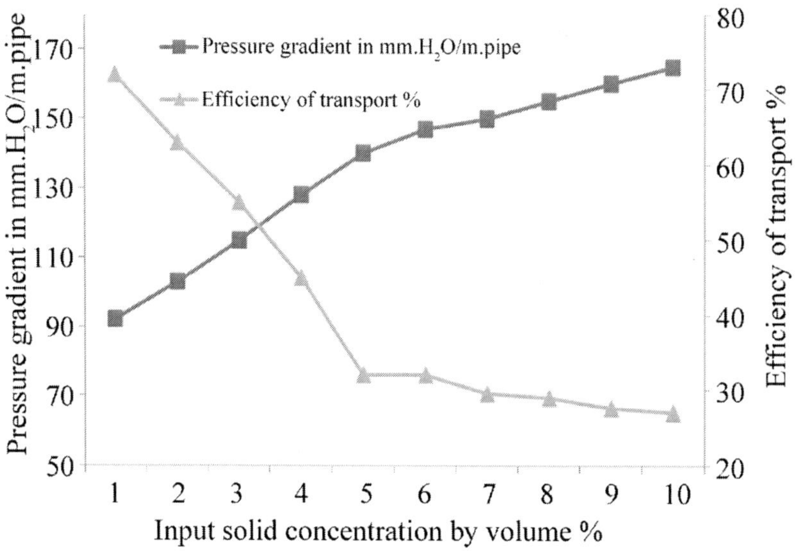

Figure 16. Variation of slurry performance characteristics for sand versus input solid concentration at Re = 37396.

2) Effects of varying input solid concentration upon efficiency of transport:

Figures from 14 to 18 indicate the effects of varying input concentration on efficiency of transport. It has been noticed that as input concentration increases the efficiency of transport decreases sharply. This remark encourages using low input concentrations to avoid sharp decrease in efficiency; hence avoid fast blocking of the pipe.

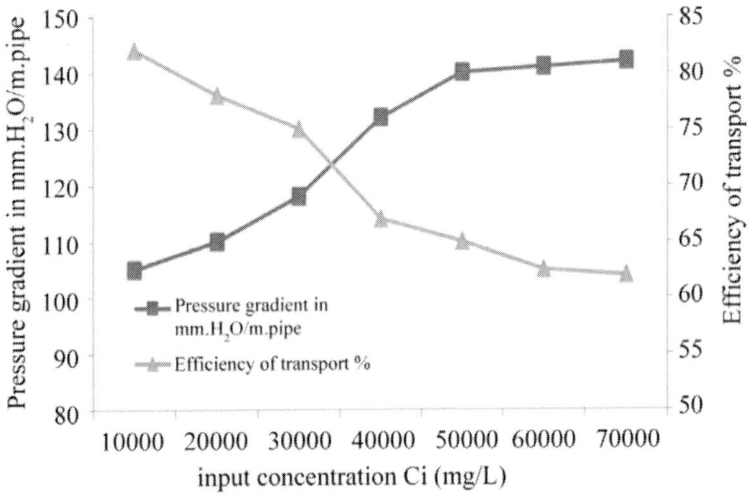

Figure 17. Variation of slurry performance characteristics for iron oxide versus input solid concentration at Re = 37396.

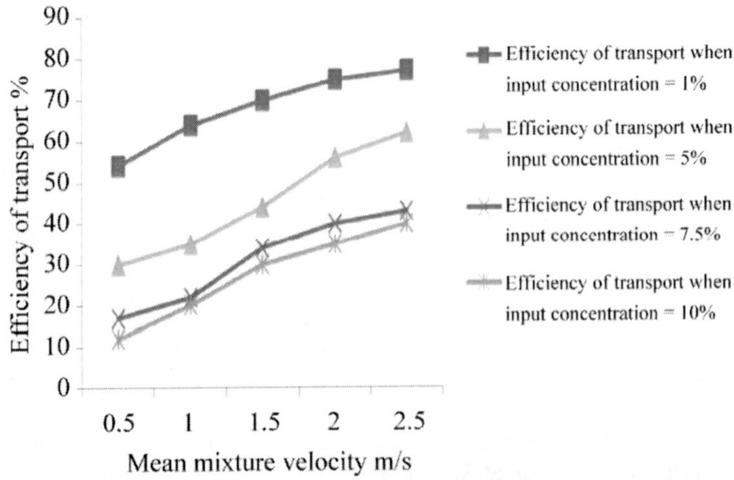

Figure 18. Effect of varying input solid concentration on the efficiency of transport of sand slurry of grain size diameter d_s = 1 mm. versus mean mixture velocity.

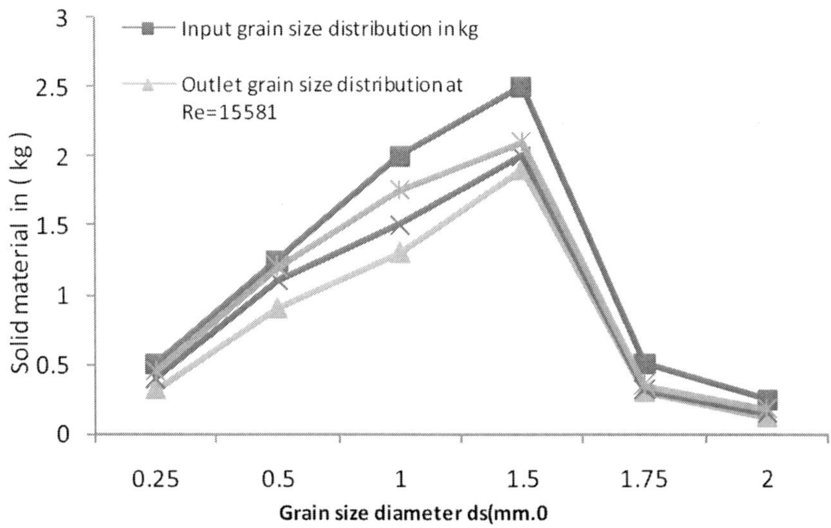

Figure 19. Effect of Reynolds number's on grain size distribution at pipe outlet at total input concentration of 0.119% by volume of iron oxide.

4.6. Effect of Reynolds Number on Grain Size Distribution

Figure 19 shows the effect of Reynolds number on grain size distribution at pipe outlet for concentration of 0.119% by volume.

From this figure we notice that:

1) High values of Reynolds number increases the ability to transport large grain sizes.

2) Transporting fine grain size does not need high Reynolds number.

5. CONCLUSIONS

The main conclusions that can be extracted from the results and observations of this work are mentioned below:

1) The input concentration plays an important role in the slurry transportation process. Increasing input concentration causes an increase in the pressure drop and

a decrease in efficiency of transport, probably due to agglomeration and flocculation of solid particles.

2) Increasing the grain size increases the pressure loss and decreases efficiency of transport.

3) As the Reynolds number increases, transport efficiency increases with an increase in pressure drop, and subsequently input power increases as well.

4) The process of transportation can be improved by using fine materials rather than course materials.

5) The specific gravity of solid materials plays an important role in efficiency of transport and also in pressure gradient. As specific gravity increases, the tendency to settle down increases, which can lead to decrease in efficiency of transport and increase in pressure drop.

6. RECOMMENDATIONS

1) The present work was carried out to investigate and deduce the dimensionless groups that affected the transporting process, but the functional dependence was still unknown. So it's recommended to use the data summed up here and define the governing equations.

A. I. Abd Al Aziz, H. I. Mohamed 2) It is also recommended to carry out experiments to investigate how the slurry transport affects the character istics of the pumps.

REFERENCES

1. D. M. Newitt, J. F. Richardson, M. Abbot and R. B. Turtle, "Hydraulic Conveying of Solids in Horizontal Pipes," Transactions of the Institution of Chemical Engineers, Vol. 33, 1955, pp. 93-113.
2. R. Durand, "The Hydraulic Transport of Coal and Solid Materials in Pipes," Proceedings of Colloquium on the Hydraulic Transport of Coal, National Coal Board, London, 1952, pp. 39-55.
3. R. C. Worster, "Proceedings of Colloquium on Hydraulic Transport of coal, National Coal Board, London, 1952.
4. K. Wagner, "Investigations on the Influence of the Grain Size Distribution for Horizontal Hydraulic Conveying of Solids in the Heterogeneous Regime," Thesis, Karlsruhe, 1982.
5. M. Weber, "Improved Durand–Equation for Multiple Application," International Symposium on Slurry Flows, 1986, pp. 61-66
6. R. M. Turain, T. F. Yuan, et al., "Pressure Drop Correlation Pipeline Flow of Solid Liquid Suspensions," AICHE Journal, Vol. 17, No. 4, 1971, pp. 809-817. doi:10.1002/aic.690170409
7. H. C. Hou, "Investigation of Optimal Grain Distribution for Transport with High Concentration," Hydraulic Transport of Solids in Pipes, city, 29-31 October 1986, pp. 177- 183.

8. M. S. Lee, V. Matousek, C. K. Chung and Y. N. Lee, "Pipe Size Effect on Hydraulic Transport of Jumoonjin Sand: Experiments in a Dredging Test Loop," Terra et Aqua-number 99, Hamburg, June 2005, pp. 1-10.

NOMENCLATURES

C_c: Delivered concentration of coarse material, volume %

C_D: Drag coefficient for free falling particle $= \frac{4gd(\rho_s - \rho)}{3\rho\vartheta_\infty^2}$ dimensionless.

C_f: Delivered concentration of fine material, volume %.

C_i: Input concentration of solids, volume %.

C_o: Output discharge concentration of solids, volume %.

D: inside pipe diameter in meter [m]

D_s: Diameter of solid particle assumed spherical, in meter. [m]

F: Fraction of fine material.

(1 – F): Fraction of coarse material.

F_L: Coefficient for deposit velocity.

$\lambda = \frac{\Delta PD}{2\rho\upsilon^2 L}$: Friction factor for pipe flow of slurry, dimensionless.

$\lambda_w = \frac{\Delta PD}{2\rho\upsilon^2 L}$: Friction factor for pipe flow in absence of solids, dimensionless.

G: Gravitational constant, m/s^2.

i: Head loss for mixture.

i_W: Head loss for water.

k: Constant dimensionless.

L: Pipe length in meter.

S: Ratio of solid to liquid densities dimensionless.

v: Mean mixture velocity in m/s.

υ_∞: Terminal deposit velocity for particles m/s.

$\left[\frac{\Delta P}{L}\right]$: Pressure gradient measured along pipeline in mmwater column per meter pipe length.

$\text{Re} = \dfrac{\rho v D}{\mu}$: Reynolds number for pipe flow, dimensionless.

$Fr = \dfrac{\upsilon}{\sqrt{gD}}$: Froude number for pipe flow, dimensionless.

GREEK LETTERS

ε: Shape factor for suspended particles, dimensionless.

$\eta = \dfrac{C_o}{C_i} \times 100\%$: Efficiency of transport.

ψ: Particle size distribution function, dimensionless.

CHAPTER 14

Solid-State Synthesis and Effect of Temperature on Optical Properties of CuO Nanoparticles

C. C. Vidyasagar Y. Arthoba NaikEmail author T. G. Venkatesha R. Viswanatha

Department of Chemistry, School of Chemical Sciences, Kuvempu University, Jnanasahyadri, Shankaraghatta-577 451, Karnataka, INDIA

ABSTRACT

Modulation of band energies through size control offers new ways to control photoresponse and photoconversion efficiency of the solar cell. The *P*-type semiconductor of copper oxide is an important functional material used for photovoltaic cells. CuO is attractive as a selective solar absorber since it has high solar absorbance and a low thermal emittance. The present work describes the synthesis and characterization of semiconducting CuO nanoparticles via one-step, solid-state reaction in the presence of Polyethylene glycol 400 as size controlling agent for the preparation of CuO nanoparticles at different temperatures. Solid-state mechanochemical processing, which is not only a physical size reduction process in conventional milling but also a chemical reaction, is mechanically activated at the nanoscale during grinding. The present method is a simple and efficient method of preparing nanoparticles with high yield at low cost. The structural and chemical composition of the nanoparticles were analyzed by X-ray diffraction, field emission scanning electron microscopy and energy-dispersive spectrometer, respectively. Optical properties and band gap of CuO nanoparticles were studied by UV-Vis spectroscopy. These results showed that the band gap energy decreased with increase of annealing temperature, which can be attributed to the improvement in grain size of the samples.

Keywords: Band gap CuO Polyethylene glycol 400 Semiconductors Solid-state reaction

INTRODUCTION

Work demand for energy is expected high and finding sufficient route to produce clean energy is an ever more pressing problem. Science had identified the key research challenges if solar energy is to provide a significant fraction of our energy needs. The huge gap between our present use of solar energy and its enormous undeveloped potential defines a grand challenge in energy research. One of the most attractive methods currently being developed is "dye sensitization" in solar cells in order to increase the efficiency of conversion of solar radiation into electrical energy. Further, the most serious problem of fossil fuel as the energy source is the harmful effect produced on the environment due to the release of hazardous gases into the atmosphere. To avoid both environmental pollution and the future energy crisis, much attention has been devoted to the development of photovoltaic solar cells. The development of photovoltaic solar cells was initiated first by the French Scientist Edmund Bequerel [1]. During the last two decades, a substantial body of research has been directed toward the synthesis of metal nanoparticles in efforts to explore their special properties and potential applications. Nanocrystalline semiconductor particles have drawn considerable interest in recent years because of their special properties such as a large surface-to-volume ratio, increased activity, special electronic properties and unique optical properties as compared to those of the bulk materials [2-4]. The large surface area to volume ratio can contribute to some of the unique properties of nanoparticles. The oxides of transition metals are an important class of semiconductors. Among various semiconductors, Cu and their oxides have attracted a great deal of attention in both fundamental research and technical application because of their catalytic, optical, and electrical conducting properties [5,6]. Copper is an important material because of its high electrical and thermal conductivities. CuO is known as p-type semiconductor exhibiting narrow band gaps (E_g = 1.2 eV) and which makes it a promising material for gas sensors, magnetic storage media, solar energy transformation, electronics, semiconductors, varistors, and catalysis [7]. It has therefore been studied together with other copper oxides, especially with respect to its applications as a photothermally active and photoconductive compound [8]. Thus, great efforts have been made to study the preparation of nanosized CuO in the past two decades. Conventional methods for the preparation of CuO powders include one step solidstate reaction at room temperature, thermal decomposition of copper salts and mechanical milling of commercial powders [9-11]. Solid-state mechanochemical processing is an effective, useful and simple processing technique widely applied to the synthesis of quasicrystalline, nanocrystalline and crystalline materials, which is not only a physical size reduction process in conventional milling but also a chemical reaction that is mechanically activated at the nanoscale during grinding. The present method is a simple and efficient method of preparing nanoparticles with high yield at low cost. Thus, there is a renewed interest in understanding the fundamental physical properties of CuO, as well as for upgrading its performance in applications. PEGs are prepared by polymerization of ethylene oxide and are commercially available over a wide range of molecular weights from 300 g/mol to 10,000,000 g/mol. While PEG with different molecular weights find use in different applications and have different physical properties (e.g., viscosity). They are used in industry as surfactants, including food, cosmetics, toothpastes and pharmaceutics; in biomedicine, as dispersing agents. PEG 400 is less toxic and soluble in water, acetone, alcohols and benzene [12,13]. In this work, to prepare low-cost advanced materials for photovoltaic energy production, we report a simple one-step, solid-state reaction in presence of PEG 400 [H-(O-CH_2-CH_2)n-OH] as size controlling agent for the preparation of CuO nanoparticles at different temperature. The process is carried out at room temperature. Band gap energy of all the synthesized nanoparticles was calculated using UV-vis spectroscopic data and synthesized the materials as in our pervious paper (C. C. Vidyasagar, et al. Powder Tech. 2011).

MATERIALS AND METHODS

Materials

Copper dichloride and sodium hydroxide were procured from Merck (Mumbai, India). PEG 400 (Polyethylene glycol 400) was procured from s.d.fine- CHEM Ltd. (Mumbai, India). All the chemicals were of analytical grade and used as received for the experiments. Double distilled water was used for the preparation of the solutions.

Characterization of nanoparticles

X-ray powder diffraction (XRD) patterns were recorded on a Philips X'Pert X-ray diffractrometer using Cu-Kα radiation. The size distribution and morphology of the samples were analyzed by Field emission scanning electron microscope (FESEM) fitted with an Energy-dispersive X-ray spectroscopy (EDAX) [Model: Nova Nano SEM600-FEI]. The absorbance of nanopowder samples were measured on a UV-Vis spectrophotometer [Model: USB 4000, Ocean Optics, USA].

Synthesis of CuO nanoparticles

In a typical synthesis, 5 g of $CuCl_2$ and 3 g of NaOH were ground separately for 5 min in an agate mortar. Thus obtained fine powder of $CuCl_2$ and NaOH were mixed with 6 ml of PEG 400 and subjected to further grinding for 30 min. The resultant paste was thoroughly washed with double distilled water and ethyl alcohol to remove the PEG 400. Finally, the product was dried at 70°C in an oven and the resultant solid was subsequently annealed at 400°C, 600C and 800°C for 2 h.

RESULTS AND DISCUSSIONS

X-ray diffraction measurements were carried out to study the crystal structure and crystalline quality. Figure 1 represents the XRD pattern of synthesized CuO nanoparticles calcined at 400°C, 600°C and 800°C, respectively. All the diffraction peaks could be indexed to crystalline monoclinic structure. The major peaks located at 2θ values correspond to the characteristic diffraction of monoclinic phase of CuO (JCPDS NO- 80-1268). No other peaks were observed belonging to any impurity such as $Cu(OH)_2$ and Cu_2O, indicating high purity of CuO nanoparticles. CuO particles annealed at higher temperature shows shift in (110), (002) and (200) planes compared to lower annealed temperatures. As the annealing temperature increases, the peak intensity of CuO nanoparticles increases with a decrease in the full width at half maximum (FWHM), which indicates *a* possible change in the grain size of CuO [12,14]. The diffraction peak (110) was narrower than (002) peak, and in turn (002) peak was narrower than (200) peak. This clearly indicates the presence of asymmetry in the crystallite shape. The height of c-axis of the crystal is bigger than the basal diameter (crystal axis, a_1, a_2) [15]. The intensity of (110), (002) and (200) peaks gradually increases with increase in annealing temperature. However, the peak intensity of the annealed samples at 800C° was found to be increased. The lattice parameters for CuO nanoparticles were calculated from the XRD data. The lattice constants *a*, *b* and *c* for CuO annealed at 400C° were found to be 0.47153 nm, 0.33774 nm and 0.51247 nm, respectively. As for CuO annealed at 600C°, the lattice constant a decreases to 0.47141 nm, *b* increases to 0.34206 nm and c increases to 0.51264 nm. For CuO annealed at 800C°, *a* decreases to 0.46392 nm, *b* decreases to 0.33934 nm and *c* decrease to 0.50409 nm. Figure 1 illustrates the changes in (002) and (200) peaks position of the nano powders. As shown in Fig. 1, when the calcinations temperature changes from 400C° to 600C°, the peaks slightly shift towards lower angles. On the other hand the strongest shift towards higher angles has been observed in the sample annealed at 800C°. This can be seen from the enlarged view of the inset in Fig. 1. These shifts depend directly on the changes of the lattice constants [22]. It can be seen that the shifts of the peaks were detected as the lattice constants changed, which results in the changes of lattice constants along *a*, *b* and *c*-axis. As the peaks shift towards lower angle, the lattice constants increase and as it shifts towards higher angle, the lattice constants decrease [20]. The increases in peak intensity at higher annealing temperatures are attributed to the increase in crystallinity.

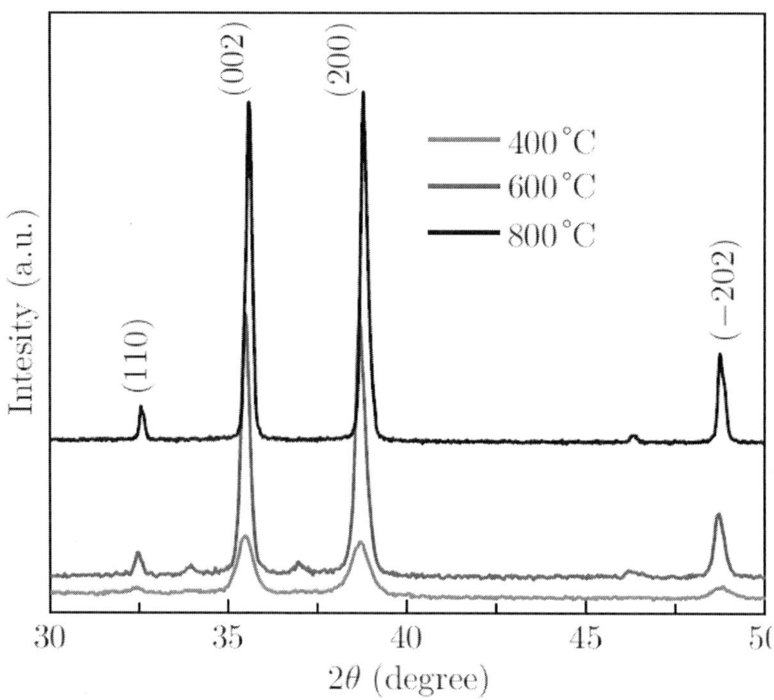

Figure. 1 XRD patterns of CuO samples prepared at different temperatures.

Figure 2 shows the FESEM micrographs of all annealed samples and EDAX patterns of CuO. It was clearly seen that the samples annealed at 800°C revealed the uniform and plate-like particles. From Fig. 2(a) it can be seen that size of CuO nanoparticles annealed at 400°C are in the range of 21-25 nm. As the annealing temperature increases from 400°C to 800°C, the crystal size of CuO was also found to be increased from 65 nm to 454 nm)

(Fig. 2(b) and 2(c)). With increasing in the annealing temperatures, the nucleation rate of the particles increases more rapidly. This is due to the increase in supersaturation of the reaction products, which accelerates the crystal core forming reaction within a short time. Under these conditions the controlling step of the reaction is transferred from grain growth to crystal nucleus formation. With the temperature continuing to rise, the phenomenon of "nuclear-aggregation" caused by the rapid formation of crystal nucleus is obvious, which results in aggregation among the crystal nucleus [12,16,17]. The rate of particle aggregation is a major factor that controls the morphology and structure (crystalline) of the final product [18]. With increasing the calcination temperature particle agglomeration was observed in FESEM images. Literature also indicates that with increase in calcination temperature the crystallinity of the particle increases as well as the surface area decreases. Variation in calcination temperature brought about changes in the crystallinity [21]. The micrographs clearly indicated that the crystallinity has increased with an increase in annealing temperature, which is in agreement with XRD results. EDAX spectra of CuO particles at different temperatures are shown in Figs. 2(d), 2(e) and 2(f), which indicates the high purity of synthesized materials where only Cu and O are present. Carbon signal are due to the Gold coating and carbon film supporting the specimen in FESEM observation [12]. Surfactants (PEG 400) would change nanoparticle's shape, size and surface properties to different extent depending upon their molecular structure, for example, nature of head group, length of hydrophobic tail and type of counter ions. The termination of the nanoparticle growth and size is controlled by the diffusion and the attachment rates of surfactants (PEG 400) on the nanoparticle surface [23, 24].

Figure 3 shows the UV-Vis absorption spectra of the samples annealed at different temperatures. The exciton absorption is at about 650-900 nm. The CuO particles obtained from the higher annealing temperatures exhibits the relatively lower absorption in the visible range. As shown in Fig. 3, the increasing annealing temperature of CuO particles is considered to be the main factor for the intense decrease in absorption [19]. The crystallinity of the particles in our experiment was improved when the substrate temperatures increased from 400°C to 800°C (Fig. a, b and c). With increase in the annealing temperature from 400°C to 800°C, the optical absorption edge slightly shifted towards longer wavelength, which may be attributed to the increase in grain size at higher annealing temperatures. The optical band gap (E_g) of the CuO was determined by using the formula, $E_g = hc/\lambda$, where h is plank's constant, c is velocity of light and λ is wavelength [12]. The crystallinity of the CuO improved when the temperature increased from 400°C to 800°C. It is obvious that the conduction type of CuO is dependent on temperature. From UV-Vis results, the shift in the band gap may be attributed to the effect of temperature on the crystallinity of CuO

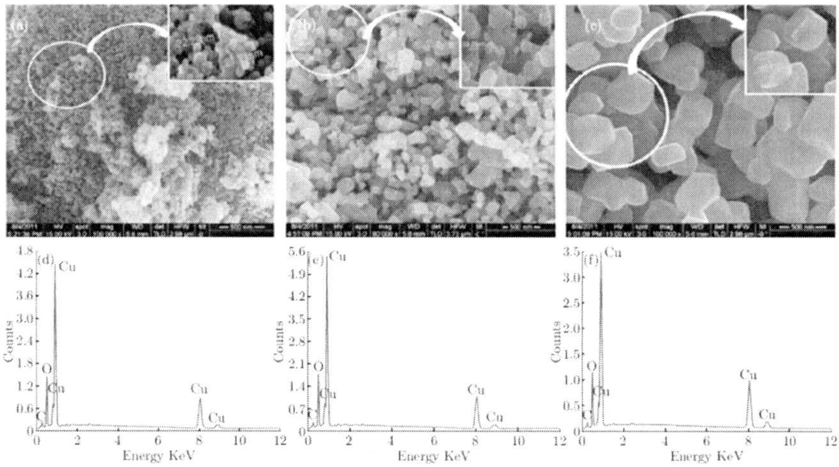

Figure. 2 FESEM images and EDAX spectra of CuO samples prepared at different temperatures. (a) and (d) 400°C, (b) and (f) 600°C, (c) and (e) 800°C.

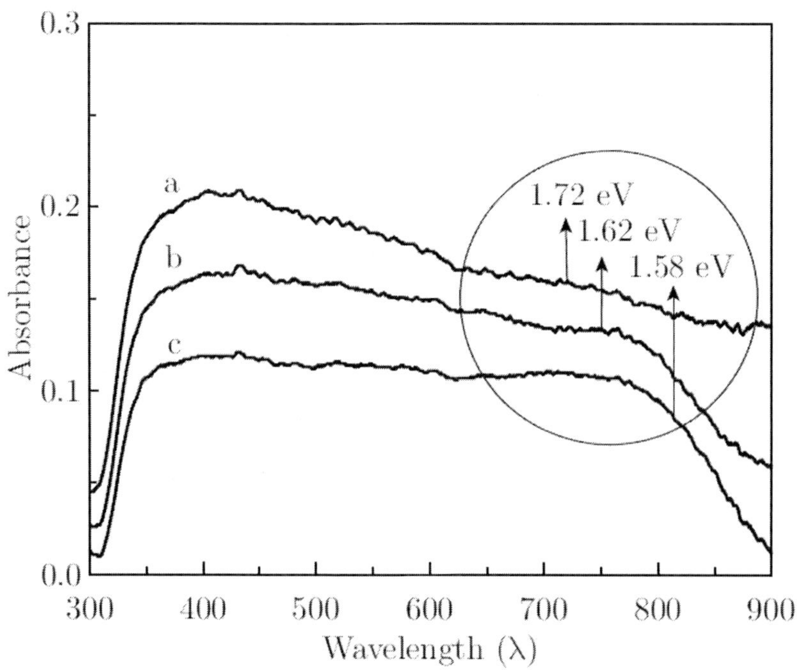

Figure. 3 UV-Vis spectra of CuO samples prepared at different

samples. On the other hand, the crystallinity become better when E_g value red shifted from 1.72 to 1.58 eV. The changes in the optical band gap of CuO samples with increase in temperature are due to the change of crystallinity at higher temperature. The observed results show that the annealing temperature is very important in the band gap assignment.

CONCLUSION

In summary, nanocrystalline CuO particles with a monoclinic structure have been synthesized through a simple, rapid, low-cost and one-step solid state reaction method using PEG 400 as a surfactant. The improved crystallinity was observed when annealed at higher temperatures. The grain size can between 21-450 nm. The intensity of peaks has been increased with increase in annealing temperature, as well as the optical band gap edge shifts towards longer wavelength region which may be attributed to the decreasing band gap of the CuO nanoparticles at higher temperatures.

ACKNOWLEDGMENTS

Authors are grateful to UGC for financial support and DST, New Delhi. I would like to thank Prof. G. U. Kulkarni, Chemistry and Physics of Material Unit, Jawaharlal Nehru for Advanced Scientific Research, Bangalore, for helping FESEM/EDAX analysis. The authors wish to thank Dept. of Chemistry, Kuvempu University, for providing laboratory facilities to carry out this work.

REFERENCES

1. Sambandam Anandan, Sol. Energy Mater. Sol. Cells 91, 843 (2007). http://dx.doi.org/10.1016/j.solmat.2006.11.017
2. Henglein, Chem. Rev. 89, 1861 (1989). http://dx.doi.org/10.1021/cr00098a010
3. Agfeldt and M. Gratzel, Chem. Rev. 95, 49 (1995). http://dx. doi.org/ 10.1021/cr00033a003
4. Hui Wang, Jin-Zhong Xu, Jun-Jie Zhu and Hong-Yuan Chen, J. Cryst. Growth 244, 88 (2002). http://dx.doi.org/10.1016/S0022-0248(02)01571-3
5. Julia Hambrock, Ralf Becker, Alexander Birkner, Jurij Weiß and Roland A. Fischer, Chem. Commun. 68–69 (2002). http://dx.doi.org/10.1039/b108797e
6. Bong Kyun Park, Sunho Jeong, Dongjo Kim, Jooho Moon, Soonkwon Lim and Jang Sub Kim, J. Colloid Interface Sci. 311, 417 (2007). http://dx.doi. org/10 .1016/j.jcis.2007.03.039
7. Masoud Salavati-Niasari and Fatemeh Davar, Mater. Lett. 63, 441 (2009). http://dx.doi.org/10.1016/j.matlet.2008.11.023

8. Chih-Hung Lo and Tsing-Tshih Tsung, J. Vac. Sci. Technol. B 23, 2394 (2005). http://dx.doi.org/10.1116/1.2122787
9. Junwu Zhu, Haiqun Chen, Hongbo Liu, Xujie Yang, Lu. Lude and Xin Wang, Mater. Sci. Eng. A 384, 172 (2004). http://dx.doi.org/10.1016/j.msea.2004.06.011
10. Claude Carel, Mona Mouallem Bahout and Jean Gaude, Solid State Ionics 117, 47 (1999).
11. 0Wang Wenzhong, Zhan Yongjie and Wang Guanghou, Chem. Commun. 727 (2001). http://dx.doi.org/10.1039/B008215P
12. C. Vidyasagar, Y. Arthoba Naik, T. G. Venkatesh and R. Viswanath, Powder Tech. 214, 337 (2011). http://dx.doi.org/10.1016/j.powtec.2011.08.025
13. Tetsuya Kida, Takanori Oka and Masamitsu Nagano, J. Am. Ceram. Soc. 90, 107 (2007). http://dx.doi.org/10.1111/j.1551-2916.2006.01402.x
14. Wang Dong, Z. Q. Chen, D. D. Wang, J. Gong, C. Y. Cao and Z. Tang, et al., J. Magn. Magn. Mater. 332, 3642 (2010). http://dx.doi.org/10.1016/j.jmmm.2010.07.014
15. Masoud Sa lavati-Niasari, Fatemeh Davar and Mehdi Mazaheri, Mater. Lett. 62, 1890 (2008).
16. Siqingaowa, Zhaorigetu, H. Yao and Garidi, Front. Chem. China 3, 277 (2006). http://dx.doi.org/10.1007/s11458-006-0036-7
17. Fan Zhang and Junling Yang, Int. J. Chem. Kinet. 1, 18 (2009).
18. Z. C. Michael Hu, T. Michael Harris and H. Charles Byers, J. Colloid Interface Sci. 198, 87 (1998). http://dx.doi.org/10.1006/jcis.1997.5290
19. T. Prem Kumar, S. Saravanakumar and K. Sankaranarayanan, Appl. Surf. Sci. 257, 1923 (2011). http://dx.doi.org/10.1016/j.apsusc.2010.09.027
20. J. C. Fan and Z. Xie, Mater. Sci. Eng. B 150, 61 (2008). http://dx.doi.org/10.1016/j.mseb.2008.02.014
21. T. H. Mahato, Beer Singh, A. K. Srivastava, G. K. Prasad, A. R. Srivastava, K. Ganesan and R. Vijayaraghavan, J. Hazard. Mater. 192, 1890 (2011). http: //dx.doi.org/10.1016/j.jhazmat.2011.06.078
22. Majid Ebrahimizadeh Abrishami, Seyed Mohammad Hosseini and Ahmad Kompany, J. App. Sci. 11, 1411 (2011). http://dx.doi.org/10.3923/jas.2011.1411.1415
23. Mehta, S. K. & Chaudhary, Savita. Sci. Topics Retrieved, January 20 (2012).
24. Chang-Woo Kwon, Tae-Sik Yoon, Sung-Soo Yim, Sang-Hyun Park and Ki-Bum Kim, J. Nanopart. Res. 11, 831 (2009). http://dx.doi.org/10.1007/s11051-008-9451-7

CHAPTER 15

Driving the Diffusion of Cesium Atoms of Graphite Monolayers on Metal Surfaces into Rhenium (1010)

A. K. Orujov

Baku State University, Baku, Azerbaijan

ABSTRACT

It has been shown experimentally for the first time that diffusion of cesium n rhenium takes place at irradiation of atoms flux heated up to T ≤ 1800 K Re(10$\bar{1}$0)-C. Analysis of thermal desorption spectra showed that cesium forms two high-temperature phases during adsorption of cesium on graphite monolayer in rhenium heated up to 1800 K. The first, intercalated phase starts filling immediately and grows up to maximum concentration. The second, diffusion phase appears with slight delay and grows incessantly during adsorption. It is for the first time that the Fick Equation has been applied for such kind of diffusion. Correspondences $\tau_{diff} = f(T)$ and average displacement L = $f(T,t)$ have been obtained at diffusion of surface cesium in rhenium in the interval of temperatures 1600 K < T < 2200 K. At conditions of $\nu = 5 \times 10^{13}$ cm$^{-2}\cdot$s^{-1}, U = +2000 V, T = 1700 K, effectiveness of cesium diffusion δCs in the system of Re(1010)-C was identified with application of suggested methods, which showed δCs = 0.45. Activation energies of Cs atoms diffusion in rhenium from rhenium were found and values E_{1p} = 5.65 ± 0.05 eV and E_{p1} = 5.85 ± 0.05 eV were obtained respectively. Effect of electrical field on intercalation and cesium diffusion in Re-C was observed.

Keywords: Surface Ionization, Graphite Monolayer, Intercalation, Diffusion

1. INTRODUCTION

The discovery of intercalation [1],[2],paved the way for the studying of superefficient diffusion of K, Cs, Na, Ba, Sm and In atoms in metals Ir and Re coated with two-dimensional graphite monolayer. The phenomenon of potassium, cesium, samarium, barium and indium atoms diffusion across the graphite monolayer in iridium and potassium in rhenium was discovered and studied during the studies [3],-[7]. It was interesting whether cesium would diffuse through the graphite monolayer in rhenium. For the purposes of studying the cesium atoms diffusion in rhenium, facets of rhenium (1010) had to be obtained, conditions enabling the formation of graphite monolayer on its surface needed to be examined and the conditions of intercalation of the graphite monolayer cesium atoms on the surface of rhenium needed to be studied, the latter of which has been thoroughly investigated in [1],[2]. As shown in the above studies, during the adsorption of Cs on Re – C, intercalated cesium leaves only at extremely high temperatures T ~2200 K and intercalated cesium prevents the dissolution of graphite islets, communicating to the islets a unique thermal stability, which we think is very questionable. Because a graphite monolayer on the surface of iridium is destroyed and carbon atoms are desorbed at T ~1900 K, while graphite monolayer on the surface of rhenium is degraded and carbon atoms are desorbed at temperatures ~5 - 10 K above the temperature of obtaining graphite monolayer. In this study, we present only a few cases to understand the sequence of processes of intercalation and diffusion. The purpose of this paper is to investigate the possibility of highly efficient diffusion of Cs atoms in rhenium with a monolayer graphite film.

2. EXPERIMENTAL SET-UP AND RESEARCH METHODS

The research was done in a static mass spectrometer of sector type with 90° inclination angle of the ion trajectories. The mass-spectrometer resolution by peak heights was about 300. The mass-spectrometer has two types of ion sources: 1) source with ionization by electrons (volumetric ionization), 2) source of surface ionization with emitter (surface ionization). The atoms or molecules falling on the emitter surface after ionization by PI are accelerated passing through the collector slit with suppressor grid and enter into mass analyzer. One of the most convenient methods commonly used in the experiment is the method of thermal desorption (TDS) flash. A sample (Figure 1.(1)) was placed in a high vacuum chamber before the entrance slit of mass-spectrometer equipped with device for detecting partial pressure or ion current. Coating an adsorbate on the sample surface can be done in different ways. In this case, it is obtained by adsorption. After obtaining the coating, thermal flash is created by selecting the sample's temperature increase mode. Thus, desorption process takes place, characterized by desorption speed and leading to increase of partial pressure on ΔP device or increase of surface ionization (SI) I^+ current. As the temperature of sample rises, the rate of particles desorption also increases, which leads to an increase in its instantaneous density. Knowing this increase, surface can be diagnosed, information on the nature and concentration of the adsorbed particles can be obtained and desorption kinetics can be studied. Experiments were performed in ultra-high vacuum with a residual pressure of 10^{-10} Torr. Thin strips of 50 × 1.5 × 0.03 mm^3 textured rhenium with facet (1010) on the surface and work function of 5.1 eV were used. Graphite monolayer was formed by exposing rhenium warmed up to 1600 K in benzene vapor or spraying from graphite rod (Figure 1.(9)). Surface carbon on rhenium was studied by thermionic emission and dissociation of CsCl molecules. Graphite monolayer is formed in benzene vapor on rhenium ribbon after saturation of rhenium Re (1010)

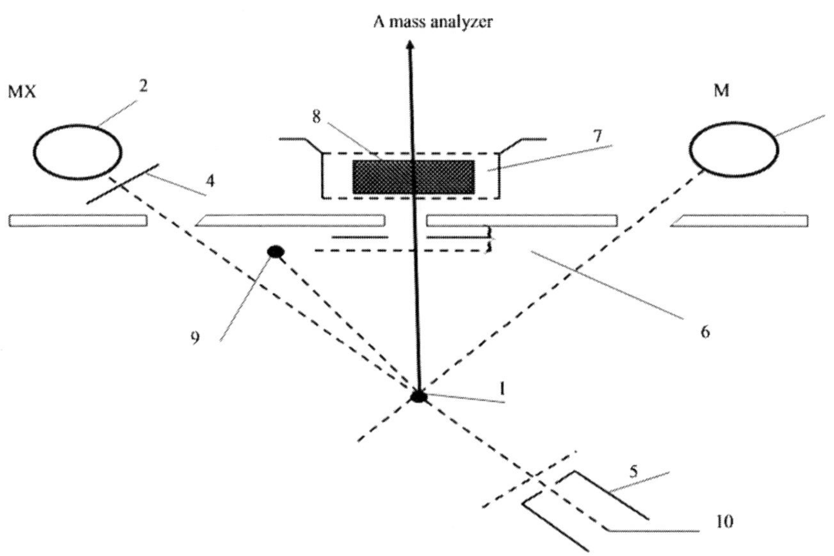

Figure 1. Scheme of ion sources. 1: emitter, rhenium ribbon, 2 - 3: source of molecules or atoms, the Knudsen camera, 4: spoiler, 5: electron gun, 6: manifold with a suppressor grid, 7: box of volume ionization, 8: ions, 9: graphite rod, 10: cathode electron gun.

carbon heated up to T ≤ 1800 K. This is consistent with the results of [8],, which are confirmed by Auger electron spectroscopy [1],[2]. Cesium on the surface of Re(10$\bar{1}$0-C (with work function ~4.5 eV) was sprayed from a Knudsen cell, while its state in the adsorbed layer was studied by method of thermal desorption spectroscopy (TDS) in the form of ionic flash with mass-spectrometer registration of Cs ion current that with ~100% efficiency transformed into cesium Cs^+ ions during desorption by mechanism of surface ionization, increasing the sensitivity by ~10 times. To register cesium, the flash was produced at the speed of −100 deg/s to 2250 K rhenium tape temperature and Cs ions current was further recorded on the constant T temperature ($T_{пл\cdot Re}$ = 3453 K). At such high temperatures, graphite islets in the adsorbed layer of rhenium are destroyed (dissolved), carbon atoms evaporate, while intercalated and diffused cesium is desorbed in the form of Cs^+ions. Neutral component of the flow desorbing during the flash was ionized in the source with electron impact before entrance into the mass-analyzer. Figure 2 separately shows the scheme of the volumetric ionizing chamber.

3. A BRIEF THEORY OF THE KINETICS OF DIFFUSION BETWEEN THE SURFACE AND VOLUME OF RHENIUM

It is well known that diffusion is an irreversible and non-equilibrium process leading to the establishment of equilibrium distribution of concentrations within phases and Equation of chemical potentials of the system's components. The most important laws of diffusion are the Fick laws [9]. The first law of Fick is like this:

$$j = -D\overline{\nabla} n \tag{1}$$

where j stands for diffusive flux, D—for diffusion coefficient, $\overline{\nabla} n$ — concentration gradient; the "minus" sign means that the flow is directed towards reducing the concentration;

Fick's second law (for one-dimensional diffusion in the direction of x)

$$\frac{\partial n}{\partial t} = \frac{\partial}{\partial x}\left(D\frac{\partial n}{\partial x} \right) \tag{2}$$

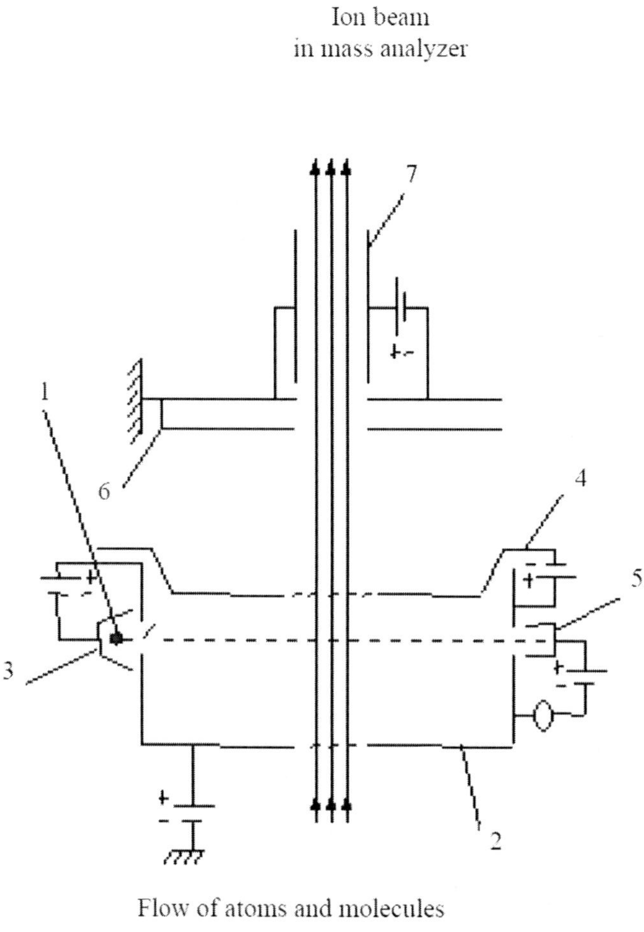

Figure 2. Scheme of ion source with electron ionization (volumetric ionization source). 1: cathode of electron impact, 2: box, 3: focusing electrode, 4: ion extracting electrode, 5: Faraday cup, 6: protective diaphragm, 7: condenser.

In the case of D's independence on the n concentration, the law looks like this:

$$\frac{\partial n}{\partial t} = D\frac{\partial^2 n}{\partial x^2} \tag{3}$$

The diffusion coefficient D shows the number of diffusing substance (in mass units) through unit area under a concentration gradient, equaling to unity; has dimensions cm^2/sec. It is determined by properties of the environment, the type of diffusing particles, depends on the temperature T and the diffusion activation energy E_a in accordance with the Arrhenius Equation

$$D = D0\ exp(-E_a/kT) \tag{4}$$

In this Equation, D_0 is numerically equal to the diffusion coefficient at the temperature gravitating to infinity. In general, the movement of impurity atoms or ions in the crystal lattice is irregular with overcoming the potential energy barriers. The solution of Equations (2) and (3) has a different appearance depending on the boundary conditions. In particular, boundary conditions for the diffusion from an ongoing source are:

$$n(0,t) = n0;\ n(x,0) = 0 \tag{5}$$

Considering this, the solution is:

$$n(x,t) = n_0\left(1 - \text{erf}\frac{x}{2\sqrt{Dt}}\right) \tag{6}$$

$$\text{erfz} = \frac{2}{\sqrt{\pi}}\int_0^y e^{-z^2}\,dz \tag{7}$$

where for the Gauss error function (abbreviation erf comes from error function), erfcz = 1 – erfz complementary error function (error function complement) is used as well.

At the diffusion of adsorbed atoms from intercalated zones with constant surface concentration of metal, the boundary conditions are well observed for our case

$$n(0,t) = 0;\ n(x, 0) \tag{8}$$

Considering these initial conditions, we obtain an expression for the concentration $n(x, t)$

$$n(x,t) = n_0 \text{erf}\frac{x}{2\sqrt{Dt}} \tag{9}$$

Comparing the literature data for different initial conditions of diffusion, we see that the value of $L = \sqrt{Dt}$, which has the dimension of length and is called diffusion length, is included in all the final results of the Fick Equation. The physical meaning of this parameter is the average distance, which the diffusing particles overcome in the direction of alignment of the concentration gradient during the time t. The diffusion length is not to be confused with the length of particle's mean free path, which is defined as the average distance traveled by the particle between two successive scattering events. If we assume that atoms diffuse in the form of monovalent positive ions from the intercalated area, then Einstein relationship connecting diffusion coefficient of charged particles with their mobility may be applied in conditions of thermodynamic equilibrium

$$\mu = \frac{eD}{kT} \tag{10}$$

The second Fick Equation taking into account both the diffusion and migration components in the absence of recombination is as follows:

$$\frac{\partial n}{\partial t} = D\frac{\partial^2 n}{\partial x^2} \pm \mu_{\pm} E\frac{\partial n}{\partial x} \tag{11}$$

where the sign "– " refers to the case of positive charges movement (with mobility μ_+), and "+"—that of negative charges (mobility μ_-), while E stands for the tension of electric field. Various parameters of diffusion can be calculated and assessed by varying the formulas (1)-(11).

4. EXPERIMENTAL RESULTS AND THEIR DISCUSSION

It has been shown experimentally that the diffusion of cesium in rhenium occurs at irradiation of atom flows warmed up to T ≤ 1800 K Re($10\bar{1}0$)-C. Figure 3 shows the thermal desorption spectra of Cs taken for different times of exposure. An extremely effective diffusion of Cs atoms in Re($10\bar{1}0$) has been discovered, stimulated by graphite monolayer on its surface. Analysis of the TD-specters has revealed that similar to the adsorption of K atoms on the Re-C [8],, during adsorption of cesium on monolayer graphite in rhenium heated up to 1800 K, cesium also forms two high-temperature phases. The first intercalated phase starts to be filled immediately and grows up to the maximum concentration. The second diffusion phase appears with some delay and

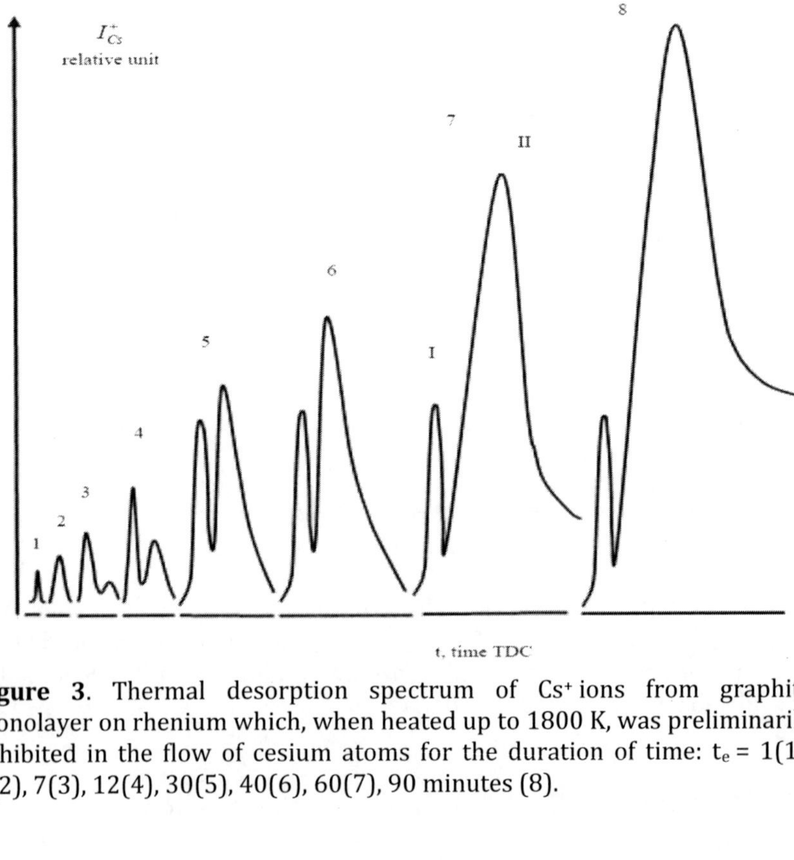

Figure 3. Thermal desorption spectrum of Cs^+ ions from graphite monolayer on rhenium which, when heated up to 1800 K, was preliminarily exhibited in the flow of cesium atoms for the duration of time: t_e = 1(1), 3(2), 7(3), 12(4), 30(5), 40(6), 60(7), 90 minutes (8).

grows continuously during adsorption. The peak's high occupancy, which in some cases exceeded monatomic cesium concentration $\sim 5\times 10^{14}$ cm^{-2} by 10 - 100 times on the surface of metals pointed at the diffusive nature of this peak and the thermal removal of cesium required a very long time (tens of hours). Some differences in the TDS of Cs^+ ions in the systems of Re(10$\bar{1}$0)-C and Ir(111)-C were observed. The diffusion phase of rhenium Cs^+ in comparison with a similar phase of iridium was stretched under the same conditions of preliminary irradiation of Ir(111)-C, and Re(10$\bar{1}$0)-C. The above differences in the forms of TDS and effectiveness of cesium atoms diffusion in Re(1010) and Ir(111) are apparently related to differences of their crystallographic structures and diffusion lengths. Another difference of cesium TDS peaks from Ir(III)-C and Re(1010)-C was in the initial desorption temperatures of the diffusion phase. If in the case of Ir(III)-C, beginning of TDS corresponded with desorption temperature of T ≥ 1900 K carbon atoms, i.e. temperature of the graphite structure destruction, in the case of Re(10$\bar{1}$0)-C it conformed with delivery temperature of monolayer graphite in Re, i.e. temperature of carbonization.

5. AVERAGE DISPLACEMENTS AND DIFFUSION LENGTH OF CESIUM ATOMS IN RHENIUM

Suppose τ_{diff} is an average lifetime of the *Cs* atom in rhenium lattice with respect to the diffusion hopping and τ_{exp} is observation time. Then, during the observation time, the *Cs* atom will make $n = \tau_{exp}/\tau_{diff}$ diffusion jumps with chaotic nature of movements, move in the given direction at a $L = \sqrt{D\tau_{exp}}$ distance or $L = \lambda\sqrt{n}$, where $\lambda = \sqrt{D_0\tau_0}$ is the length of one diffusion jump, which is usually of roughly atomic size or approximately coincides with the metal's crystal lattice period. It is known that rhenium with 1,373 Å atomic radius and Re1 + 1.33 Å, Re7 + 0.56 Å ionic radii crystallizes in a hexagonal close-packed lattice (a = 2.760 Å, c = 0.458 Å). Suppose that $\tau_{diff} = \tau_0 \exp E_{diff}/kT$. Let's compute the diffusion length for three different exposure times t_1 = 10 min = 600 s, t_2 = 2 *h* = 7200s and t_3 = 10 h = 36,000 s at several temperatures of rhenium, i.e., the average distance in angstroms, for which the *Cs* atoms (atomic radius of cesium is 2.67 Å, radius of the monovalent ion is 1.67 Å) shift deep into rhenium and the average lifetime of Cs atoms (Table 1) in the crystal lattice of rhenium. Comparison of the obtained results showed that the average displacements of diffusion for cesium atoms in *Re*(10$\bar{1}$0) is less than in Ir(III).

It can be seen that even at a sufficiently high temperature of T = 2200 K and longer exposure times of t = 10 h cesium penetrates rhenium tape by a small distance of 480 Å, which makes ~175 atomic layers of rhenium, while 30 mkm thick rhenium tape contains ~10^5 atomic layers. Consequently, almost all diffused cesium is probably located in a thin surface layer of rhenium. It can be said with a high probability that the above differences in the TDS spectra are associated with the rhenium and iridium crystal lattice structure, as well as the dimensions of atoms. The mentioned experimental data show that doping of iridium, rhenium with K, Cs, Na, Ba, Sm, In, etc. atoms may lead to the change of 5d-zone of transition metals' electron density.

Dissolution of surface *Cs* in rhenium. Activation energy of E diffusions of surface *Cs* in rhenium was found in special experiments. For this purpose, a two-dimensional layer of cesium intercalated with an initial concentration of N was formed under the graphite monolayer on rhenium, then T was increased and reduction of this concentration of N was monitored (by flash) through diffusion of cesium in rhenium. It is obvious that the

Table 1. Correspondences $\tau_{diff} = f(T)$and average displacement of $L = f(T,t)$ at the diffusion of surface cesium in rhenium.

T, K	τ_{ag}, s	L_1(Å), for t_1 = 600 s	L_2(Å) for t_1 = 7200 s	L_3(Å) for t_1 = 36,000 s
1600	100,730	0.2	0.7	1.7
1700	8793	0.72	2.5	5.6
1800	1007	2.2	7.9	16.5
1900	145	5.6	19.4	43.5
2000	25	7.2	46.8	104.7
2100	5	30	104	234
2200	1.2	62	214	478

reduction of dN concentration of surface Cs during the time dt is associated with diffusion flow v and average lifetime of adsorbed atoms *Cs* with respect to dissolution of

$tp1 = t_0 \exp Ep_1/(kT)$ in rhenium with ratio

$$dN = -vdt = -Ndt/t_{p1} \tag{12}$$

Integration of the relationship (12) gives us

$$N = N \exp(-t/t_{n1}), \tag{13}$$

where from we can deduce the activation energy of surface Cs dissolution in rhenium:

$$E_{p1} = k\text{T}\, ln(t/t_0\, ln(N_0/N)) \tag{14}$$

It appears that $E_{1p} = 5.65 \pm 0.05$ eV.

Discharge of Cs from the bulk of rhenium to its surface. Let's consider diffusion of Cs from the bulk to the surface of rhenium. Let's say, N_1 is Cs concentration in the plane lagging behind the surface by the distance of a single diffusion jump. Then the flow of Cs coming to the surface and associated with average lifetime of atom with respect to the diffusion on surface

$t_{pl} = t_0 \exp E_{pl}/(kT)$, will equal to

$$v = N_1/t_1p = N_1/t_0 \exp[E_1p/(kT)] \tag{15}$$

We will register the flow of Cs atoms coming to metal surface by I ions of Cs^+. From the condition of conserving the $v_1n = v_{\partial ec}$ flow at 100% efficiency of ionization and S ion collection area, we get

$$I = ev_{\partial ec}S = eN_1S/t_0 \exp[E_1p/(kT)]. \tag{16}$$

Suppose that during the experiment, the $v_{1n}{}^{t}exsp << N_1$ correlation takes place. Then the flow

v_{1n} will be constant and

$$I = B \exp[-E_1p/(kT)]. \tag{17}$$

Experimentally obtained dependence $\ln I = f(1/T)$ was linear and the activation energy of diffusion of Cs atoms of rhenium was found to be E_{p1} = 5.85 ± 0.05 eV. As can be seen from this correlation $E_{1p} < E_{p1}$, potential barrier for the diffusion of cesium atoms in the case of rhenium is also asymmetric, but the difference between energies of diffusion in Re and from Re was smaller than in the case of cesium atoms diffusion in Ir. In the simplest case, when the diffusion coefficient does not depend on the concentration of diffusible component, activation energy E_a can be deduced from the slope of lg$D$ dependence on 5040/T according to the Equation obtained by taking the logarithm of correlation (4). The results of these calculations will be given in the following materials.

6. THE EFFECTIVENESS OF DIFFUSION AND INFLUENCE OF THE ELECTRIC FIELD ON INTERCALATION AND DIFFUSION

The proportion of flux or efficiency of δ diffusion diffusing in Re at irradiation of Re(1010)-C with Cs flow was identified by two independent methods:

1) By surface ionization (PI) of cesium atoms in Re via the formula

$$\delta(T) = 1 - \frac{I(T)}{I_0} \tag{18}$$

where I_0 and I(T) are cesium ion currents from a clean surface of rhenium with graphite monolayer respectively.

2) From comparison of the number of incident atoms of cesium on Re-C at this temperature for a certain time t to the number of diffusing cesium atoms in rhenium during this time $\delta = \frac{v_{\partial iff.}}{v_{\partial iff.} + v_{\partial es.}}$. Both methods gave almost identical results. Under the conditions v = 5 × 1013 sm^{-2}·s^{-1}, U = +2000 V, T = 1700 K, the effectiveness of cesium diffusion δ_{Cs} in Re (1010)-C was determined via the above methods, which gave δ_{Cs} = 0.45. This value δ_{Cs} is more than δ_{Cs} of cesium in the system Ir(III)-C (δ_{Cs} = 0.25). Thus, the

formation of graphite monolayer on the surface of rhenium due to the huge increase in the lifetime of adsorbed cesium atoms under graphite islets as compared with those of rhenium with an open surface enormously increases the efficiency of surface diffusion of cesium in rhenium, which grows by about $\sim 10^8$ times from 1.5×10^{-9} to 0.45, approaching one. The effect of electric field on the intercalation and diffusion of cesium in Re-C similar to the one detected for potassium in Re-C [7], was observed. The intercalation and diffusion of cesium, just like that of potassium, occurs only in case of a positive potential of iridium tape. The noticeable intercalation and diffusion of cesium atoms started at $U \geq +150$ V and increased with growth of voltage up to 4000 V roughly by linear law. At the potential of 600 V - 700 V, the intercalated phase reaches saturation and practically does not rise with growing voltage. The density of diffusion phase increases linearly with growth of U up to 4000 V (maximum voltage of power supply source). This effect is explained by decoration of the edges of graphite islets with K or Cs ions. At pre-irradiation of Re(1010)-C heated to high-temperature by the large flow (10^{13} - 10^{14} $cm^{-2} \cdot s^{-1}$) of cesium atoms, a significant number (hundreds of monolayers) of cesium atoms diffused into rhenium during a long time (an hour, 2 hours, etc.). This followed from the analysis of TDS and also from measuring of PI current flow of cesium atoms diffusing from the volume at high temperatures T = 2200 K. Therefore it is proposed to use rhenium as an effective source of cesium atoms and ions. For example, after exposure rhenium ribbon heated up to T = 1700 K Re(1010)-C in a stream of Cs atoms with density $\nu = 1 \times 10^{14}$ $cm^{-2} \cdot s^{-1}$ for a time of t = 1 hr at T = 2200 K emitted a steady stream of Cs^+ ions with density $\nu = 1 \times 10^{12}$ $cm^{-2} \cdot s^{-1}$ for the duration of t = 30 min and at T = 2150 K, rhenium ribbon emitted steady stream of Cs^+ ions with density $\nu = 5 \times 10^{11}$ $cm^{-2} \cdot s^{-1}$ for the duration of t >1 hour. Prolonged contact with the atmosphere of rhenium ribbon saturated with cesium had no effect on the reemitting of cesium from rhenium and iridium [10],tape again under high vacuum at high temperatures. This fact of prolonged existence of cesium within the volume of highly heated metal screen can be used in practice and requires development of a process model.

REFERENCES

1. Tontegode, A.Y. (1991) Progress in Surface Science, 38, 201 429.
2. Tontegode, A.Y. and Rutkov, E.V. (1993) Progress of Physical Sciences (Russian), 163, 1053.
3. Tontegode, A.Y. and Yusifov, F.K. (1993) FTT, 35, 987-993.
4. Tontegode, A.Y. and Yusifov, F.K. (1994) Surface, 4, 20-24.
5. Orudjov, A.K, Dashdamirova, A.O. and Yolchiyeva, A.K. (2010) Physics of Metals and General Metallurgy, 109, 366-368.

6. Orudjov, A.K. and Dashdamirov, A.O. (2007) Diffusion of Na Atoms in Ir with Irradiation of Ir-C Flux of Na Atoms. VII International Scientific Conference "Solid Body Chemistry and Modern Microand Nano-Systems", Kislovodsk, 17- 22 September 2007, 95-96.
7. Orudjov, A.K. (2013) Physics of Metals and General Metallurgy, 114, 63-67.
8. Orudjov, A.K., Nurullayev, Y.G. and Dashdamirov, A.O. (2005) News of Baku University, Series of Physical and Mathematical Sciences, 3, 140-149.
9. Bockstein, B.S. and Yaroslavtsev, A.B. (2005) Diffusion of Atoms and Ions in Solids. M: MISA, 362.
10. Orudjov, A.K. (2011) Physics of Metals and General Metallurgy, 111, 626-627.

CHAPTER 16

Production of Dissimilar Metals Materials by the Method of Solid-State Joining

B. V. Borts, I. M. Korotkova, O. O. Lopata, V. I. Sytin, V. I. Tkachenko, I. O. Vorobyov

National Science Center "Kharkov Institute of Physics & Technology", Kharkov, Ukraine

ABSTRACT

Mechanisms taking place while hot rolling in vacuum of dissimilar materials joined in solid phase are described. It is shown that at joining of materials in solid phase redistribution of atoms from one material into another occurs on the interface of joining. On the base of calculation and expe- rimental results it is concluded that the ultimate strength of the interface of joint is always higher than the ultimate strength of less durable material. Pair zirconium-stainless steel SS (Type AISI 321) and carbon steel (Type C22E) those are used in nuclear power are investigated.

Keywords: Vacuum Hot Roll Bonding, Solid Phase, Ultimate Strength, Interface

1. INTRODUCTION

Problems of service life and of safe operation of constructions, machines and units of nuclear power stations pose the tasks of strength and reliability increase for units and structural elements encourage development of new materials-dissimilar metals bonded. There is the necessity in materials with high strength and low density, weld ability with heterogeneous materials, wear resistance and electroconductivity, high-temperature resistance and corrosion resistance. None of existing materials [1] , can't guarantee the needed complexity of properties. And only "laminar" multilayer allows to produce absolutely new material, properties of which differ from properties of its components. Production of safe and long-lived composition materials for unconnected joints of structural elements produced of materials with different properties is very topical task.

Together with fusion welding of materials, solid-state joining is important [2] method because it provides wide possibilities especially in specific conditions of the use. This method allows to produce [3] the necessary structures and properties of metals in the area of their joining: increased static strength, fatigue strength at high and low temperatures, corrosion resistance in corrosive media, corrosion-mechanical strength, serviceability under high heat loads and under different irradiations.

2. MECHANISMS THAT PROCEED AT SOLID-STATE JOINING OF HETEROGENEOUS

During vacuum hot rolling activation of surfaces occurs at the expense of their shearing plastic deformation that promotes the fracture of oxide film on joined surfaces [4]. On pure (juvenile) surfaces activation [5] centers are produced [2] associated with the escape of dislocations on surface. If dislocations that provide the elastic defor- mation produce dipoles of wedge disclination and appear on metal surface into the area of physical contact then they are the center of disturbance. Each disclination may represent [6] effective way of fluxes of heterogeneous materials atoms directed towards each other. Mutual transport of atoms and energy transfer occurs by these channels that balances all the system. Further deformation of material causes the advance of metal atoms that have increased energy and displacing by produced channels of defected structure on long distance. In experi- ments atoms of metals were detected that were displaced into crystalline lattice of other material on depth up to 20 mkm. On interface of two metals the flux of heterogeneous atoms sets up directed in opposite directions. Due to this in the narrow range (1 - 5 mkm) random rearrangement of two joining lattice occurs and the zone with mixed distribution of atoms of two materials forms on the interface of solid phase joint.

Investigation of joint of heterogeneous materials in solid phase by X-ray microanalysis had showed the pre- sence of solid zones on boundaries of joints niobium-copper and niobium-zirconium. Even transition of one material into the other is observed on distance of several micrometers (Figure 1).

The performed investigations had showed that contacting surface layers of metal atoms may be considered as special two-dimensional solid solution. Dissolution of foreign atoms in crystalline lattice causes its distortion, variation of interatomic bonds and energy threshold for joining of metal atoms.

It is impossible to explain the joining of contacting metals only by mutual diffusion. It is refuted by the fact that joint may be obtained also between mutually insoluble metals (copper-lead, for instance).

Atom of metal belonging to one crystalline lattice may start the penetration in other under two conditions. The first condition—atoms of two metals must be approached to distances comparable with parameters of crystalline lattice. The second condition is the formation of declined dipoles and transfer of additional energy to atoms of metal, penetrating from one crystalline lattice to the other. After penetration of atoms the bonds of definite type form between them and atoms of the matrix of other bonded samples. After this the diffusion may proceed in usual manner.

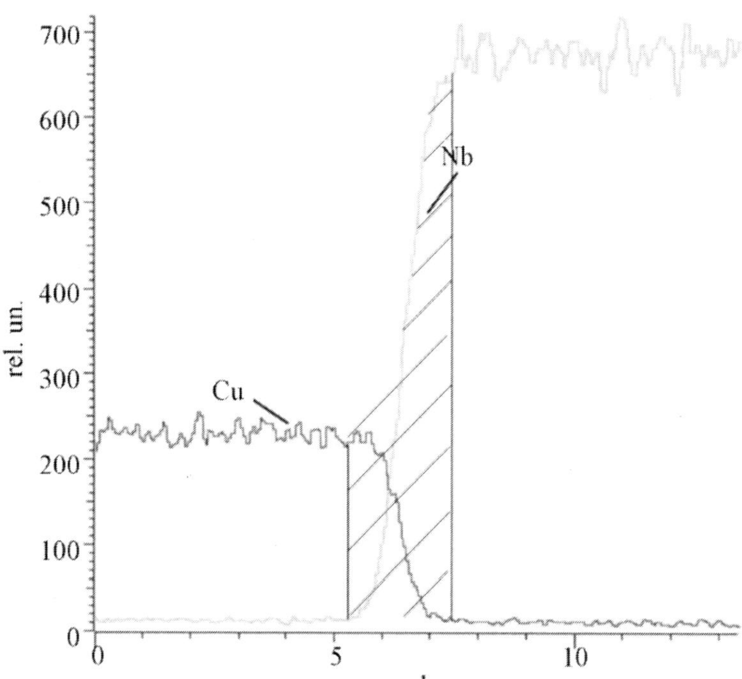

(a)

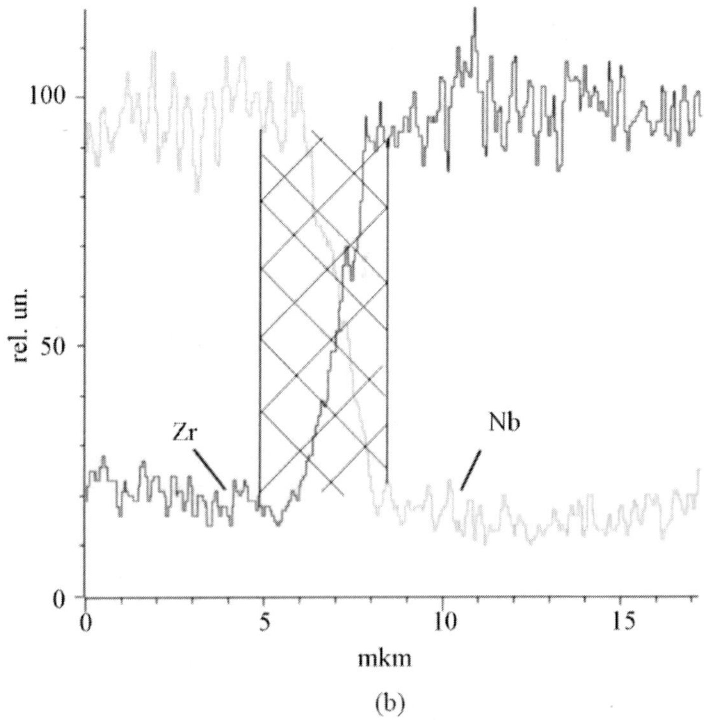

Figure 1. Spectra of micro X-ray spectrometric analysis near the interface of adjoint: (a) Niobium-copper; (b) Niobium-zir- conium.

3. DETERMINATION OF RELATIVE ULTIMATE STRENGTH OF THE INTERFACE OF JOINT

Theoretical model for ultimate strength of interface of heterogeneous materials joining in solid phase is de- scribed in details in [7]. In the presented paper we will consider only the final conclusions and statements. On joining of heterogeneous metals in solid phase the range of their strength variation will be in the limits $[\sigma_{Sp}^{M} \sigma_{S0}^{T}]$ where σ_{Sp}^{M} –yield strength of less hard metal; σ_{Sp}^{M} –yield strength of more hard metal. Expression for relative ultimate strength of joining is.

$$\bar{\sigma}_B\left(t^*\right)\Big|_p = \frac{kT}{E_A}\frac{\sigma_{S0}^{T}}{\sigma_{Sp}^{M}}\frac{\sigma_{S0}^{T}+\sigma_{Sp}^{M}}{\sigma_{S0}^{T}-\sigma_{Sp}^{M}}\ln\left(\frac{\sigma_{S0}^{T}}{\sigma_{Sp}^{M}}\right) \equiv \frac{kT}{E_A}Q_p\left(\frac{\sigma_{S0}^{T}}{\sigma_{Sp}^{M}}\right) \tag{1}$$

where k—Boltzmann's constant; E_A—activation energy of plastic deformation of joined materials; T—temper- ature of rolling; Q_p—relative coefficient of strength of joined metals interface.

Introducing the coefficient of the strength range expansion $g = \sigma_{S0}^{T}/\sigma_{S0}^{M}$ we may represent expression (1) as a plot. Family of curves for dependencies of ultimate strength of interface of monometals joining on parame- ters of expansion of strength range g was plotted.

Figure 2 shows the curve 1 that characterizes the strength of the interface of monometals joining. With the increase of limits of plasticity range of two different metals the strength characteristics of bimetals pair increases in comparison with monometallic one (see Figure 2, curves 2-4).

Such increase is observed until the range of bond strength will be determined by more hard metal. In this case the relative ultimate strength of sample bonding interface is determined by the ultimate strength of more hard material. Therefore the ultimate strength of interface of heterogeneous metals is considerably higher than the ul- timate strength of soft metal and higher than the ultimate strength of multilayer material produced of soft mo- nometals. The ultimate strength of heterogeneous metals interface is considerably higher than the ultimate strength of hard material but is lower the ultimate strength of multilayer material of hard monometals.

Figure 3 shows the tension diagram that shows that ultimate strength of joining interface is higher than the strength of less strength material. In this case the rupture of pair Stainless steel SS (Type AISI 321) and carbon steel (Type C22E) occurred by carbon steel under tension 430 MPa that corresponds to the ultimate strength of carbon steel.

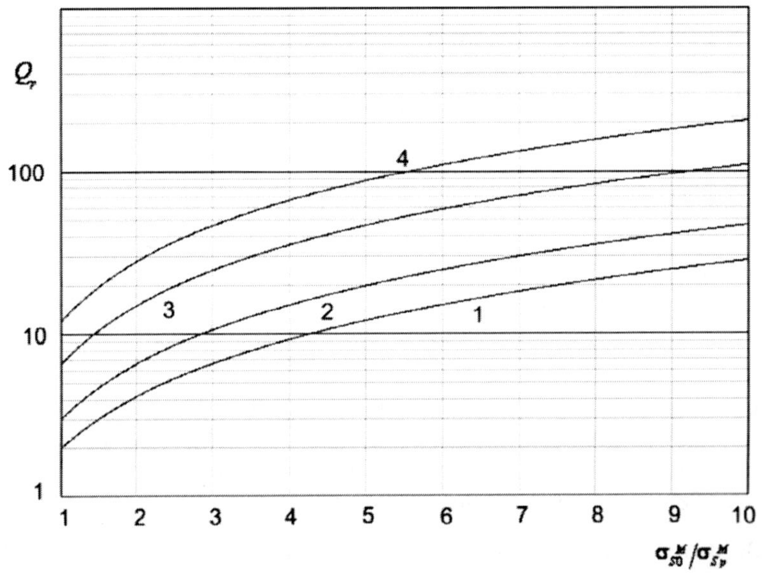

Figure 2. Dependence of relative strength coefficient of interface of joined metals on ratio of their yield strength $\sigma_{S0}^{M} / \sigma_{Sp}^{M}$ at different coefficients of yield strength range expansion: $g = 1$ (1); $g = 1.5$ (2); $g = 3$ (3); $g = 5$ (4).

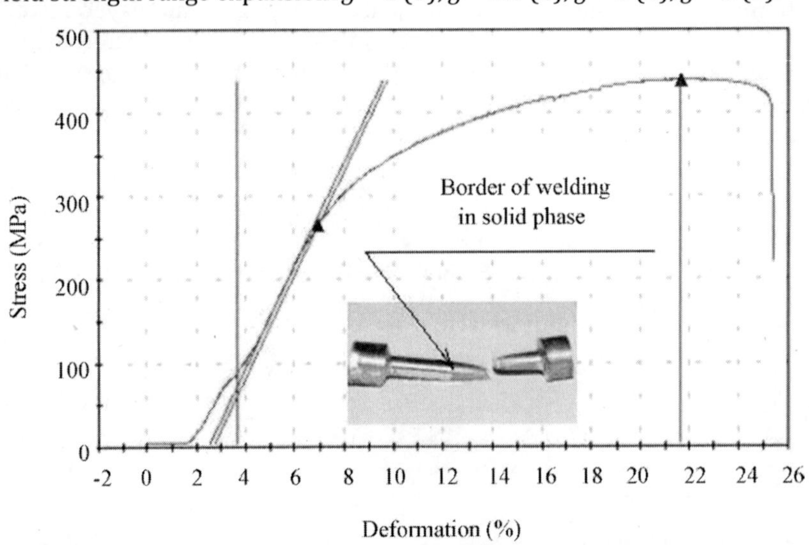

Figure 3. Tensile stress-deformation diagram of SS-C22E sample at 20°C. The inset below shows the point of specimen fracture in steel C22E.

4. DISSIMILAR METALS SOLID-PHASE BONDED FOR USE IN NUCLEAR ENGINEERING

4.1. Joint Zirconium-Stainless Steel

To increase the service life and safe operation of assemblies and units of structures of reactor WWER-1000 high strength continuous joints of alloy zirconium-1% niobium with stainless steel 12Cr18Ni10Ti, were produced by solid-state joining by high temperature rolling in vacuum.

Influence of working environment on the properties of multilayer material is investigated. Method of autoc- lave in liquid medium at temperatures and pressures simulating operation conditions was used. At autoclave testing two-times distilled water close by its composition to the coolant of first circuit of reactor WWER-1000 was used. Corrosion testing was carried out at temperature 350°C and under the pressure 16.5 MPa. Distilled water has considerably strong corrosion influence on materials due to the absence of inhibitors in this water.

Corrosion is the process of chemical or electrochemical interaction of metals with corrosion medium. Para- meters of corrosion properties of solid phase-bonded materials components at autoclave were following: char- acteristic of polished metallographic specimen surface, namely, continuity, homogeneity of the film color, cracks and spallation; mechanical properties (strength and plasticity) of rupture specimens after different times of exposure in autoclave at operational regimes.

For investigation of corrosion micro metallographic specimens were polished and studied after testing during 50, 100, 500 and 1000 hours without preliminary etching structure of solid-phase bonded samples for monitor- ing the changes of surface of pair components.

Figure 4 and Figure 5 show the surfaces of the solid-phase bonding Zr1Nb-Ni-SS and Zr1Nb-Nb-SS before and after corrosion treatment.

For inseparable joint Zr1Nb-Ni-SS after rolling in initial state the transient zone of depth less 5 mkm appears in nickel from side bordering on zirconium. After 50, 100, 500 and 1000 hours of testing the depth of transient zone remains invariable. Probably, the temperature of corrosion testing (350°C) is insufficient for mutual diffu- sion of the solid-phase bonding elements on the interface of their mating.

With the increase of testing time some pitting corrosion occurs on the surface of nickel and also grain struc- ture of nickel and zirconium manifests. Stainless steel remains invariable. Oxides of blue color are only ob- served around precipitates of remained α-phase and carbides.

Study of mechanical properties of multilayer materials Zr1Nb-Nb-SS and Zr1Nb-Ni-SS were carried out at temperature 20°C in initial state and also after corrosion treatment in conditions of autoclave testing. Results of performed tests are shown on Figure 6.

As it is seen from presented data all compositions in initial state (after rolling) have rather high strength prop- erties. The higher values of ultimate strength 445 - 465 MPa are observed for composition ZrNb-Ni-SS. The fracture of specimen of multilayer material Zr1Nb-Ni-SS proceeds by interface zirconium-nickel. Multilayer material Zr1Nb-Nb-SS break down by interlayer of niobium.

(a) (b)

Figure 4. State of the surface of the solid-phase bonding Zr1Nb-Ni-SS: (a) In initial state; (b) After corrosion treatment during 100 hours.

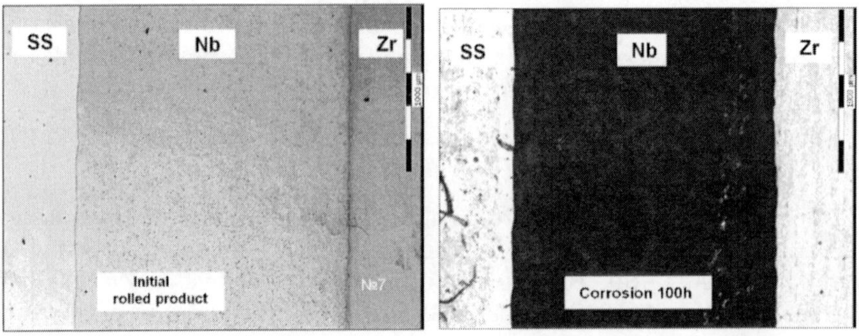

(a) (b)

Figure 5. State of the surface of the solid-phase bonding Zr1Nb-Nb-SS: (a) In the initial state; (b) After corrosion treatment during 100 hours.

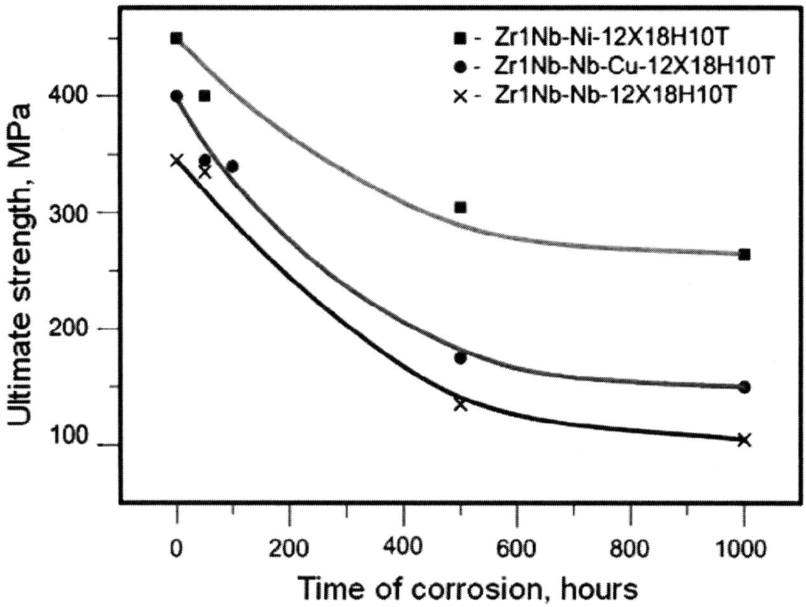

Figure 6. Dependence of ultimate strength of investigated com- position on time of corrosion testing.

After corrosion testing in water at operational temperature 350°C and pressure 16.5 MPa considerable de- crease of investigated compositions strength is observed. For Zr1Nb-Nb-12Cr18Ni10Ti the value of ultimate strength decreases from 340 - 350 MPa in initial state to 100 - 110 MPa after corrosion treatment during 1000 hours. Metallographic study had showed that during autoclave discontinuities as cracks on interface of interlay- ers with zirconium and steel were formed in the result of electrochemical corrosion. Separation into layers proceeds along specimen perimeter in the places of contact with corrosive medium. This causes the considerable decrease of joint strength and preliminary break down of the solid-phase bonding.

For Zr1Nb-Ni-SS decrease of strength is also observed after corrosion testing during 1000 hours (σ_b= 225 - 280 MPa) that can be explained by the formation of transient zone on interface zirconium-nickel.

4.2. Laminar Multilayer Material Stainless Steel SS (Type AISI 321) and Carbon Steel St20 (Type C22E)

This material was produced in two manners: by solid-state joining of steel C22E (St20) and stainless steel Cr18Ni10Ti directly or through the interlayer of nickel. The temperature of rolling was (980°C ± 20°C) at relative reduction (28% ± 3%).

Temperature dependencies of mechanical characteristics of the solid-phase bonding steel 20-steinless steel (see Figure 7(a) and Figure 7(b)).

Testify the high strength of solid-state joining. Introduction of nickel as barrier interlayer doesn't influence considerably on values of ultimate strength and of yield strength of this material.

Metallographic analysis of the solid-phase bonding structure was performed with the use of metallographic complex (microscope GX-51 and analyzer of image IA-32). Figure 8 shows the microstructure of pair steel C22E-SS.

After vacuum rolling any additional phases were not detected on interface of joint steel C22E-stainless steel Cr18Ni10Ti. Microstructure of main components on interface differs from layers that are on depth. Near interface on distance of 20 - 30 mkm from it new recrystallized grains have appeared with size 10 - 12 mkm without signs of deformation in the grain body. In steel C22E decarburized zone (see Figure 8(a)) had appeared with columnar structure of ferrite with transverse dimensions 55 - 60 mkm and not thick zone of finer equiaxial grains of ferrite with dimensions 35 - 40 mkm, passing to ferrite-perlite structure characteristic for low-carbo- nized steel (see Figure 8(a)).

After annealing ferritic zone in steel C22E increases in multilayer material rolled without interlayer. Dimen- sions of this zone depend on combination of temperatures and time of exposure. So, after annealing at 500°C during 10 hours substantial structural changes in comparison with the state after rolling are not observed neither in stainless steel nor in steel C22E (see Figure 8(b) and Figure 8(c)). High strength properties are preserved (σ_b = 400 - 440 MPa at 20°C). After lengthy exposure at temperatures 500°C and 600°C during 50 hours chemical

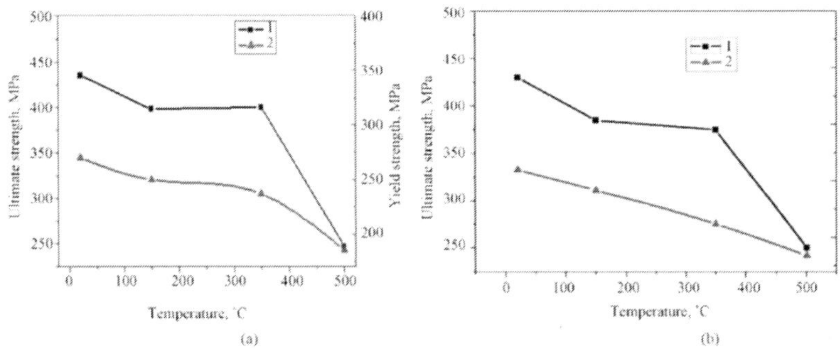

Figure 7. Temperature dependencies of mechanical characteristics of the solid-phase bonding steel C22E-SS (a) and steel C22E-Ni-SS (b); 1—σ_b; 2—σ_{02}.

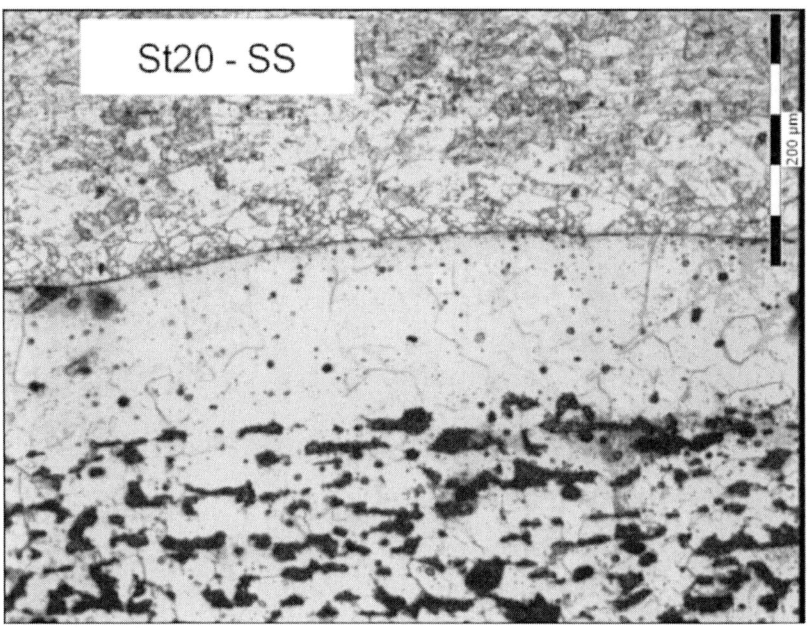

(a)

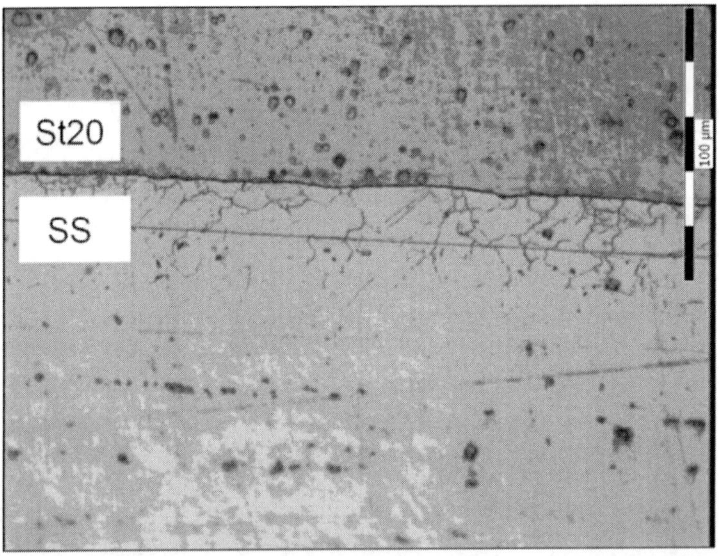

(b)

(c)

Figure 8. Microstructure of components of the solid-phase bonding steel C22E-SS (×200): (a) On interface of layer joining; (b) Interface after annealing at 500°C during 10 h (×500); (c) Interface after annealing at 500°C during 10 h after durable etching (×200).

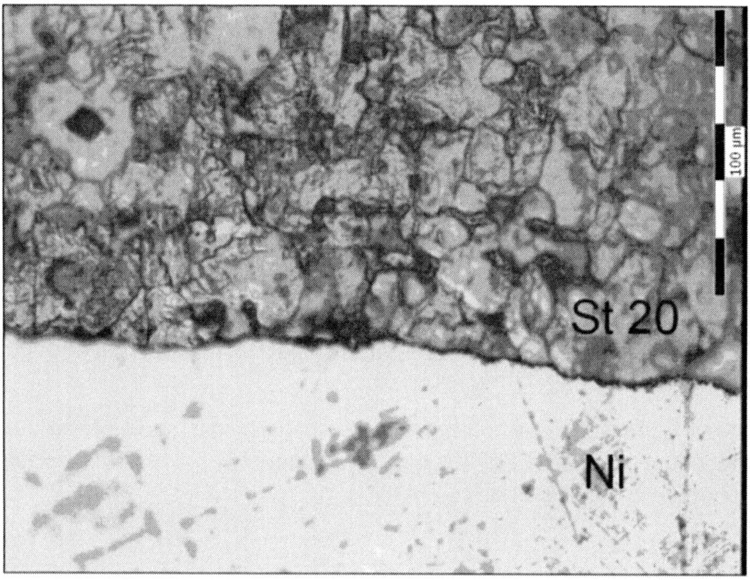

Figure 9. Microstructure of the solid-phase bonding C22E- Ni-SS on interface steel C22E -Ni after annealing at 500°C, 10 h (×200).

and structural inhomogeneities are detected both in steel C22E and in stainless steel 12Cr18Ni10Ti. Zone of fer- rite increases more than twice; from the side of stainless steel narrow light band appears with dimension of ~ several micrometers, related with interdiffusion of multilayer material elements.

For the delay of carbon diffusion from steel C22E into the stainless steel the barrier interlayer of nickel was introduced into the solid-phase bonding. Microstructure of the solid-phase bonding steel C22E-Ni-SS is shown on Figure 9.

Analysis of specimen structure both after rolling and after thermal treatment had showed the absence of structural changes and mutual transport of elements on the interface of the solid-phase bonding steel C22E- Ni-SS.

5. CONCLUSIONS

1) The main mechanism responsible for material solid-state joining on rolling is the onset of rotational plas- ticity in the crystal and its fragmentation with the formation of joint disclinations.

2) It is possible to obtain solid-state joining by the method of hot rolling in vacuum. The ultimate strength of inhomogeneous metals interface exceeds essentially the ultimate strength of less strength metal.

3) This method of joining may be used for production of vacuum-proof and durable adapters, zirconium-steel, titanium-steel, low-carbon and stainless steel.

4) Zone of solid-state joining in multilayers the solid-phase bonding produced by the method of hot rolling in vacuum represents a new material with improved characteristics.

REFERENCES

1. Kundu. S. and Ghosh, M. (2005) Diffusion Bonding of Commercially Pure Titanium to 304 Stainless Steel Using Copper Interlayer. Materials Science and Engineering: A, 407, 154-160. http://dx.doi.org/10.1016/j.msea.2005.07.010
2. Karakozov, E.S. (1976) Joining of Metals in Solid Phase. Metallurgiya, Moscow, 263 p.
3. Thomas, K. (1994) Roll Welding. ASM Hand Book, 6, 312-314.
4. An, J., Liu, Y.B., Lu, Y. and Sun, D. (2001) Hot Roll Bonding of Al-Pb Bearing of Alloy Strips and Hot Dip Alumi- nized Steel Sheet. Materials Characterization, 47, 291-297. http://dx.doi.org/10.1016/S1044-5803(01)00190-5
5. Hwang, Y.M., Hso, H.H. and Lee, H.J. (1995) Analysis of Sandwich Sheet Rolling by Stream Function Method. In- ternational Journal of Mechanical Sciences, 37, 297. http://dx.doi.org/10.1016/0020-7403(95)93522-8
6. Borts, B.V., Vanzha, A.F., Lopata, A.T., Neklyudov, I.M. and Shevchenko, S.V. (2005) Investigation of Processes of Welding of Multi Layers Structures of Crystals of Different Chemical Composition by Hot Rolling in Vacuum. Prob- lems of Atomic Science and Technology, No. 5. Series: Physics of Radiation Damage and Radiation Material Science 88, 196-198.
7. Borts, B.V. and Tkachenko, V.I. (2008) Phenomenological Description of Welding of Heterogeneous Metals in Solid Phase. Proceedings of XVIII International Conference on Physics of Radiation Phenomena and Radiation Material Science, Alushta, Crimea/Kharkov, 8-13 September 2008, 185-187.

CHAPTER 17

Investigating Structural Features Which Control The Dissolution of Bioactive Phosphate Glasses: Beyond The Network Connectivity

Jamieson K. Christie ✉, Richard I. Ainsworth, Nora H. de Leeuw

Department of Chemistry, University College London, 20 Gordon Street, London WC1H 0AJ, UK

ABSTRACT

We have used classical molecular dynamics simulations to characterise the structure of three compositions of silver-containing phosphate glasses with 45 mol% P_2O_5, 30 mol% CaO, and varying amounts of Na_2O and Ag_2O. These compositions all have the same network connectivity, allowing us to highlight two other structural features which will affect the glass dissolution. Firstly, the number of different phosphate chains bonded to each modifier atom was computed and it was observed that silver and sodium bind to roughly the same number of phosphate chains, despite the differences in their local environments. Secondly, the clustering of modifier cations was characterised and shown to be enhanced at low concentrations of sodium and silver, but not to exist for calcium.

Keywords: Bioactive glass; Phosphate; Glass; Molecular dynamics; Dissolution

1. INTRODUCTION

Very many different types of glasses are implanted into the body for medical reasons. Because of its amorphous structure, glass is not restricted to specific stoichiometries, as crystalline materials are. This means that glass has a wider range of possible compositions, and the properties of an amorphous implant can be tuned to optimise the efficacy of the therapy by varying the glass composition.

Glasses implanted into the body will react chemically with their local environment [1,2], and many will degrade [3,4]. Understanding the degradation is crucial [5]: the dissolution products of 45S5 bioglass are useful and promote the creation of new bone [6], by contrast, aluminosilicate glass compositions used for radiotherapy must be as durable as possible [7]. If a glass is to deliver a drug or nutrient at a specific rate, then its dissolution needs to be precisely controlled [8,9]. Much recent research work has sought to improve our understanding of glass dissolution processes and the structural and compositional features that affect them.

In principle, the factors which control glass dissolution are wide-ranging and could include the glass structure at various length scales, the diffusion of ions through the glass and their exchange between the surface and environment, changes in the glass surface structure, the solution chemistry and pH, among others. However, despite these possibilities, a good understanding of the glass dissolution can often be obtained from its bulk atomic structure alone, which is useful because the bulk structure is accessible from experiment and computer simulation.

The most important structural parameter which affects the dissolution of a glass is the *network connectivity*[10,11]. The network connectivity (NC) is defined as the average number of bridging oxygen (BO) atoms bound to a network-forming cation, where a BO atom is defined as an oxygen atom which is chemically bound to two network polyhedra. Oxygen atoms which do not connect two network polyhedra are called non-bridging oxygen (NBO) atoms. The NC depends critically on composition because the inclusion of network-modifying cations such as sodium or calcium typically breaks T–O–T bonds (where T is a network former, phosphorus in these glasses) causing the formation of NBO and decreasing the network connectivity [12].

High values of NC typically indicate a well-connected glass network which is not prone to dissolution and is not bioactive. Lower values indicate a fragmented, disconnected network which has many reactive sites and is both prone to dissolution and bioactive. 45S5 bioglass, which is highly bioactive and bonds to both bone and soft tissue, has a NC of 1.9; likewise, bioactive phosphate glasses which have 45 mol% P_2O_5[3] have NCs of 1.8. Whilst bioactivity is difficult to define and is certainly not a precise function of NC, Hill [11] identified that compositions with NC as high as 2.6 can still be bioactive, but Tilocca identified that compositions with NC greater than 3.0 are bioinactive and very durable [12,13].

It was clear from its first uses [10,11] to describe glass structure, that the NC cannot capture all of the structural features which affect the dissolution. Recently, we and others have shown through computer simulation other features of the glass structure, particularly present at length scales beyond the first neighbour shell, which are also pertinent. We briefly outline some here:

i. Clustering of network-modifying cations, such as Na, Ca, Ag, Sr, Y. These cations are not necessarily distributed evenly throughout the glass [13–16], and cluster together, which decreases bioactivity [12].
ii. Mesoscale segregation of network. The NC can vary spatially throughout the material: for example, fluorinated 45S5 bioglass separates into silicate-rich and silicate-poor regions due to fluorine's preferential bonding to the network modifiers in the glass [17,18], which does not happen for fluorinated phosphate-based glass [19].
iii. Modifier-chain bonding. For phosphate-based bioactive glasses with < 50 mol% P_2O_5, the dissolution is controlled by the bonding of the phosphate chains to the modifier atoms [20], such that small changes in the Na/Ca ratio, which do not alter the NC, lead to order-of-magnitude changes in the dissolution rate [3,21].

In this work, we use an example multicomponent system, i.e., a silver-containing phosphate-based bioactive glass, with 45 mol% P_2O_5 and varying amounts of Na_2O, Ag_2O and CaO. In addition to the interest from the glass structure perspective, silver-containing glasses are known to be antibiotic [8,22]. Substituting Ag_2O for Na_2O leads to a decrease in glass dissolution rate [22] for glasses with 50 mol% P_2O_5. There is no evidence for mesoscale segregation of this system, and so we will concentrate on the other two factors, namely: clustering of modifier cations, and the nature of network-modifier-chain bonding and its changes with composition. These properties are accessible through the use of molecular dynamics (MD) simulation techniques, which have been used not just to study glass properties, but also to connect the atomic structure of glass to its use as biomedical implants [12,13,15,16,20,23–25]. In this work, we use classical MD, in which the interatomic forces are approximated by an empirical expression. Although this approximation can introduce errors, it also reduces the computational expense sufficiently to allow us to model systems containing thousands of atoms, which are large enough to look at modifier-chain bonding and clustering. Smaller, more accurate, models created with first-principles MD, in which the forces are computed from a quantum-mechanical representation of the electronic structure, are not large enough to allow this.

2. METHODS

Classical molecular dynamics simulations were performed using the DLPOLY code [26] using a formal-charge, polarizable interatomic force field previously used by us to describe silver-containing [27] and other [20,28] phosphate-based bioactive glasses, in simulations which have given glass structures in agreement with experiment.

The simulation methodology used is identical to that in Ref. [27], which we outline briefly here. Models containing about 2000 atoms were prepared by randomly and independently placing atoms into a cubic periodic box of the appropriate density, subject to the constraint that no atom be closer to another than within ~ 85–90% of the expected interatomic separation. The densities of these compositions were extrapolated from the experimentally measured density of the Ag-free composition [29], using our knowledge of how Ag_2O substitution for Na_2O affects the density of very similar glass compositions [27]. After a short zero-temperature relaxation, the simulation was run in an NVT ensemble for 50 ps at 2500 K to equilibrate the liquid. The temperature was then reduced in units of 100 K, and at each temperature, the simulation was run for 50 ps, corresponding to a cooling rate of ~ 2 K/ps. At 300 K, the simulation was run for 250 ps, the last two-thirds of which forms the production run. All data stated are averaged over snapshots taken from this run. Simulated compositions and densities are given in Table 1.

Table 1. The simulated compositions (mol%) and their densities.

Composition	P_2O_5	Na_2O	Ag_2O	CaO	Density (g cm^{-3})
A5	45.0	20.0	5.0	30.0	2.707
A10	45.0	15.0	10.0	30.0	2.854
A15	45.0	10.0	15.0	30.0	3.001

Although this cooling rate is substantially faster than that used to prepare glasses experimentally, simulated cooling rates of this order of magnitude have been used to prepare accurate structural models of glasses in agreement with experimental results using classical molecular dynamics simulations [28,30].

3. RESULTS

Pictures of the models at 300 K are given in Fig. 1.

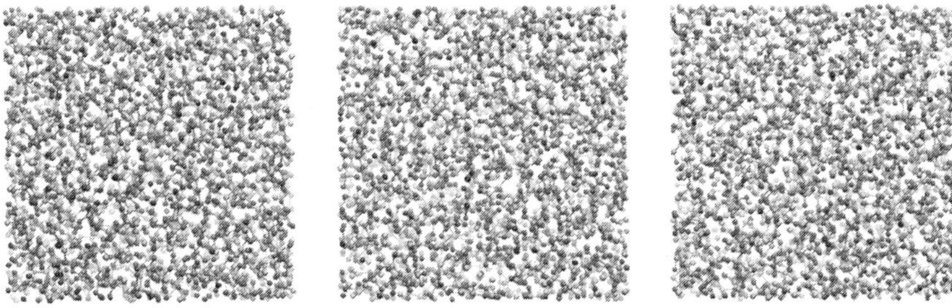

Figure. 1. Views of the A5 (left), A10 (middle) and A15 (right) compositions. The colours are red (oxygen), tan (phosphorus), dark blue (sodium), light blue (calcium) and pink (silver).

Because the local silver environment in phosphate-based glasses has been studied before [27], as has the structure of Ag-free phosphate glasses [20,28], we mention them only in the context of those structural features which most profoundly affect the dissolution, which have not yet been characterised for these glasses.

The modifier–oxygen bond lengths and coordination numbers are given in Table 2. The coordination numbers are broken down into bonding to bridging and non-bridging oxygen atoms. Although all modifier atoms prefer to bond to NBO, when more than one are present, they "compete" to satisfy their preferred bonding environment [31]. This means that the NBO/BO ratio often differs around different modifier atoms, affecting the bioactivity [15,20].

Oxygen–modifier–oxygen bond-angle distributions for the three compositions are given in Fig. 2. Although modifier atoms typically have rather broad bond-angle distributions, reflecting the wide range of possible bonding environments, there are often differences between the different modifiers, particularly in the amount of "intra-tetrahedral" bonding, in which a modifier atom bonds to two atoms from the same phosphate PO_4 tetrahedron [16,20].

As mentioned in the introduction, different modifier atoms bond to different numbers of phosphate chains. The oxygen atoms in the first coordination shell of the modifier atoms are also bonded to PO_4 phosphate tetrahedra which are parts of these phosphate chains. The distribution of numbers of phosphate chains bonded to each type of modifier atom was calculated and is given in Table 3.

Table 2. Bond lengths and coordination numbers to all oxygen atoms (CN), to bridging oxygen atoms only (CN(BO)), and to non-bridging oxygen atoms only (CN(NBO)). Cutoffs are 3.2 Å for all modifier–oxygen pairs.

	Ag–O	Na–O	Ca–O
A5			
Bond length (Å)	2.30	2.35	2.33
CN	5.88	6.34	6.37
CN (BO)	1.15	1.18	0.36
CN (NBO)	4.73	5.16	6.01
A10			
Bond length (Å)	2.29	2.36	2.33
CN	5.59	6.33	6.32
CN (BO)	0.89	1.21	0.40
CN (NBO)	4.69	5.12	5.92
A15			
Bond length (Å)	2.27	2.37	2.33
CN	5.46	6.17	6.23
CN (BO)	0.89	1.28	0.40
CN (NBO)	4.57	4.89	5.83

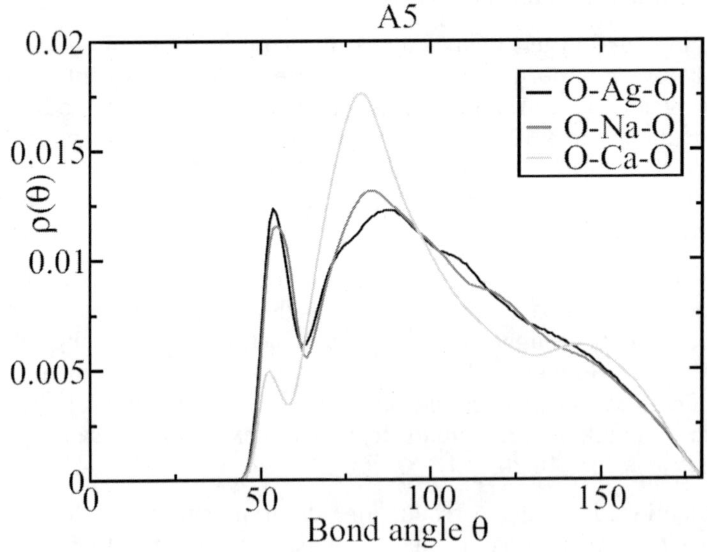

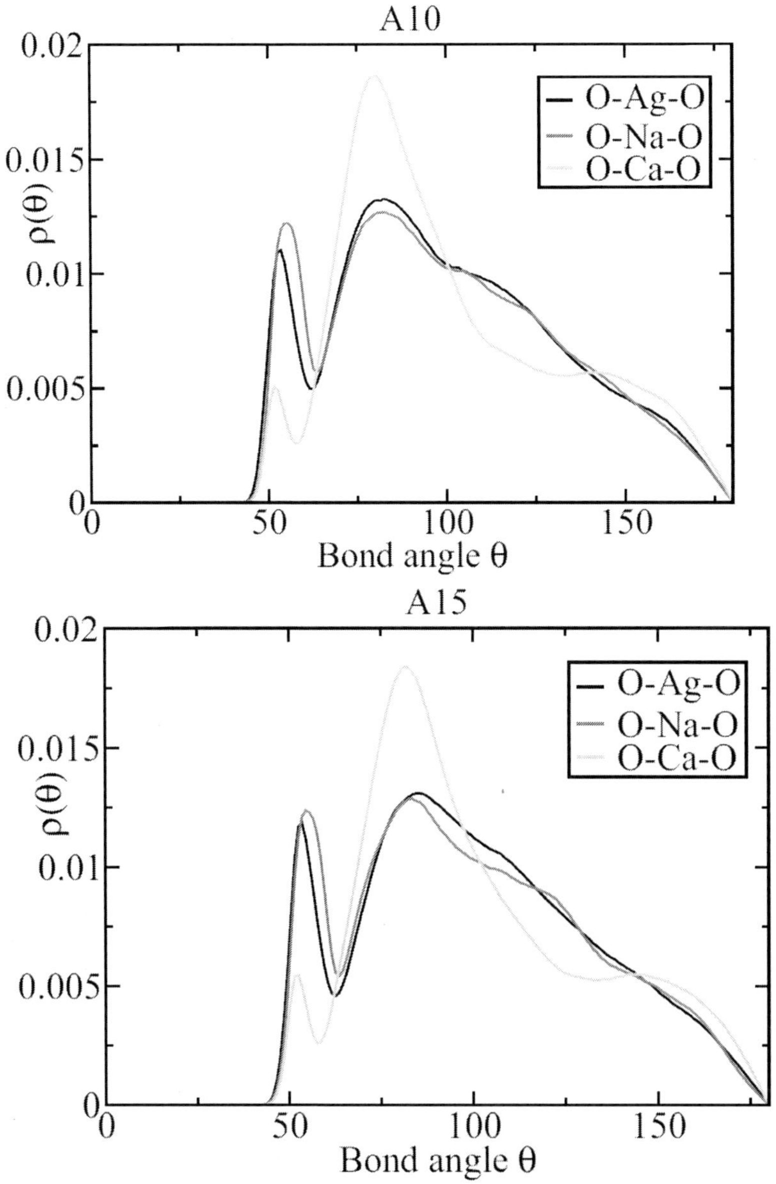

Figure. 2. Oxygen–modifier–oxygen bond-angle distributions for the A5 (top), A10 (middle) and A15 (bottom) compositions. The colours are black (O–Ag–O), red (O–Na–O), green (O–Ca–O).

Table 3. The distribution (in %) of numbers n of phosphate chains bound to different modifier atoms for the glass compositions studied.

	n (%)							Average
	0	1	2	3	4	5	6	
A5								
Ag	0	0	16	44	38	2	0	3.26
Na	0	0	14	49	35	3	0	3.25
Ca	0	0	5	26	41	26	2	3.94
A10								
Ag	0	2	15	52	28	3	0	3.21
Na	0	0	16	45	35	2	0	3.16
Ca	0	0	4	27	52	16	1	3.83
A15								
Ag	0	3	14	49	32	4	0	3.01
Na	0	2	25	47	25	2	0	3.21
Ca	0	0	7	33	44	16	2	3.73

Clustering of modifier cations was assessed by comparing the observed number of atoms N_{obs} in a coordination sphere to that which would be expected if the modifier cations were distributed homogeneously throughout the sample N_{hom} and computing the ratio $r = N_{obs}/N_{hom} = (\mathrm{CN}+1)/(\frac{4}{3}\pi r_c^3 \rho)$, where CN is the M–M coordination number of the central modifier atom M, r_c is the cutoff radius, and ρ is the number density of modifier atoms in the model [13,14,32]. In this work, r_c is set to 5 Å, as in Ref. [14]. Clustering ratios are reported in Table 4. Values of $r > 1$ imply that clustering is occurring, with a ratio $r = 1$ implying that the atoms are distributed as if randomly.

Table 4. The clustering ratios r for the different modifier atoms M.

M	A5	A10	A15
r(Ag)	2.0	1.5	1.3
r(Na)	1.1	1.2	1.5
r(Ca)	1.0	0.9	1.0

4. DISCUSSION

As outlined in the introduction, in this work we are interested in the structural features which affect the dissolution of phosphate glasses, other than the network connectivity. To remove any influence of changes in the NC from our results, the glass compositions chosen here have been constructed such that they all have the same NC.

The local atomic environment of silver in related phosphate glasses has already been discussed at length [27], as has the structure of Ag-free phosphate glasses [20,28], and so we confine this discussion to those structural features which are likely to affect the glass dissolution.

In glasses with < 50 mol% P_2O_5, the structure is made up of finite-length unbranched chains of PO_4 phosphate tetrahedra, bound to the modifier atoms, in this case, sodium, calcium and silver. For Ag-free glasses, we have shown [20] that Na binds to fewer chains than Ca does, and that increasing the Ca/Na ratio of the glass will make the glass less prone to dissolution, as seen experimentally [3], because of the increased amount of cross-linking between different phosphate chains. We know from experiment [22] that substituting Ag_2O for Na_2O reduces the dissolution rate for glasses with 50 mol% P_2O_5.

The silver ion has a single charge, as does the sodium ion, and the effects of changing the Ag/Na ratio are likely more subtle. We see from Table 3 that silver and sodium typically bond to roughly the same number of phosphate chains (3.0–3.3 for Ag, 3.2–3.3 for Na).

Although the two ions have the same charge, Ag^+ is slightly larger, with an ionic radius of 1.29 Å compared to 1.16 Å for Na^+. The field strength of an ion is defined [31] as the ionic charge divided by the square of the interatomic distance, and the distance will be larger for a larger ion. Ag^+ therefore has a field strength between that of sodium and calcium, and will therefore [31] be bonded to a number of BO that is less than that of sodium but more than that of calcium, as we confirm from Table 2. This could imply that a silver ion is bound to more phosphate chains than a sodium ion, because BO will connect two phosphate tetrahedra, which are part of the same chain by definition. This will mean that for a given number of tetrahedra chemically bonded to a modifier atom, there could be fewer chains bonded to a silver ion than to a sodium ion.

However, two effects mitigate this trend. Firstly, silver has a smaller coordination number than sodium (Table 2), reducing the potential number of phosphate chains in its first coordination shell. Secondly, sodium and silver have similar amounts of "intra-tetrahedral" bonding, that is, bonding to two oxygen atoms from the same PO_4 tetrahedron, which is visible from the similar intensities of the peaks at ~ 60° in the O–Ag–O and O–Na–O bond-angle distributions (Fig. 2). This implies a similar number of phosphate chains in the coordination shell of these two ions.

We have previously identified that the field strength of the modifier ions affects the dissolution of the glass containing them, and that modifier atoms with higher field strengths will likely bind to more phosphate chains, e.g., substituting a lower-field-strength modifier by a higher-field-strength modifier will increase glass durability [3,20,22]. It is clear from this study, though, that this is not always the only or dominant effect.

Clustering of modifier cations is known to inhibit bioactivity [12] by strengthening the glass network, although the exact mechanism is unknown. Clustering can be characterised from simulation through the calculation of the clustering ratio *r*.

From Table 4, we see that calcium is distributed homogeneously throughout the glass, with *r* very close to 1 for all compositions. Silver and calcium both cluster slightly ($r > 1$) but seem to cluster together more at lower concentrations: silver has its highest clustering ratio for the A5 composition, and sodium has its highest clustering ratio for the A15 composition, which has the lowest amount of sodium. This phenomenon has also been observed when yttrium clusters in aluminosilicate glasses [13].

These clustering ratios are not much larger than one, and hence are unlikely to be detrimental to the bioactivity of the glass, as similar ratios have been reported for compositions known to be bioactive [14,15], but this clustering could affect other properties of these glasses, for example, the dynamics of silver ion release and hence the glass's antibacterial properties.

5. CONCLUSION

To conclude, we have looked at compositions of Ag-containing phosphate glasses with identical network connectivities in order to characterise the effect of structural features on the bioactivity and dissolution. By doing so, we have highlighted some structures – in this case, phosphate–modifier bonding and modifier clustering – which affect the dissolution of these bioactive glasses.

It is rather challenging to disentangle the effects of these structures on the dissolution rate. The experimental data [22] that we have, are on glasses with 50 mol% P_2O_5 which has a NC of 2.0 and therefore a polymeric structure, so it is possible that our simulated data are not directly comparable. For our simulated compositions, we observe that silver and sodium ions are bound to about the same number of phosphate chains which would imply that the change in dissolution has a different cause. Each of these two ions clusters spatially, and this is enhanced at low concentration, but the precise connection between changes in the clustering ratio and dissolution is unknown. This underlines the fact, as mentioned in the introduction, that the dissolution of these glass compositions can depend on several properties of the glass.

Therefore, although these features may not form a complete picture of the glass dissolution, in order to optimise these glasses for a specific dissolution rate, for example for a specific therapy, they must be taken into account.

ACKNOWLEDGEMENTS

We acknowledge the EPSRC (grant no. EP/J008095/1) for funding. Via our membership of the UK's HPC Materials Chemistry Consortium, which is funded by EPSRC (grant nos. EP/F067496 and EP/L000202), this work made use of the facilities of HECToR and Archer, the UK's national high-performance computing service.

REFERENCES

1. L.L. Hench, H.A. PaschallJ. Biomed. Mater. Res., 7 (1973), p. 25
2. L.L. Hench, J. WilsonScience, 226 (1984), p. 630
3. J.C. KnowlesJ. Mater. Chem., 13 (2003), p. 2395
4. B. Zberg, P.J. Uggowitzer, J.F. LófflerNature Mater., 8 (2009), p. 887
5. Hoppe, N.S. Güldal, A.R. BoccacciniBiomater., 32 (2011), p. 2757
6. I.D. Xynos, A.J. Edgar, L.D.K. Buttery, L.L. Hench, J.M. PolakJ. Biomed. Mater. Res., 55 (2001), p. 151
7. E.M. Erbe, D.E. DayJ. Biomed. Mater. Res., 27 (1993), p. 1301
8. S.P. Valappil, D.M. Pickup, D.L. Carroll, C.K. Hope, J. Pratten, R.J. Newport, M.E. Smith, M. Wilson, J.C. KnowlesAntimicrob. Agents and Deliv., 51 (2007), p. 4453
9. D.M. Pickup, R.J. Newport, J.C. KnowlesJ. Biomat. App., 26 (2012), p. 613
10. Z. StrandBiomater., 13 (1992), p. 317
11. R.G. HillJ. Mater. Sci. Lett., 15 (1996), p. 1122
12. TiloccaProc. R. Soc. A, 465 (2009), p. 1003
13. J.K. Christie, A. TiloccaChem. Mater., 22 (2010), p. 3725
14. J.K. Christie, J. Malik, A. TiloccaPhys. Chem. Chem. Phys., 13 (2011), p. 17749
15. J.K. Christie, A. TiloccaJ. Mater. Chem., 22 (2012), p. 12023
16. J.K. Christie, A. TiloccaJ. Phys. Chem. B, 116 (2012), p. 12614
17. G. Lusvardi, G. Malavasi, F. Tarsitano, L. Menabue, M.C. Menziani, A. PedoneJ. Phys. Chem. B, 113 (2009), p. 10331
18. J.K. Christie, A. Pedone, M.C. Menziani, A. TiloccaJ. Phys. Chem. B, 115 (2011), p. 2038
19. J.K. Christie, R.I. Ainsworth, N.H. de LeeuwBiomater., 35 (2014), p. 6164

20. J.K. Christie, R.I. Ainsworth, D. Di Tommaso, N.H. de LeeuwJ. Phys. Chem. B, 117 (2013), p. 10652
21. Ahmed, M. Lewis, I. Olsen, J.C. KnowlesBiomater., 25 (2004), p. 491
22. Ahmed, D. Ready, M. Wilson, J.C. KnowlesJ. Biomed. Mater. Res., 79A (2006), p. 618
23. Tilocca, N.H. de LeeuwJ. Phys. Chem. B, 110 (2006), p. 25810
24. Tilocca, A.N. Cormack, N.H. de LeeuwChem. Mater., 19 (2007), p. 95
25. Tilocca, A.N. Cormack, N.H. de LeeuwFaraday Discuss., 136 (2007), p. 45
26. I.T. Todorov, W. Smith, K. Trachenko, M.T. DoveJ. Mater. Chem., 16 (2006), p. 1911
27. R.I. Ainsworth, J.K. Christie, N.H. de LeeuwPhys. Chem. Chem. Phys., 16 (2014), p. 21135
28. R.I. Ainsworth, D. Di Tommaso, J.K. Christie, N.H. de LeeuwJ. Chem. Phys., 137 (2012), p. 234502
29. D. Carta, D.M. Pickup, J.C. Knowles, I. Ahmed, M.E. Smith, R.J. NewportJ. Non-Cryst. Sols., 353 (2007), p. 1759
30. K. Vollmayr, W. Kob, K. BinderPhys. Rev. B, 54 (1996), p. 15808
31. A.K. Varshneya, Fundamentals of Inorganic Glasses, Society of Glass Technology (2006)
32. R.N. Mead, G. MountjoyJ. Phys. Chem. B, 110 (2006), p. 14273

CHAPTER 18

Influence of Groundwater Hypothetical Salts on Electrical Conductivity Total Dissolved Solids

S. A. M. Al Dahaan[1], Nadhir Al-Ansari[2], Sven Knutsson[2]

[1]Department of Geology, Faculty of Science, University of Kufa, Kufa, Iraq

[2]Lulea University of Technology, Lulea, Sweden

ABSTRACT

A relationship between electrical conductivity (EC) and total dissolved solids (TDS) was tested for solutions of same salinity levels with respect to different artificial salts with their combinations. Results showed remarkable jumping at the order of the artificial salt sequence specially that of the magnesium type. A computer model is designed with an input of EC and TDS. The output will be the possible prevailing artificial salts. The accuracy of the model was tested by using the groundwater data of Safwan-Zubair area south of Iraq and it proved to be significant at 95% matching. The 5% unmatched results are due to the possibility of having more than one type of prevailing salt.

Keywords: Electrical Conductivity, A Computer Model, Artificial Salts, Groundwater, Total Dissolved Solids

1. INTRODUCTION

Salinity is a measure of the amount of salts in the water, while total dissolved solids (TDS) as salinity parameters are often calculated using laboratories test [1]. When salts are dissolved in water, their ions dissociated and increased both the amounts of dissolved solids in the solution and their conductivity [2]. Electrical conductivity gives an indication of the amount of total dissolved substitution in water [3]. The estimations of total dissolved solids (TDS) content are based on electrical conductivity (EC) measure- ments [4].

Electrical conductivity (EC) for groundwater is the ability of 1 cm^3 water to conduct an electric current at 25°C and is measured in micro Siemens per centimeter, so it depends on the total amount of soluble salts (TDS) as charged particles [5] [6] [7] [8].

$$\mathrm{TDS}(\mathrm{ppm}) = 0.64 \times \mathrm{EC}(\mu\mathrm{S/cm}) = 640 \times \mathrm{EC}(\mathrm{dS/m}) \quad (1)$$

Groundwater conductance is a function of type of present ions, types of dissolved constituents and temperature [9]. Electrical conductivity is an indirect measurement of salinity, and it is temperature dependent and good indicator of the total salinity, but it does not provide any information about the ionic composition within the water sample [10] [11]. The response of the conductance value to temperature changes is somewhat varies for different salts and many concentrations. In dilute solutions however, for most ions, an increase of 1°C can increase conductance by about 2% - 3% [12]. It also increases with the increase of the total dissolved salts [13]. The variation of conductivity gives important information on the evolution of water quality. Electrical conductivity is a measurement to estimate the amount of total dissolved solids by factor of 0.55 - 0.90 for converting conductivity into total dissolved solids [14]. The spatial distribution of EC is controlled by several factors and practices, which may cause salinity variation [8] [15] [16]. Some of these factors are the depth of the collected samples, concentration and type of concentration, mobility of groundwater, valence, temperature of water, type of soil or rock leaching, the long term flow with high rates of discharge, and the distance between the recharge and discharge area [13]. The internationally accepted standard unit for reporting EC of water is deciSiemens per meter (dS/m). Note that this standard unit was adopted recently. An older, equivalent unit often appears in water quality reports from the 1980s or earlier which is: milliohms per centimeter. Although the term "mhos" may at first appear strange, it was chosen by early researchers for reasons that involve physics. EC, as its name implies, is a type of conductivity—the opposite of resistivity, measured in ohms. Hence, for EC, researchers adopted the term "mho"- "ohm" written backwards. Here is a quick summary of the various EC units you might encounter when reading papers from the literature:

$$1\,\mathrm{dS/m} = 1\,\mathrm{mmho/cm} = 1000\,\mu\mathrm{mho/cm} \quad (2)$$

Units used for measuring electrical conductivity of water are MicroSiemens per centimeter µS/cm, millisiemens per centimeter (mS/cm) and DeciSiemens per meter dS/m. Groundwater contains different of both ionic and uncharged species in various amounts and proportions that constitute the dissolved solids. Thus it is not clear whether specific conductance measurements can be used to obtain accurate estimates of TDS. The electrical conductivity for absolute pure water is 0.055 µS/cm, distilled water 0.5 µS/cm, power plant boiler water 1.0 µS/cm, deionizer water 0.1 - 10 µS/cm, good city water 50 µS/cm, drinking water 0.5 - 1 mS/cm, ocean water 53 mS/cm, 10% NaOH 355 mS/cm and 10% H_2SO_4 is 432 mS/cm [17].

Total dissolved solids term TDS describes all solids, commonly mineral salts that are dissolved in water [18] [19]. There is a close relation between TDS and the electrical conductivity [20]. As more salts are dissolved in water; the value of electric conductivity becomes higher. The majority of solids, which remain in the water after filteration are dissolved ions. Water of excellent purity without salts has a very low electrical conductivity (lennetech.com). Total dissolved solids TDS are differentiated from total suspended solids TSS, because the latter cannot pass through a filter of two micrometers and yet are called suspended in solution [21]. However, when the concentration of salt reaches a certain level, electrical conductivity is no longer directly related to salts concentration because of ion pairs weaken each other's charge, so that above this level, higher TDS cannot result equally higher electrical conductivity. Electrical conductivity can be converted to estimate total dissolved solids by using the following equation [13] :

$$\mathrm{TDS(ppm)} = 0.64 \times \mathrm{EC}(\mu\mathrm{S/cm}) = 6.4 \times \mathrm{EC\,mS/cm} = 640 \times \mathrm{EC(dS/m)} \quad (3)$$

Electrical conductivity for water is directly related to the concentration of dissolved ionized solids in water. Ions from the dissolved solids in water are creating the ability of water to conduct an electrical current. It can be measured by using a conventional conductivity meter or TDS meter within about 10% accuracy, when correlated with laboratory TDS measurements.

Relationship of total dissolved solids and specific conductance for groundwater can be approximated by the following equation (epa.gov):

$$\mathrm{TDS} = \mathrm{keEC} \quad (4)$$

where TDS is expressed in mg/L or ppm and EC is the electrical conductivity in microSiemens per centimeter or µS/cm at 25°C. Correlation factor ke is between 0.55 - 0.8.

Rainwater has TDS about of 20 ppm or less. Fresh water of lakes, rivers, and ground- water is more variable, with TDS ranging from 20 - 1000 ppm.

Brackish water is, by definition, water with TDS exceeding 1000 ppm and ranging as high as that of seawater, at about 35,000 ppm.

Total dissolved solids TDS is the total amount of solids remaining when a water sample evaporates to dryness [22]. Dissolved solids are those that pass through a filter with 2.0 μm or smaller pores. A simple method for determining the concentration of dissolved solids is to filter the water, evaporate the filtrate and weight the residue. The TDS represents a total summation of ionic concentrations of cations and anions. It is measured by the ppm or mg/l units [9].

The aim of study is to present a model clarifying the effect of the prime salt at same salinity level on the values of the electrical conductivity. Also the effects of the presence of difference artificial salts at same salinity level on the electrical conductivity. Finally this model is programmed with input of electrical conductivity and total dissolved solids to predict the type of the artificial salt.

2. MATERIALS AND METHODS

Preparation of different artificial salts at same level of salinity in five replicates was carried out. Salinity levels were taken from 500 ppm with 500 ppm increment up to 3000 ppm. Then the salinity levels were increased by 1000 ppm from 3000 ppm up to 7000 ppm. The electrical conductivity was measured by conductor meter at 25°C for prime salts and their combinations [23].

The statistical test (F, t) is applied for the predicted of TDS and the input TDS [24].

3. RESULTS AND DISCUSSIONS

The obtained results for prime salt of chloride type started from top NaCl to KCl to bottom $CaCl_2$(Figure 1). At same TDS salinity Level, the EC values increase from bottom to top of the sequence. This means that for the same EC value the TDS increase from top to bottom. All of the prime salts were below the mixed water type number 4. The $MgCl_2$ jumped out of the sequence. The $MgCl_2$artificial prime salt jump from the chloride natural sequence to the sulphate sequence (Figure 2). The position is at the end of the sulphate replacing the $MgSO_4$ position which is the second jump type. Here the value of the TDS at same level of EC will be from top to bottom increase. The sulphate sequence is of higher TDS than the chloride sequence at given level of EC. The

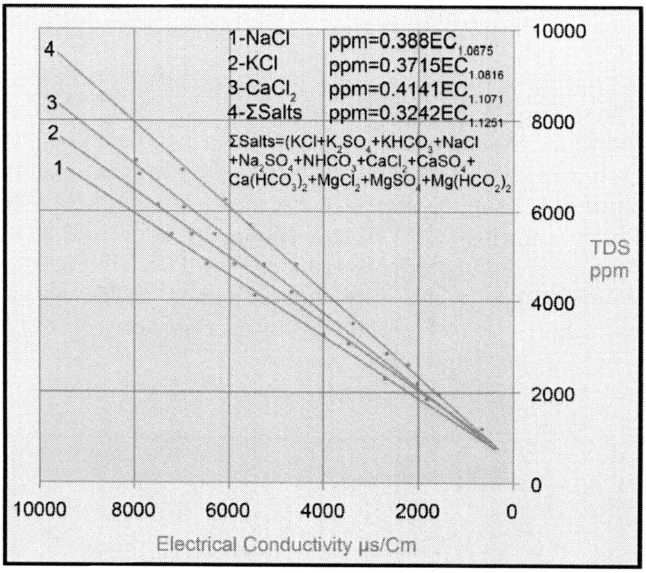

Figure 1. EC and TDS for prime salts 1, 2, 3 and first stage mixed salts 4.

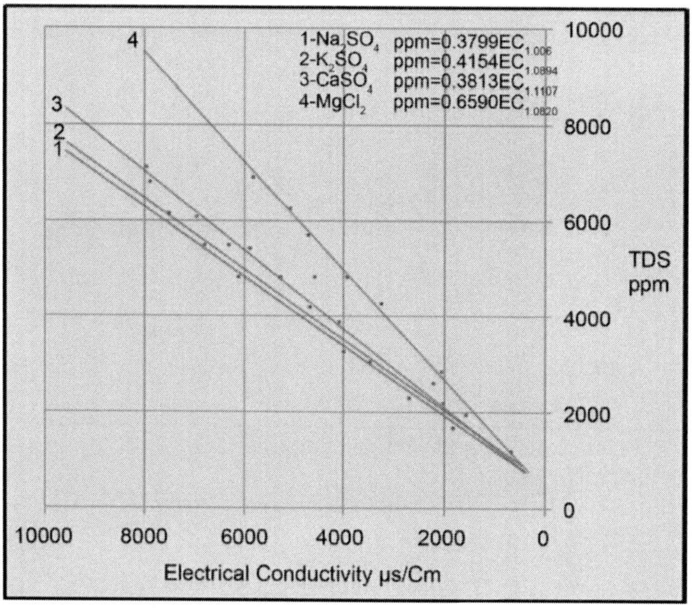

Figure 2. Prime salts EC and TDS second stage.

bicarbonate sequence (Figure 3) is characterizes by the $KHCO_3$ top passes toward $Ca(HCO_3)_2$ and then jumping to $NaHCO_3$ and followed to the $Ca(HCO_3)_2$. The $MgSO_4$ jumped from the sulphate to the bicarbonate sequence following the $NaHCO_3$ and replacing the $Mg(HCO_3)_2$. The bottom of the sequence is $Mg(HCO_3)_2$.

The flow chart of such jump in the sequence of salts is shown in Figure 4. Also all the artificial salts of the chloride type Na, K, Ca and Mg were mixed with the sulphate and bicarbonate artificial salts combination. The obtained results showed that the EC value is decreased from the mixing of Na_2SO_4 to K_2SO_4 passed to $NaHCO_3$, $KHCO_3$ and ending by $MgSO_4$. This is the same result of the obtained individual sequence. The TDS will increase from top to the bottom and all of the sequence is higher in TDS values than the sulphate at same level of EC. There is an overlap between the $KHCO_3$ and the $MgCl_2$ in position (**Figure 5**).

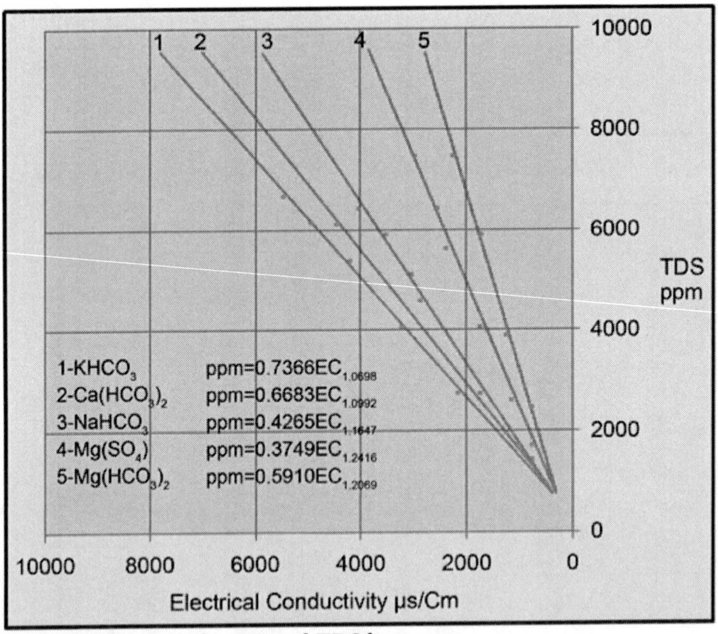

Figure 3. Prime salts EC and TDS last stage.

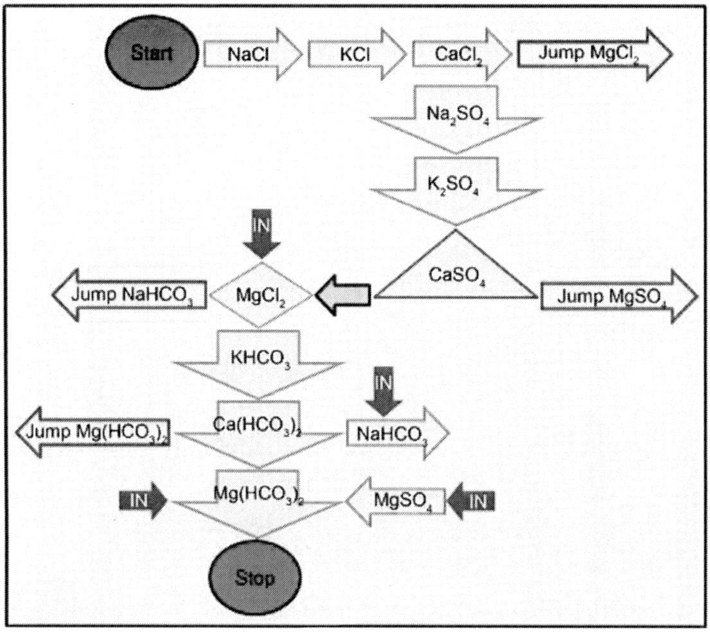

Figure 4. Flow chart for the jump of EC at same salinity TDS with different water type.

4. COMPUTER MODEL

The mixed salt relationship presented in Figures 1-5, and type curve number 4 is the first point considered in the model. The calculated ECC by this type curve will be compared with the actual EC. There will be three conditions:

1) ECC < EC:

This is the condition of chorded artificial salt combination. Then Chloride Water Type will be in action.

2) ECC > EC:

This is the condition of sulphate and bicarbonate artificial salts. The SOCH subroutine will be in action.

3) ECC = EC:

This is the ideal case of more than one prevailing artificial salt combination. The limits are taken for ±5% relative difference. This mean matching of 95%.

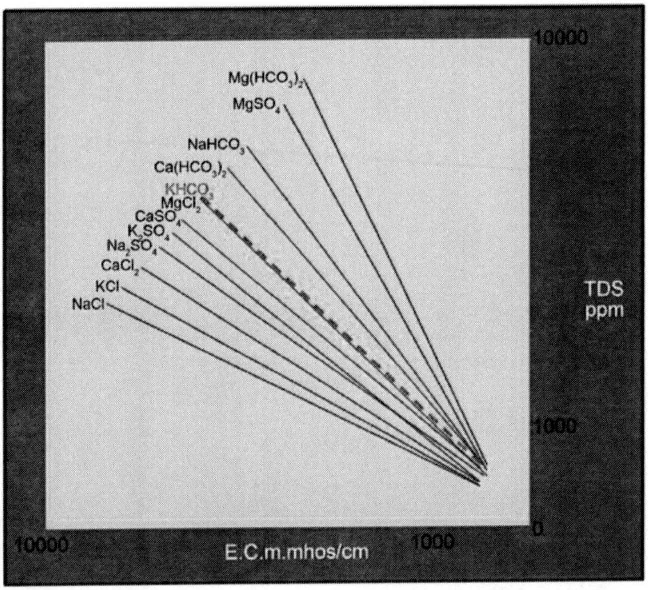

Figure 5. Prime artificial salts behavior with EC and TDS relationship.

5. CONCLUSION

The prime salt of chloride group is of higher EC value than the sulphate water group followed by the bicarbonate of same salinity. This sequence is changed when salt combinations are considered. For chloride and sulphate groups, the EC value decreases from sodium to potassium and to calcium in a solution of same salinity. This order is changed for bicarbonate group as the sodium bicarbonate salt is at the end of the sequence. The presence of the magnesium anion salts decreases the EC rapidly. The $MgCl_2$ follows the $CaSO_4$ and the $MgSO_4$ follows the $NaHCO_3$ and this sequence is ended by the $Mg(HCO_3)_2$. For the case that the magnesium is the prevailing cation, then the EC value is less than expected. There are significant differences in the EC values of the prime salts in a solution of same salinity in the mixed state. This fact is due to the behavior of the prime salts in a solution related to their stages of development and interaction. The mixed stat position is between chloride and the sulphate water groups, while the bicarbonate is still far away from the mixed condition. Thus it is not possible to use any relation of a given basin to another basin without taking into consideration the prevailing water type and salinity level. A computer model is built for the prediction of the type of the hypothetical salt. The input data are the EC and TDS. The calculated electrical conductivity ECC is according to the mixed condition of no prevailing salt. The relation between the actual and the calculated electrical conductivity will determine the type of prevailing salt.

REFERENCES

1. Siosemarde, M., Kafa, F., Pazira, E., Sedghi, H. and Ghaderi, S.J. (2010) Determine of Constant Coefficients to Relate Total Dissolved Solids to Electrical Conductivity. International Journal of Environmental, Chemical, Ecological and Geophysical Engineering, 4, 457-459.
2. Hölting, B. and Coldewey, W.G. (2012) Hydrogeologie-Einführung in die Allgemeine und Angewandte. Springer Spektrum, Berlin, 439 p.
3. Yilmaz, E. and Koc, C. (2014) Physically and Chemically Evaluation for the Water Quality Criteria in a Farm on Akcay. Journal of Water Resource and Protection, 6, 63-67.https://doi.org/10.4236/jwarp.2014.62010
4. Hubert, E. and Wolkersdorfer, C. (2015) Establishing a Conversion Factor between Electrical Conductivity and Total Dissolved Solids in South Africa Mine Waters. Water SA, 41, 490-500. https://doi.org/10.4314/wsa.v41i4.08
5. Todd, D.K. (1990) Groundwater Hydrology. John Wiley and Sons, New York, 535 p.
6. Todd, D.K. (2007) Groundwater Hydrology. 2nd Edition, 3rd Reprint, John Wiley and Sons, Inc., India.
7. State Water Resources Control Board (2016) Groundwater Information Sheet. Salinity, Division of Water Quality, GAMA Program, USA. http://www.waterboards.ca.gov/gama/docs/coc_salinity.pdf
8. United States Geological Survey (2016) Sources of Dissolved Solids in Brackish Groundwater. http://water.usgs.gov/ogw/gwrp/brackishgw/sources.html
9. Boyd, C.E. (2000) Water Quality an Introduction. Kluwer Academic Publisher, USA, 330 p.
10. Hem, J.D. (1985) Study and Interpretation of the Chemical Characteristics of Natural Water. 3rd Edition, U.S. Geological Survey Water-Supply Paper, Vol. 2254. Washington DC.
11. Hem, J.D. (1991) Study and Interpretation of the Chemical Characteristics of Natural Water. Geological Survey Water-Supply Paper 1973, U.S. Government Printing Office, Washington DC, 363 p.
12. Boyd, C.E. (1999) Water Quality: An Introduction. Kluwer Academic Publishers Group, The Netherlands.
13. Detay, M. (1997) Water Wells—Implementation, Maintenance and Restoration. John Wiley and Sons, London, 379 p.
14. Werkema, D.D. (2002) Geoelectrical Response of an Aged LNAPL. Plume: Implications for Monitoring Natural Attenuation. Unpublished PhD Dissertation, Western Michigan University, 136 p.
15. Ledesma-Ruiz, R., Pastén-Zapata, E., Parra, R., Harter, T. and Mahlknecht, J. (2015) Investigation of the Geochemical Evolution of Groundwater under Agricultural Land: A Case Study in Northeastern Mexico. Journal of Hydrology, 521, 410-423.https://doi.org/10.1016/j.jhydrol.2014.12.026
16. Al-Shidi, F.K.R. (2014) Study the Quality of Groundwater of Al-Zoroup Area in Mhdah State, the Sultanate of Oman. MSc Thesis, United Arab Emirates University, College of Science, 160 p. http://scholarworks.uaeu.ac.ae/cgi/viewcontent.cgi?article=1010&context=all_theses
17. Atekwanaa, E.A., Atekwanaa, E.A., Roweb, R.S., Werkema Jr., D.D. and Legalld, F.D. (2004) The Relationship of Total Dissolved Solids Measurements to Bulk Electrical Conductivity in an Aquifer Contaminated with Hydrocarbon. Journal of Applied Geophysics, 56, 281-294. https://doi.org/10.1016/S0926-9851(04)00057-6

18. Advanced Purification Engineering Corp. (2016) Water Quality, Total Dissolved Solids in Water. http://www.freedrinkingwater.com/water-education2/72-tds-water.htm
19. Safe Drinking Water Foundation (2016) TDS AND pH. http://www.safewater.org/PDFS/resourcesknowthefacts/TDS_AND%20_pH.pdf
20. Mackereth, F.J.H., Heron, J. and Talling, J.F. (1978) Water Analysis, Scientific Publication No. 36. Fresh Water Biological Association, 120 p.
21. Dezuane, J. (1997) Handbook of Drinking Water Quality. 2nd Edition, John Wiley and Sons.
22. Drever, J.I. (1997) The Geochemistry of Natural Water, Surface and Groundwater Environments. 3rd Edition, Prentice Hall, USA, 436 p.
23. Wood, W.W. (1976) Guidelines for Collection and Field Analyses of Groundwater Samples for Selected Unstable Constituents. Techniques of Water-Resources Investigations of the United States.
24. Meddis, R. (1975) Statistical Handbook for Non-Statisticians. McGraw-Hill Book Company, UK, 161 p.

CHAPTER 19

Effect of Concentration on the Optical and Solid State Properties of ZnO Thin Films Deposited by Aqueous Chemical Growth (ACG) Method

S. L. Mammah[1,2], F. E. Opara[2], F. B. Sigalo[2], S. C. Ezugwu[3], F. I. Ezema[3]

[1]Department of Science Laboratory Technology, School of Applied Sciences, Rivers State Polytechnic, Bori, Nigeria

[2]Department of Physics, Faculty of Science, Rivers State University of Science and Technology, Port Harcourt, Nigeria

[3]Department of Physics and Astronomy, University of Nigeria, Nsukka, Nigeria

ABSTRACT

Thin films of Zinc Oxide (ZnO) having different concentrations were deposited using the Aqueous Chemical Growth (ACG) method. The films were characterized using Rutherford Back Scattering (RBS) spectroscopy for chemical composition and thickness, X-Ray Diffraction (XRD) for crystallographic structure, a UV-VIS spectrophotometer for the analysis of the optical and solid state properties which include spectral absorbance, transmittance, reflectance, refractive index, direct band gap, real and imaginary dielectric constants, absorption and extinction coefficients and a photomicroscope for photomicrographs. The average deposited film thickness was 100 nm. The results indicate that the values of all the optical and solid state properties investigated vary directly with concentration except transmittance which is the reverse. Thus, the optical and solid state properties of ZnO thin film deposited by the Acqueous Chemical Growth method can be tuned by deliberately controlling the concentration of the precursors for various optoelectronic applications including its application as absorber layer in solar cells.

Keywords: Component; Formatting; Style; Styling; Insert

1. INTRODUCTION

Zinc Oxide is an amphoteric oxide which exists in the form of white powder (zinc white) at room temperature [1]. Though zinc oxide exists naturally in the earth crust (zincite), it is the synthetic ZnO produced in laboratories that are usually used in industries [2].

Crystalline zinc oxide is thermochromic [3]. Apart from the ability of ZnO to react with acids, it also reacts with bases to produce soluble zincates [4-6]. Zinc oxide has high stability at room temperature and decomposes into zinc vapour and oxygen at about 1975°C [7]. Though zinc oxide can exist in either of hexagonal wurtzite, cubic zinc blende and cubic (rock salt) structure, the hexagonal wurtzite structure is the most common at room temperature. The different polymorphs of ZnO do not possess the property of symmetry inversion. This and other properties such as ionic bonding account for the strong piezoelectricity of both hexagonal and zinc blende ZnO [8,9].

The hexagonal wurtzite structure has the point group 6 mm (Hermann-Mauguin notation) or C_{6v} (schoenflies notation) and a space group of $P6_3me$ or C^4_{6v}. Its lattice constants are a = 3.25 Å and C = 5.2 Å [10].

The many applications to which ZnO can be put has made it to be a material for constant scientific study. The diverse applications of ZnO are derived from its unique properties which include non-toxicity, good electrical conductivity, high luminous transmittance, good substrate adherence, good optical behaviour and stability in plasma atmosphere [11,12].

ZnO has been tested and found to behave as a semiconductor having a wide bulk direct band gap of about 3.37 ev at room temperature [13] ZnO has been used for the fabrication of light emitting diodes, varistors, photo detectors, piezoelectric cantilever, gas sensors, buffer layer in solar cells and in photonic crystals [14-21].

As a result of the strategic importance of ZnO to humanity, various deposition methods have been used for its production in the form of thin films. Among these methods are the chemical bath deposition (CBD) [22-26].

Successive Ionic Layer Adsorption and Reaction (SILAR) method [27,28].

Spray pyrolysis method [29].

Electro deposition method [30,31].

The ACG method is a novel method for the deposition of ZnO thin films; hence the literature on it is sparse.

Indeed F. I. Ezema (2004) [26] reported that "Thin films of ZnO have been used but its preparation techniques have been restricted to sputtering, vacuum evaporation, chemical vapour deposition, spray prolysis, molecular beam epitaxy, sol gel and pulse laser".

Among the advantages which the ACG method have over the other methods are simplicity, low cost, reproducibility, availability of materials, environmental friendliness (non toxicity of residue), non requirement of surfactant, templates and complexing agents, low temperature requirements, ability to produce nanostructures, high purity (absence of surfactants), suitability for large scale production etc. [32,33].

This paper reports the effect of concentration of precursors on the optical and solid state properties of ZnO thin films deposited using the ACG method with $Zn(NO_3)_2 6H_2O$ as precursor material. The spectral analysis of the optical and solid state properties was carried out using a Unico UV-2102 PC spectrophotometer.

2. EXPERIMENTAL DETAILS

ZnO thin films were deposited on clean glass slides using Aqueous Chemical Growth method.

Equimolar concentrations of hydrated zinc nitrate ($Zn(NO_3)_2 \cdot 6H_2O$) and hexamine in 80 ml of water were used as precursors.

Hexamine was used to make the zinc nitrate alkaline. Three samples having different concentrations of the precursors were prepared as shown in Table 1 below.

Masses of samples were measured using an analytical microbalance. The masses of the components of each sample were put into a 100 ml pyrex bottle together with 80 ml of water. The contents were properly mixed using a magnetic stirrer.

The pyrex bottles were then tightly corked and care fully placed in an oven at a temperature of 90.00°C.

The average deposition time for the films was twelve hours.

The chemical reactions which resulted in the crystallization of ZnO are:

$C_6H_{12}N_4 + 6H_2O$ ® $6HCHO + 4NH_3$

$$NH_3 + H_2O \rightarrow NH_4^+ + OH^-$$

$2OH^- + Zn^{2+}$ ® $ZnO + H_2O$ [34]

3. RESULTS AND DISCUSSION

The elemental composition of sample C_2 was analyzed using Rutherford Backscattering (RBS).

The result is as shown in Figure 1. Judging from the film composition shown in the Table 2 below, we conclude that the elements contained in sample C_2 are Zn (0.050) and oxygen (0.950) while the glass substrate has the composition O (0.500), Si (0.120), Ca (0.100), Al (0.100) and Na (0.180). This is summarized in Table 2. The thickness of the film is given as 100 nm.

The X-Ray diffractogram of the ACG ZnO thin films were studied to determine its crystalline nature.

The thin films were scanned continuously between 0 and 70 at step size of 0.03 and at time per step of 0.15 s. The Figure 2 below shows the intensity of peaks versus diffraction angle 2θ for sample C_2 ZnO thin film using CuK_α radiation source having a wavelength of 1.54056 Å. The X-ray diffractogram reveal several peaks corresponding to directions of strong reflections. Some of the observed peaks are 30.00°, 32.50°, 34.42°, 45.56° and 54.87°, and 67.50°.

The 34.42° agree with the preferred orientation along the (002) plane reported by Ajuba et al. (2010) [35] and Shinde et al. (2005) [36] who used the CBD and SILAR methods respectively. Also, the 67.50° fairly agree with the (112) plane reported by Ajuba et al. (2010) [35,36].

The diffractogram reveal sharp diffraction peaks which is indicative of good crystallinity [37-41].

Table 1. Three samples of different concentrations of the precursor

Sample of mass	$Zn(NO_3)_2 \cdot 6H_2O$	Mass of Hexane	Conc. of $Zn(NO_3)_2 \cdot 6H_2O$	Conc. of Hexane	Volume of H_2O
C_3	0.60 g	0.030 g	0.025 M	0.025 M	80.00 ml
C_4	1.20 g	0.060 g	0.050 M	0.050 M	80.00 ml
C_2	2.40 g	0.112 g	0.100 M	0.100 M	80.00 ml

Table 2. Elemental composition of ZnO thin film and substrate from RBS analysis.

	Oxygen	Zinc	Silicon	Calcium	Aluminium	Sodium
ZnO thin film	0.950	0.050	-	-	-	-
Glass substrate	0.500	-	0.120	0.100	0.100	0.180

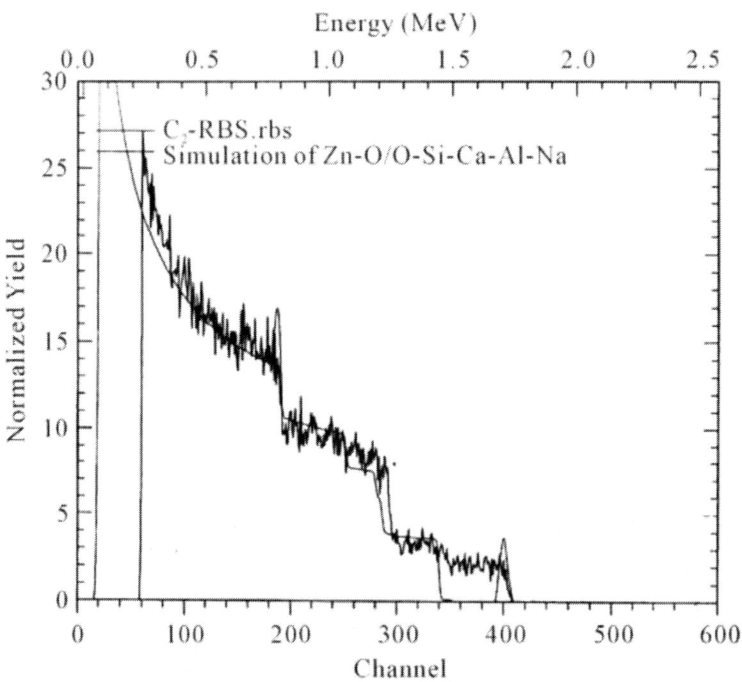

Figure 1. RBS analysis for ACG ZnO thin film (sample C_2).

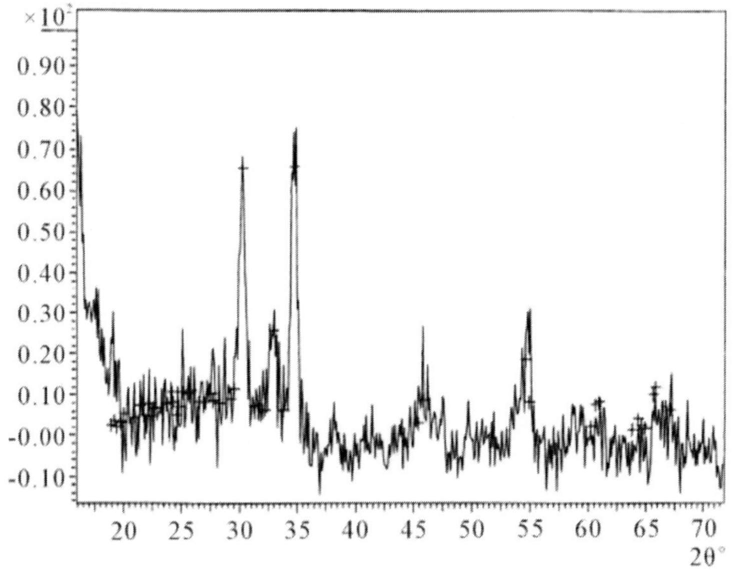

Figure 2. X-Ray diffractogram for ACG ZnO thin film (sample C_2).

The XRD pattern shows that film is crystallized in the wurtzite ZnO hexagonal P6(3) mc structure. (JCPDF CARD no. 36-1451) [42].

The mean size of the crystallite was estimated to be 12 nm using the Sherrer formula which is given as:

$$D = \frac{K\lambda}{\beta \cos \theta} \tag{1}$$

where $K = 0.9$

λ =1.541 Å is the diffraction peak angle (34.42°) and β is the Full Width at Half Maximum (FWHM) corresponding to the diffraction peak.

The spectral absorbance of films is as shown in Figure 3.

The 0.05 M concentration has an average absorbance value of 0.28 in the visible region while the 0.025 M concentration has an average absorbance value of 0.268 in the visible region.

However, the 0.05 M concentration has the lowest absorbance value of 0.21 in the infrared region.

Absorption peaks clearly occurred in film C_2 at 368 nm, 496 nm and 656 nm but were not so clearly defined in films C_3 and C_4. Ezema [26] reported absorption peaks at 3.68 nm, 449 nm and 566 nm using Chemical Bath Method. The difference in the absorption peaks may be due to deposition method. The spectral transmittance of the films is as shown in Figure 4. Transmittance decreases with concentration in the visible region. The 0.1 M, 0.05 M and 0.025 M concentrations has average transmittance of 45.5%, 53.2% and 54.60% respectively in the visible region.

However, the 0.05 M concentration has the highest transmittance of about 61.8% in the ultraviolet region. The transmittance of the films in the visible region is generally lower than the above 85% transmittance reported using Chemical Bath Deposition method [43].

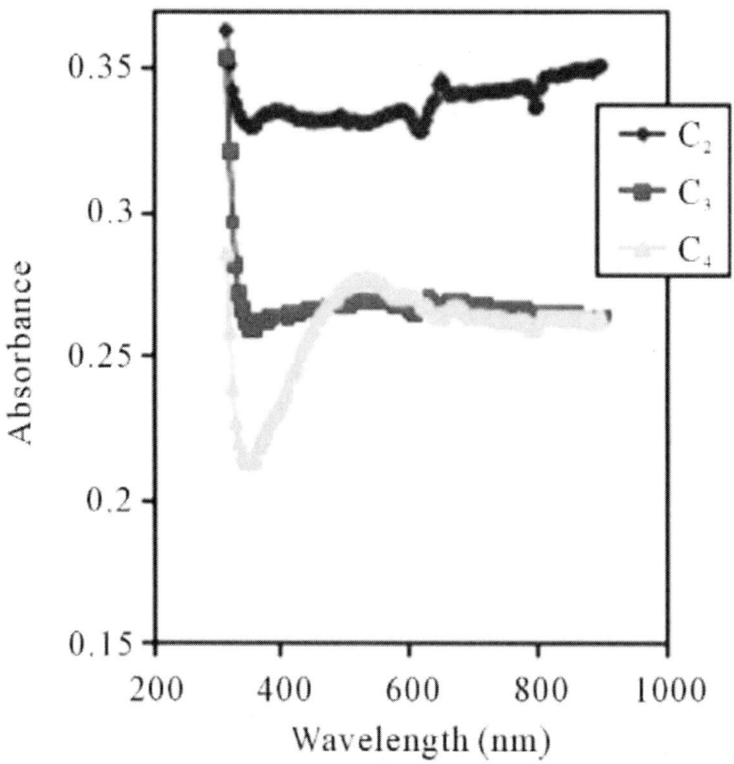

Figure 3. Absorbance vs. wavelength for ZnO thin film at different conc.

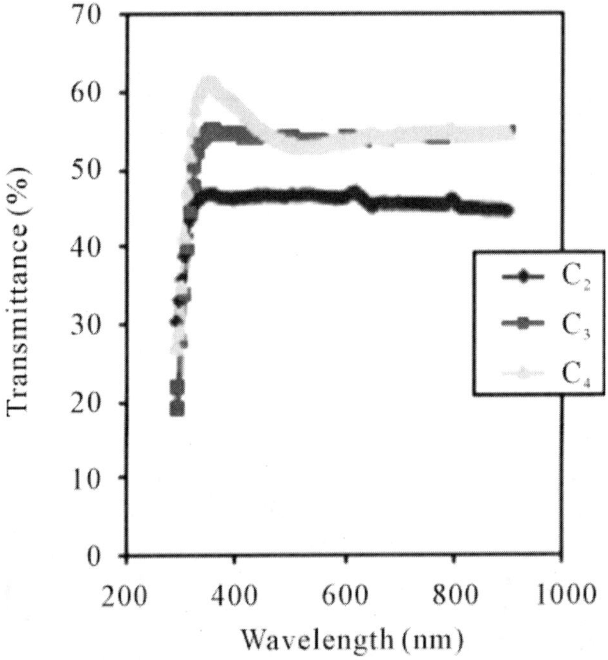

Figure 4. Transmittance vs. wavelength for ZnO thin films at different conc.

It is also much lower than the 90% 95% transmittance in the UV-VIS-NIR regions for ZnO thin films deposited using spray pyrolysis method [44,45] and Sol Gel method [46].

It has been reported that the transmittance of a film increases as its thickness decreases [26].

The average thickness of the films being reported on is 100 nm which is far smaller than the 0.203 μm and 0.069 μm thicknesses which gave average transmittances of 62% and 75% respectively that has been reported [26]. Thus, the transmittance of films deposited by ACG method are relatively lower compared to the ones deposited by other methods even though the thicknesses are smaller. The spectral reflectance of the films being reported on is as shown in Figure 5. Reflectance increases with concentration in the visible region.

The 0.1 M, 0.050 M and 0.025 M concentrations reflect an average of 20.25%, 19.38% and 18.88% of visible electromagnetic waves. These values are higher than the average of 12% to 17% reflectance of films deposited using Chemical Bath Method as reported by Ezema [26].

Thus, films deposited by ACG method are better reflectors than those deposited by Chemical Bath Method. Absorption coefficient increases with concentration in the visible region as shown in Figure 6.

The average absorption coefficients for the 0.025 M, 0.05 M and 0.1 M concentrations are 0.62, 0.65 and 0.77 respectively. However, in the UV region, 0.05 M has the lowest absorption coefficient of about 0.49.

Refractive index is fairly constant in the visible region as shown in Figure 7. The 0.1 M, 0.05 M and 0.025 M concentrations have average refractive indices of 2.3, 2.225 and 2.2 respectively. However, the 0.05 M concentration has the lowest refractive index of about 2.05 in the UV region.

It has been reported that the refractive index of ZnO films doped with Li deposited by spray pyrolysis lie between 1.60 and 2.20 at a wavelength of 500 nm [45], while Ezema [26] gave the average refractive index of the ZnO films deposited by Chemical Bath method as being between 1.64 and 1.98 and also reported observed peak values of 2.28 at 368 nm and 1.72 at 569 nm. In our own case, the refractive indices for the different concentrations were generally uniform in the visible region.

Extinction coefficient decreases with concentration and also decreases with increasing photon energy (wavelength) in the visible region as shown in Figure 8. The value of extinction coefficient for 0.1 M concentration ranges from 37.85×10^{-3} for 2 eV to 27.14×10^{-3} for 3 eV in the visible region. Similarly, the values of extinction coefficient for 0.05 M concentration ranges from 35×10^{-3} for 2 eV to 19.64×10^{-3} for 3 eV and that of the 0.025 M concentration ranges from 30×10^{-3} to about 21.43×10^{-3} at 3 eV in the visible region.

However, the 0.05 M concentration has the lowest extinction coefficient of about 13.93×10^{-3} in the ultraviolet region. In general all the concentrations have the lowest value of extinction coefficient in the higher energy region (between 3 eV and 4 eV) and the maximum value in the low energy region (between 1 eV and 2 eV). Low and high values of extinction coefficient at high and low energy regions respectively have been reported [26].

The band gap increases with concentration as shown in Figure 9. The average band gap for the 0.025 M, 0.05 M and 0.1 M concentrations are 1.16 eV, 1.22 eV and 1.4 eV respectively.

These values are lower than the direct band gaps of ZnO films deposited by Chemical Bath method [26] which lie between 1.60 eV and 1.80 eV.

Our values are also far lower than 3.29 eV band gap obtained for aluminium doped ZnO deposited using radio frequency (r.f) megnetron sputtering [18], and intrinsic band gap of 3.20 eV for ZnO films deposited by spray pyrolysis

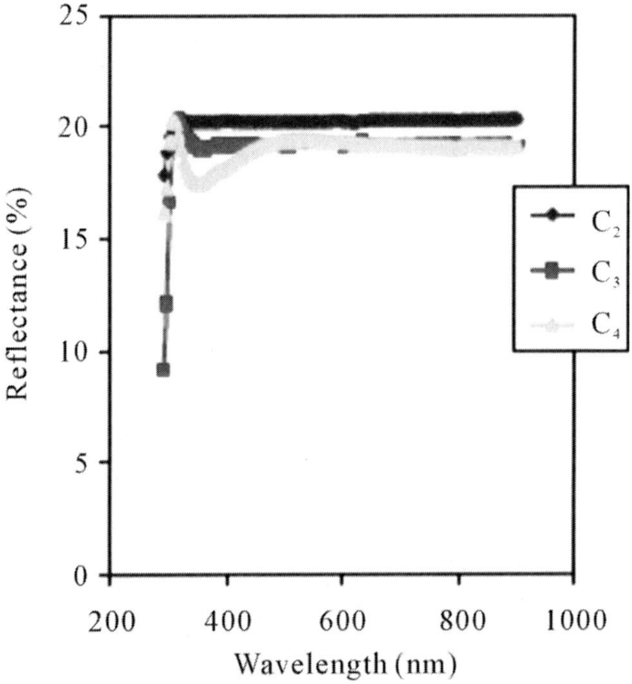

Figure 5. Reflectance vs. wavelength for ZnO thin films at different conc.

[29]. The low values of the direct band gap obtained indicate that the ZnO thin film materials prepared by ACG method are suitable for use as absorber layers in solar cells. The low values of the band gaps obtained in this work may be due to preparation conditions Ezema et al. [26].

Real dielectric constant increases with concentration in the visible region as shown in Figure 10. The 0.025 M, 0.05 M and 0.1 M concentrations have average real dielectric constants of 4.76, 4.85 and 5.15 respectively in the visible region of the electromagnetic spectrum.

However, the 0.05 M concentration has the lowest real dielectric constant of about 4.09 in the ultra-violet region. Figure 11 is the Imaginary dielectric constant that increases with concentration in the visible range.

The 0.1 M, 0.05 M and 0.025 M concentrations have average values of imaginary dielectric constant which

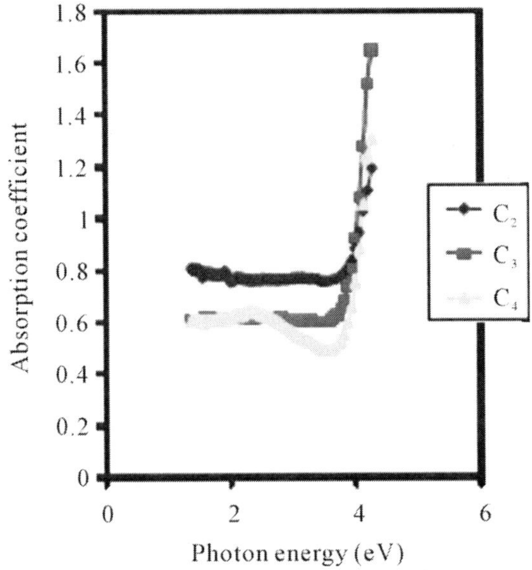

Figure 6. Absorption coefficient vs. photon energy for ZnO at different conc.

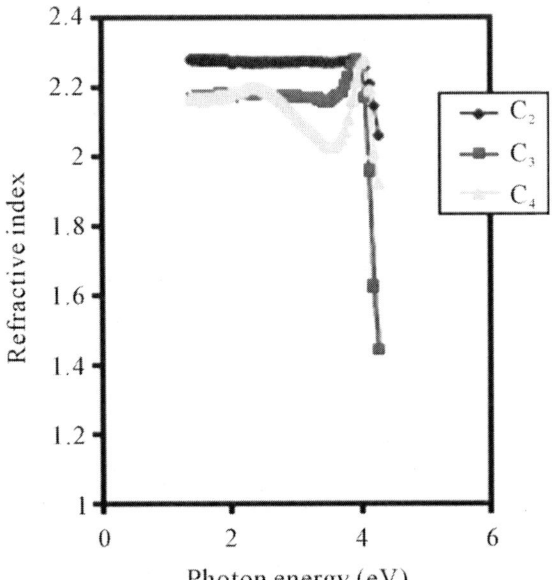

Figure 7. Refractive index vs. photon energy for ZnO thin films at different conc.

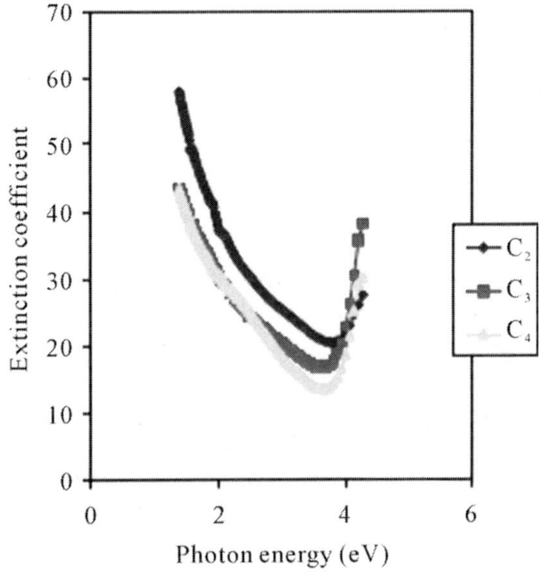

Figure 8. Extinction coefficient vs. photon energy for ZnO thin films at different conc.

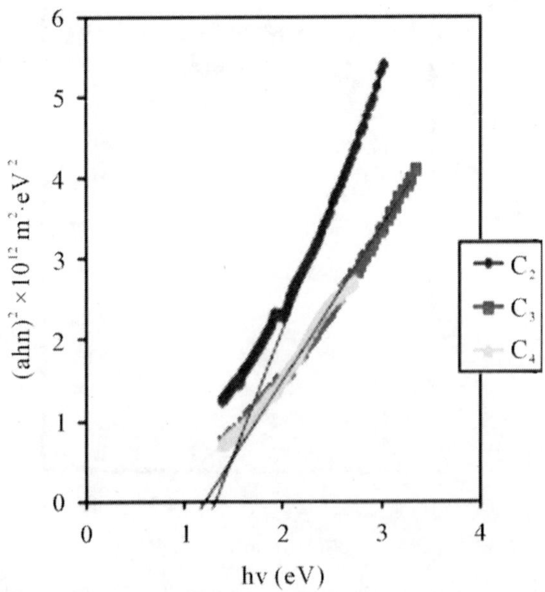

Figure 9. Direct band gap plot for ZnO thin films at different conc.

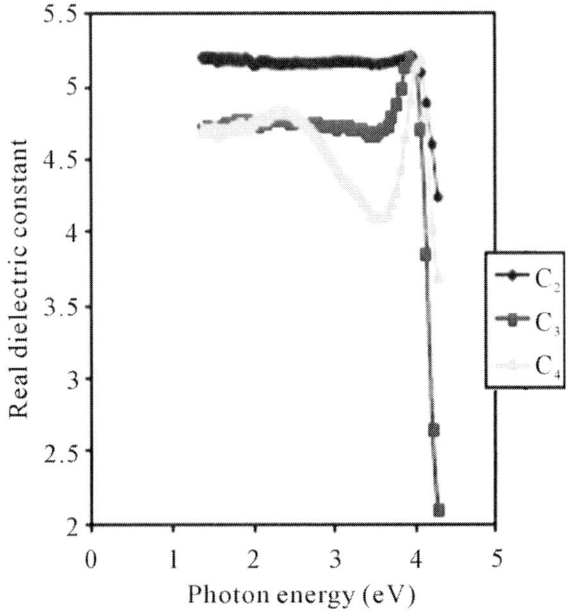

Figure 10. Real dielectric constant vs. photon energy for ZnO thin films at different conc.

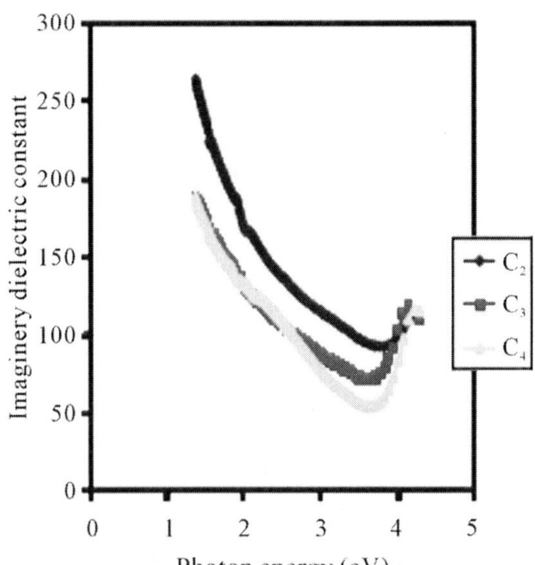

Figure 11. Imaginery dielectric constant vs. photon energy for ZnO thin films at different conc.

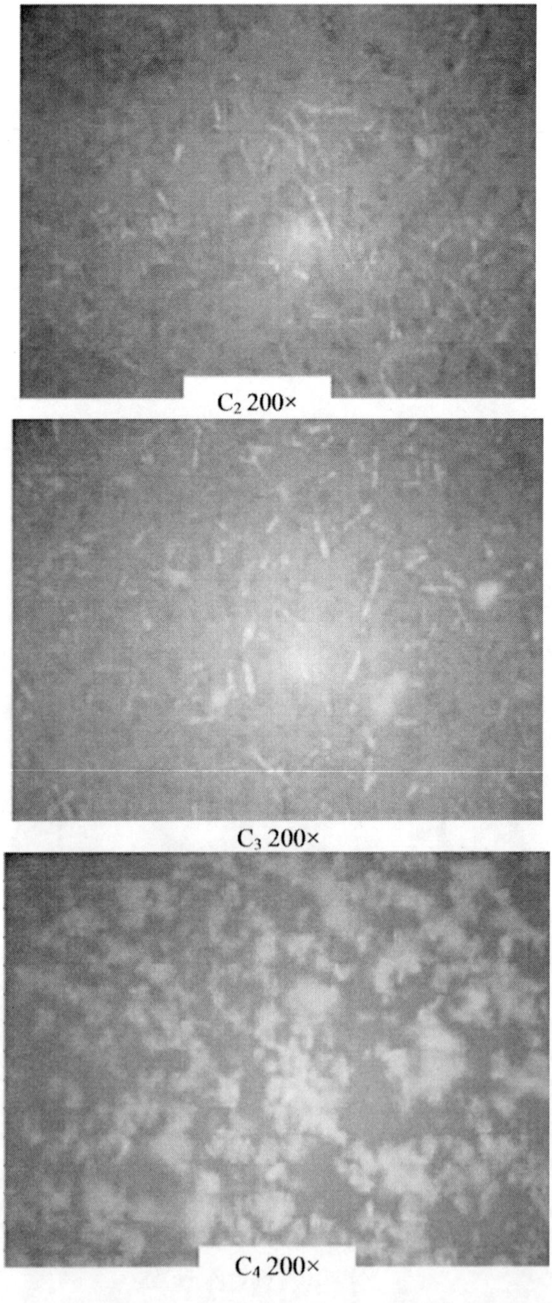

Figure 12. Photomicrographs of ACG ZnO thin films.

ranges from 180.3 × 10^{-3} for 2 eV to 118.18 × 10^{-3} for 3 eV, 143.93 × 10^{-3} for 2 eV to 78.785 × 10^{-3} for 3 eV and 143.93 × 10^{-3} for 2 eV to 89.39 × 10^{-3} for 3 eV respectively. This shows that the imaginary dielectric constant decreases with increasing photon energy (decreasing wavelength) for all concentrations in the visible region.

However, the 0.05 M concentration has the lowest imaginary dielectric constant of about 56.06 in the ultra violet region. All concentrations have their lowest imaginary dielectric constant values in the ultra-violet region.

The photomicrographs as shown in Figure 12 indicate uniform deposition of the ZnO thin films on the glass substrate. While C_2 and C_3 show stretches of rod C_4 shows clusters of thin films on the substrates.

4. CONCLUSIONS

Crystals of ZnO have been successfully grown on glass slides in the form of thin films from the aqueous solution of hexahydrated zinc nitrate and hexamine after the fashion of Lionel Vassieres [47]. The concentration of the precursors was found to vary directly with absorbance, reflectance, absorption coefficient, extinction coefficient, refractive index, direct band gap, real dielectric constant and imaginary dielectric constant. The transmittance of the thin films was found to vary inversely with the concentration of the precursors.

S. L. Mammah, F. E. Opara, F. B. Sigalo, S. C. Ezugwu, F. I. Ezema The results indicate the suitability of ZnO thin films prepared by the Aqueous Chemical Growth method for various optoelectronic applications such as absorber layer in solar cells.

REFERENCES

1. T. Kiyoshi, Y. Akihiko and S. Adarsh, "Wide Bandgap Semiconductors: Fundamental Properties and Modern Photonic and Electronic Devices," Springer, Heidelberg, 2007, p. 257.
2. C. Klingshirn, "ZnO: Material, Physics and Applications," Physical Chemistry, Vol. 8, No. 6, 2007, pp. 782-803. HUdoi:10.1002/cphc.200700002U
3. E. Wiberg and A. F. Holleman, "Inorganic Chemistry," Elsevier, Amsterdam, 2001.
4. J. W. Nicholson, Journal of Materials Science, Vol. 33, No. 225, 1998.
5. J. L. Ferracane, "Material in Dentistry: Principles and Application," Lippincott Williams & Wilkins, Philadelphia, 2001.
6. C. K. Park, M. R. Silsbee and D. M. Roy, "Setting Reaction and Resultant Structure of Zinc Phosphate Cement in Various Orthophosphoric Acid Cement-Forming Liquids," Cement and Concrete Research, Vol. 28 No. 1, 1998, pp. 141-150. doi:10. 1016/S0008-8846(97)00223-8UH

7. N. N. Greenwood and A. Earnshaw, "Chemistry of the Elements," Butterworth-Heinemann, Oxford, 1997.
8. U. Özgür, Y. T. Alivov, C. Liu, A. Teke, M. A. Reshchikov, S. Doğan, V. Avrutin, S. J. Cho, et al., "A Comprehensive Review of ZnO Materials and Devices," Journal of Applied Physics, Vol. 98, No. 4, 2005, Article ID: 041401.
9. S. Baruah and J. Dutta, "Hydrothermal Growth of ZnO Nanostructures," Science and Technology of Advanced Materials, Vol. 10, No. 1, 2009, Article ID: 013001.
10. U. Rossler, "Landolt-Bornstein, New Series, Group III," Springer, Heidelberg, 1999.
11. T. Mahalingam, V. S. John, M. Raja, Y. K. So and P. J. Sabastian, "Electrodeposition and Characterization of Transparent ZnO Thin Films," Solar Energy Materials and Solar Cells, Vol. 88, No. 2, 2005, pp. 227-235.
12. Y. R. Shinde, T. P. Gujar, C. D. Lokhande, R. S. Nane and S. H. Han, Materials Chemistry and Physics, Vol. 326, 2006.
13. A E. Ajuba, S. C. Ezugwu, P. U. Asogwu and F. I. Ezema, Chalcogenide Letter, Vol. 10, 2010, pp. 573-579.
14. T. Soki, Y. Hatanaka and D. C. Look, "ZnO Diode Fabricated by Excimer-Laser Doping," Applied Physics Letters, Vol. 76, No. 22, 2000, pp. 3257-3259.
15. Y. Lin. C. R. Gorla, S. Linng, N. Emanetoglu, Y. Tor, H. Shen and M. Wraback, 'Ultraviolet Detectors Based on Epitaxial ZnO Films Grown by MOCVD," Journal of Electronic Materials, Vol. 29, No. 1, 2000, pp. 69-74.
16. V. R. Shinde, T. P. Gujar and C. D. Lokhande, "LPG Sensing Properties of ZnO Films Prepared by Spray Pyrolysis Method: Effect of Molarity of Precursor Solution," Sensors and Actuators H, Vol. 120, No. 2, 2007, pp. 551-559.
17. A Ennaoui, S. Siebentrith, M. Ch. Lux-Steiner, W. Riedl and F. Karg, "High-Efficiency Cd-Free CIGSS Thin-Film Solar Cells with Solution Grown Zinc Compound Buffer Layers," Solar Energy Materials and Solar Cells, Vol. 67, No. 1-4, 2002, pp. 31-40.
18. Y. Chen, D. Bagnall and T. Yao, "ZnO as a Novel Photonic Material for the UV Region," Materials Science and Engineering B, Vol. 75, No. 2-3, 2000, pp. 190-198.
19. S. Liang, H. Sheng, Y. Liu, Z. Hio, Y. Lu and H. Shen, Journal of Crystal Growth, Vol. 225, No. 110, 2001.
20. M. H. Koch, P. Y. Timbrell and R. N. Lamb, "The Influence of Film Crystallinity on the Coupling Efficiency of ZnO Optical Modulator Waveguides," Semiconductor Science and Technology, Vol. 10, No. 11, 1995, p. 1523.
21. C. R. Gorla, N. W. Emanetoglu, S. Liang, W. E. Mayo, K. Lu, M. Wraback and H. Shen, "Structural, Optical, and Surface Acoustic Wave Properties of Epitaxial ZnO Films Grown on (012) Sapphire by Metalorganic Chemical Vapor Deposition," Journal of Applied Physics, Vol. 85, No. 5, 1999, pp. 2595-2603.
22. V. R. Shinde, C. D. Lokhande, R. S. Mane and S. H. Han, "Hydrophobic and Textured ZnO Films Deposited by Chemical Bath Deposition: Annealing Effect," Applied Surface Science, Vol. 245, No. 1-4, 2005, pp. 407-413.
23. A Ennaoui, M. Weber, R. Scheer and H. J. Lewerenz, "Chemical-Bath ZnO Buffer Layer for $CuInS_2$ Thin-Film Solar Cells," Solar Energy Materials and Solar Cells, Vol. 54, No. 1-4, 1998, pp. 277-286.
24. D. S. Boyle, K. Governder and P. O'Brien, Chemical Communications, Vol. 1, No. 80, 2002, p. 23.
25. M. Ortega-Lope, A. Avila-Gaecia, M. L. Albor-Aguitera and V. M. Sankez Resendiz, "Improved Efficiency of the Chemical Bath Deposition Method during Growth of ZnO Thin Films," Materials Research Bulletin, Vol. 38, No. 4, 2003, pp. 1241-1248.
26. F. I. Ezema, Journal of Research (Science), Vol. 15, No. 4, 2004, pp. 343-350.
27. V. R. Shinde, T. P. Gujar and C. D. Lokhande, "Studies on Growth of ZnO Thin Films by A Novel Chemical Method," Solar Energy Materials and Solar Cells, Vol. 91, No. 12, 1961, pp. 1055-1061.
28. A Jimenez Gonzalez and R. Suarez-Para, "Effect of Heat Treatment on the Properties of ZnO Thin Films Prepared by Successive Ion Layer Adsorption and Reaction (SILAR)," Journal of Crystal Growth, Vol. 167, No. 3-4, 1996, pp. 649-655.

29. M. L. de la Olvera, A. Maldonado, R. Asomoza and M. Melendez-Lira, "Effect of the Substrate Temperature and Acidity of the Spray Solution on the Physical Properties of F-Doped ZnO Thin Films Deposited by Chemical Spray," Solar Energy Materials and Solar Cells, Vol. 71, No. 1, 2002, pp. 61-71.
30. P. M. Izaki and T. Omi, Journal of The Electrochemical Society, Vol. 144, 1997, pp. 1949-1952.
31. D. Gal, G. Hodes, D. Lincot and H. W. Sechock, "Electrochemical Deposition of Zinc Oxide Films from NonAqueous Solution: A New Buffer/Window Process for Thin Film Solar Cells," Thin Solid Films, Vol. 361-362, 2000, pp. 79-83.
32. X. Hu, Y. Masuda, T. Olyi and K. Kato, "Fabrication of ZnO Nanowhiskers Array Film by Forced-Hydrolysis-Initiated-Nucleation Technique Using Various Templates," Thin Solid Films, Vol. 518, No. 2, 2009, pp. 621-624.
33. X. Zhang, L. Wang and G. Zhow, "Synthesis of WellAligned ZnO Nanowires without Catalysts," Reviews on Advanced Materials Science, Vol. 10, 2005, pp. 69-72.
34. S.-F. Lee, L.-Y. Lee and Y.-P. Change, Journal of Science and Engineering Technology, Vol. 5, No. 3, 2009, pp. 13-20.
35. A E. Ajuba, S. C. Ezugwu, B. A. Ezekoye, F. I. Ezema and P. U. Asogwa, Journal of Optoelectronics and Biomedical Materials, Vol. 2, No. 2, 2009, pp. 73-78.
36. V. R. Shinde, C. D. Lokhande, R. S. Mane and S. H. Han, Applied Surface Science, Vol. 245, No. 407, 2005.
37. P. Li, H. Liu, F. X. Xu and Y. Wei, "Controllable Growth of ZnO Nanowhiskers by a Simple Solution Route," Materials Chemistry and Physics, Vol. 112, No. 2, 2008, pp. 393-397.
38. H. Zhai, W. Wu, F. Lu and H. S. Wang, "Effects of Ammonia and Cetyltrimethylammonium Bromide (CTAB) on Morphologies of ZnO Nanoand Micromaterials under Solvothermal Process," Material Chemistry and Physics, Vol. 112, No. 3, 2008, pp. 1024-1028.
39. D. Yiamsawas, K. B. Savanitchakul and W. K. W. Supamonkon, Journal of Microscopy Society of Thailand, Vol. 13, No. 75, 2009.
40. S. Y. Chu and T. M. Yan, "Characteristics of Sol-Gel Synthesis of ZnO-Based Powders," Journal of Materials Science Letters, Vol. 19, No. 4, 2000, pp. 349-352.
41. D. Geeth and T. Tilagarathi, Digest Journal of Nanomaterials and Biostructures, Vol. 5, No. 1, 2010, p. 297301.
42. C. Gumus, O. M. Ozkendir, H. Kavak and Y. Ufuktepe, Journal of Optoelectronics and Advanced Materials, Vol. 8, No. 1, 2006, pp. 299-303.
43. C. Cruz-Vázquez, F. Rocha-Alonzo, S. E. Burruel-Ibarra, M. Inoue and R. Bernal, "Fabrication and Characterization of Sulfur Doped Zinc Oxide Thin Films," Superficies y Vacío, Vol. 13, 2001, pp. 89-91.
44. A Sachez-Juarez, A. Tiburcio-Silver and A. Ortiz, Solar energy Materials and Solar Cells, Vol. 52, No. 3-4, 1998, pp. 301-311. HUdoi:10.1016/S0927-0248(97)00246-8U
45. P. Pushparajah, A. K. Arof and S. Radhakrishna, "Physical Properties of Spray Pyrolysed Pure and Doped ZnO Thin Films," Journal of Physics D: Applied Physics, Vol. 27, No. 7, 1994, pp. 1518-1521. HUdoi:10.1088/0022-3727/27/7/027U
46. E. Jimenez Gonzalex and J. A. Soto Urueta, "Optical Transmittance and Photoconductivity Studies on ZnO:Al Thin Films Prepared by the Sol-Gel Technique," Solar Energy Materials and Solar Cells, Vol. 52, No. 3-4, 1998, pp. 345-353. HUdoi:10.1016/S0927-0248(97)00243-2U
47. L. Vayssieres, "On the Design of Advanced Metal Oxide Nanomaterials," International Journal of Nanotechnology, Vol. 1, No. 1-2, 2004, pp. 1-41.

CHAPTER 20

Molecular Dynamics Simulations of Bio-Active Phosphate-Based Glass Surfaces

Sergio E. Ruiz Hernandez [a,*], Richard I. Ainsworth [b], NoraH. de Leeuw[a]

[a]*School of Chemistry, Cardiff University, Main Building, Park Place, Cardiff CF10 3AT, UK*

[b]*Department of Chemistry and Biochemistry, University of California, San Diego 9500, La Jolla, California 920093-0359, USA*

ABSTRACT

Classical molecular dynamics (MD) simulations were used to study the structural changes in the surfaces of biocompatible phosphate glasses with compositions $(P_2O_5)_{0.45}(CaO)_x(Na_2O)_{0.55-x}$ (x = 30, 35, 40) to evaluate their effect on the solubility of the material. Direct comparison of the data for the three compositions highlighted the critical role that an enhancement in Na^+ concentration plays in the hydrolysis of the material, which is responsible for the release of network components into solution. The calculations also confirm that the most soluble material is $(P_2O_5)_{0.45}(CaO)_{0.30}(Na_2O)_{0.25}$, has the lowest calcium coordination number, thereby causing fewer cross links to phosphate chains.

Keywords: Molecular dynamics Dissolution Phosphate Bioactive glass

1. INTRODUCTION

Phosphate-based glasses (PBGs) in the $CaO–Na_2O–P_2O_5$ system have a number of biomedical applications [1,2]. Their ionic compositions mimic the mineral constituent of hard tissue, and they can therefore act as a template to repair soft tissue [3]. Based on their controlled ion release capabilities, they can also temporarily replace hard tissues whereas metal-doped PBGs are used in veterinary medicine to treat trace element deficiencies [1,4,5]. PBGs can also be employed to deliver drugs, from simple antibacterial ions such as Ag or Cu [1,6,7] or radioisotopes [8] to potentially more complex organic molecules for chemotherapy applications [9].

The success in medicinal utilization of this unique group of materials is closely related to their ability to dissolve completely in vivo in aqueous media, where the solubility rate can be tuned by altering the Ca^{2+}/Na^{+} ratio. It is known that the solubility of these ternary phosphate-based bioglasses decreases with an increase in the Ca^{2+}/Na^{+} ratio in the composition [10,11]. Much of our recent computational work has been dedicated to improving our understanding of glass dissolution processes and the structural and compositional features that affect them [12–14]. For example, previous theoretical studies of the bulk glass have shown that the mentioned change in solubility is caused by Ca^{2+} binding more phosphate chains than Na^{+}, thereby satisfying its bonding preference for non-bridging oxygen neighbours at the expense of sodium [12]. Also Ca^{2+} binds to more PO_4^{3-} tetrahedra and has a lower concentration of intra-tetrahedral phosphate bonds than Na^{+}, where both features help to enhance the glass's durability as more Ca^{2+} is included [12]. The same authors proposed that the clustering of modifier cations, which affect the solubility of these bioactive glasses by strengthening the glass network, should also be considered [14].

Thus far, theoretical studies on PBGs have been limited to the bulk material, but further detailed investigations must directly focus on the interface between the glass and the physiological environment, considering that it is at the surface that the bioactive processes occur. Moreover, among the wide range of factors affecting the solubility of the ternary phosphate-based bioglasses are the ion exchange between the surface and environment and variations in the glass surface structure. Classical molecular dynamics simulations (MD) with an accurate force field [13–19] have proven to be a highly suitable computational tool to obtain atomistic models of melt-derived glasses. They can provide a precise representation of the bonding and coordination environments found in the PBG surfaces, thus enabling the generation of larger statistical samples [15] than would be available from ab initio methods.

In this work we have therefore used MD to explore structural changes in the glass surfaces of compositions $(P_2O_5)_{0.45}(CaO)_x(Na_2O)_{0.55-x}$ (x = 30, 35, 40) and their influence on the material's bioactivity and solubility.

2. METHODS

2.1. Structural models

The composition of the glasses and the notation used in this work are shown in Table 1 and the protocol for obtaining the models is described in details in Ainsworth et al. [1]. Starting from the volume-optimized bulk supercell, two slab geometries with 3D periodicity, but with the top and bottom faces exposed to vacuum (Fig. 1), were obtained by elongating the c vector of the simulation cell by ~ 30 Å, producing four surface samples by exposing different sections of the bulk glass.

Table 1. Composition of the glass studied.

Notation	**Composition**	**ρ (g/cm³)**	**Number of atoms**	**Box size**
N25	$(P_2O_5)_{0.45}(CaO)_{0.30}(Na_2O)_{0.25}$	2.56	3001	34.65
N20	$(P_2O_5)_{0.45}(CaO)_{0.35}(Na_2O)_{0.20}$	2.57	2998	34.69
N15	$(P_2O_5)_{0.45}(CaO)_{0.40}(Na_2O)_{0.15}$	2.59	3001	34.71

2.2. Evaluation of interatomic potential energies

Our simulations are based on the Born model of solids [21], where ions are assumed to interact via long-range Coulomb forces and additional short-range forces given by simple parametric functions which represent electron-mediated interactions, e.g. Pauli repulsions and Van der Waals dispersion attractions between neighbouring electron charge clouds. The electronic polarizability of the ions is also accounted for via the model by Dick and Overhauser [22], in which each polarizable ion is represented by a core and a massless shell, connected by a spring. The spring constant and the distribution of the ion charge between the core and shell determine the polarizability of the ion. In our case, only the oxygen anion was considered polarizable, while cations were modelled as rigid ions. The potential model used here for the PBGs, shown in Table 2, is the formal-charge, polarizable force field to describe inter-atomic forces, which has previously been shown to accurately reproduce the structures and properties of glasses in the system P_2O_5–CaO–Na_2O, compared to experiment and ab initio methods. [1,10,12].

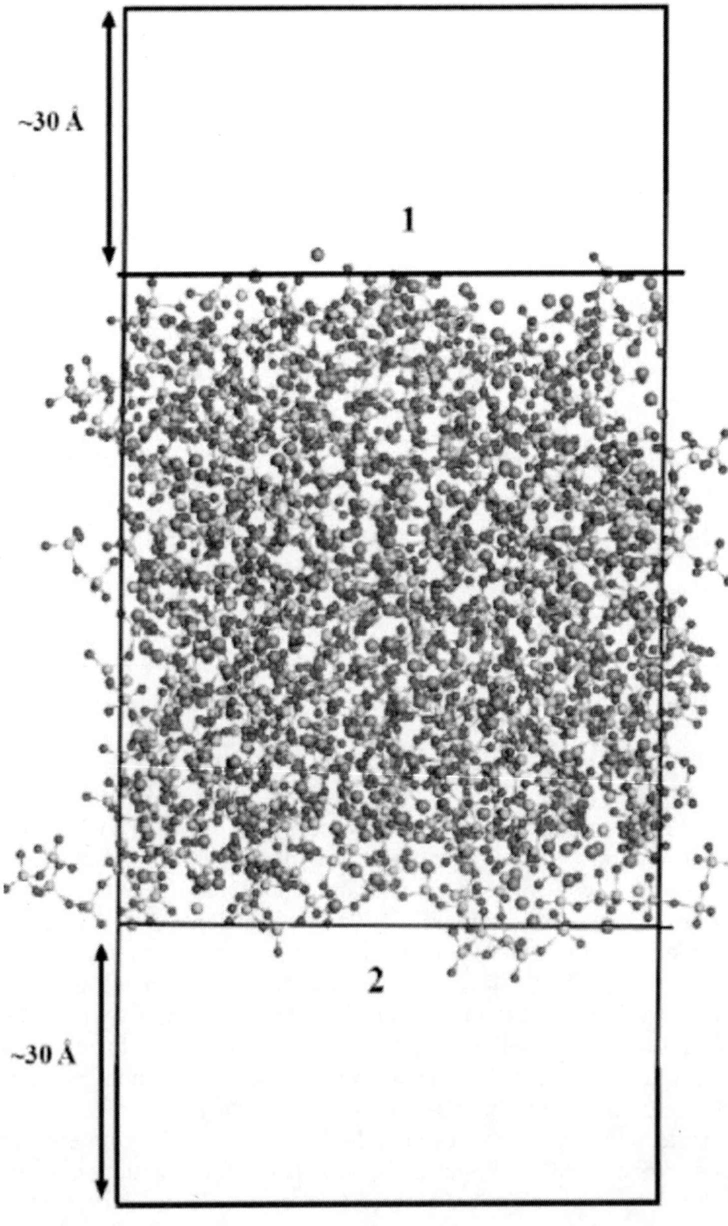

Figure. 1. Representation of the simulation cell exposing different surfaces, here labelled 1 and 2. P (pink), O (red), Ca (green) and Na (violet).

Table 2. Potential parameters used in this work.

Species	**Charge**		**Spring constant (eV Å^{-2})**
	Core	**Shell**	
P	+ 5.0		
Ca	+ 2.0		
Na	+ 1.0		
O	+ 0.85	– 2.85	74.92
	Buckingham potentialA exp (– r/ρ) – C/r^6		
	A (eV)	ρ (Å)	C (eV Å^6)
P__Os	1020.0	0.34322	0.03
Os__Os	22,764.3	0.149	27.88
Ca__Os	56,465.3453	0.193931	0.0
Na__Os	2152.3566	0.309227	0.09944
	Three-body potential		
	$(1/2)k_{3b}(\theta - \theta_0)^2$		
	k_{3b} (eV rad^{-2})		Θ_0
Os__P__Os	3.3588		109.47
P__Os__P	7.6346		141.1793

2.3. Molecular dynamics simulations

The MD runs were performed in the constant volume and temperature (NVT) canonical ensemble at 300 K using the DL_POLY code (version 2.20) [23], where all atoms were allowed to move during the simulations. An Evans thermostat [24] was used and the timestep between successive integrations of the Newtonian equation of motion was set to 0.1 fs. The motion of oxygen shells was treated by assigning them a small mass (0.2 a.u.) to simulate the fast adaptation of the electronic polarization to the change in ionic positions [25,26]. The simulation time for each production run was 1 ns, and 50 ps of equilibration.

3. RESULTS

3.1. Species distribution in the surfaces

We have considered all the atoms within 10 Å from the interface to be surface species, in agreement with previous definitions [27]. Significant reordering of the species at the interface, occurs in all three glass compositions. Figs. 2 and 3 show example topographies observed during the simulation, including some orthophosphate ions (Fig. 2b), under-coordinated phosphate units (P_{3C}) (Fig. 3a), and the formation of multiple-ring structures, some of them formed by three (3 M) or four (4 M) phosphate units (Figs. 2a, c, 3b). Small 3 M rings were not reported in previous theoretical studies of the bulk material [1,10], although they have been found before in experiment [28] and have been reported to be a very significant factor in the nucleation of apatite on silicate-based bioglass [29–31]. Many of the ring structures observed here are directly exposed at the interface and disconnected from the main phosphate network.

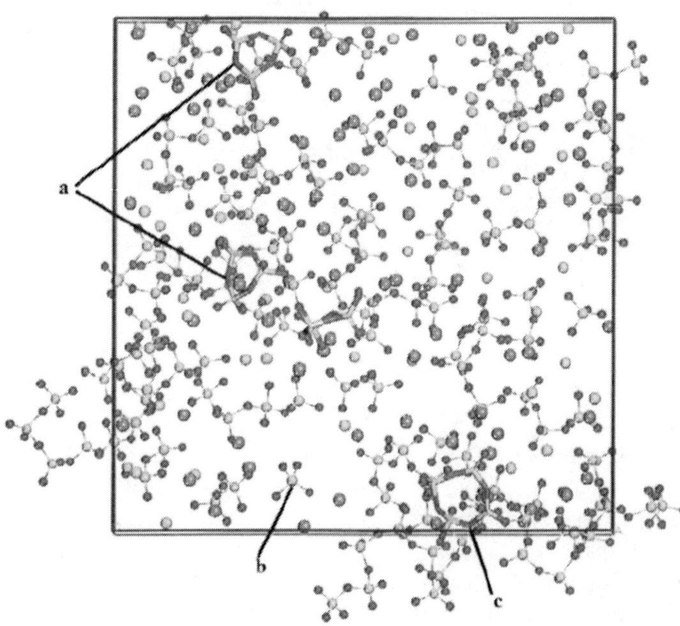

Figure. 2. Top view of a sample surface in N25 showing some features observed after relaxation. Only the top 10 Å of the slab are displayed. (a) 3 M ring (b) Q^0 phosphates (c) 4 M ring. Color code: P (pink), O (red), Ca (green) and Na (violet).

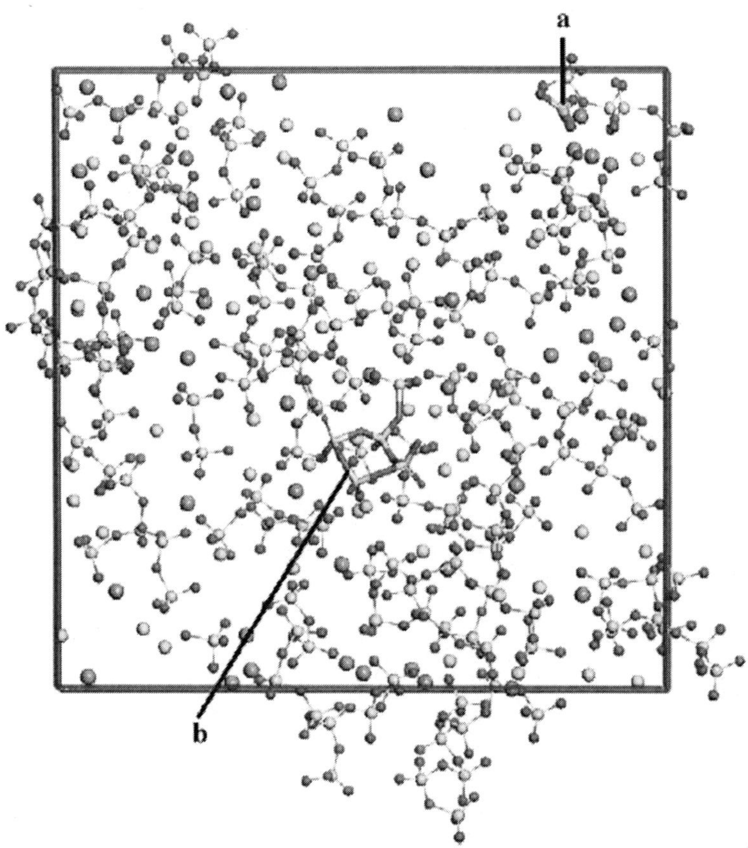

Figure. 3. Top view of a sample surface in N15 showing some features observed after relaxation. Only the top 10 Å of the slab are displayed. (a) Under-coordinated phosphorus (P_{3c}). (b) 4 M rings. Color code: P (pink), O (red), Ca (green) and Na (violet).

As occurs in silicate-based bioglasses [18], some of the highlighted structures as well as the modifier concentrations affect the glass dissolution process. The distribution of the species near the interface can be represented using the z-profiles in Fig. 4, for the three glass concentrations studied in this work, where we have defined, rather arbitrarily, the interface as the plane passing through the centre of the topmost oxygen atoms. The z-profiles represent the portion of atom x (where x = Ca, Na, O, P), that can be found at height z of the surface. There are clear compositional changes in the interface region compared to the bulk-like region ($z = 0$).

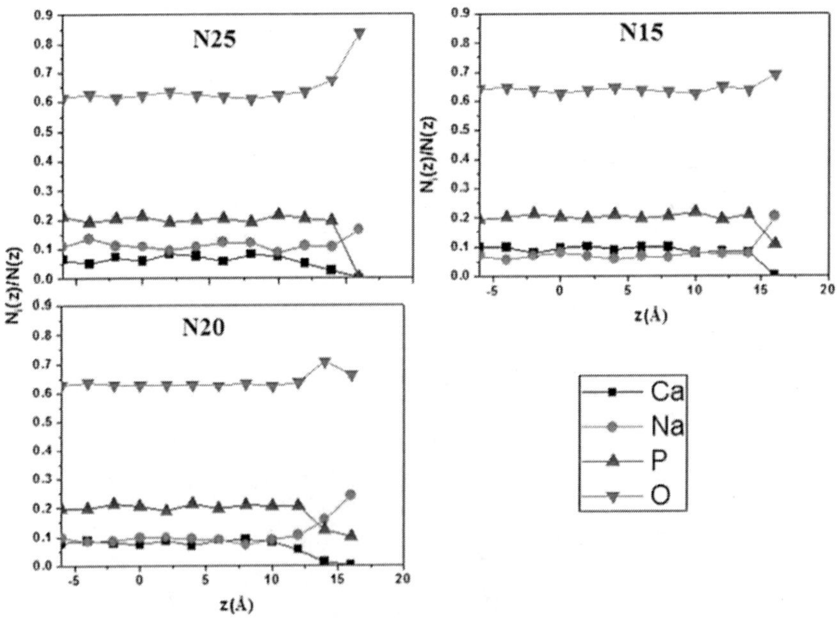

Figure. 4. z-profiles of the fraction of Ca, Na, P and O atoms in the surface slab models after the simulation. Each slab was divided in 2 Å thick slices and the number of atoms in the slice averaged over the MD and further averaged over the four surfaces. z = 0 marks the bulk-like region of the slab.

All surfaces in N20 and N15 experience an increase in sodium concentration at z > 10 Å, where in some cases the Na^+ are found as the topmost species in the surface, similar to what has been reported for the silicate-based glasses [18]. In the case of N15 at z > 10 Å, the fraction of sodium equals the calcium value in surfaces with less sodium content. For N25 surfaces, the increase is only significant at the interface z > 13 Å. In the PBGs, however, there is no drop in the silicon counterpart, i.e. phosphorous content, but instead the calcium content is less in N25 and N20. This effect is not surprising, if we consider that in these two systems there is more Na^+than Ca^{2+} in the bulk glass. Table 3 confirms that Na^+ is occupying more surface sites than Ca^{2+} in N25 and N20, whereas in N15 where the bulk composition of $Ca^{2+}$is bigger than Na^+, the latter only becomes more prominent close to the interface.

Table 3. Concentration (number density nm^{-3}) of the different species in the top two layers of the slab.

Species	N25	N20	N15
Ca^{2+}	0.83	0.78	2.21
Na^{+}	2.29	1.84	2.15
NBO	9.43	7.98	13.05

The creation of the surfaces also causes the exposure of many non-bridging oxygens (NBOs), several of them projecting out of the interface. Fig. 5 shows that many NBOs are products of the creation of the surface. The three glass surfaces, which contain similar amounts of oxygen, express higher fractions of NBOs in the top layer compared to the bulk-like region, leading to similar NBO concentrations for N25 and N20, while N15 exhibits more exposed NBOs (Table 3). However, the behaviour of the three glass compositions is slightly different in the layers near the interface. N25 exhibits a minor drop in NBOs compared to the bulk, before attaining the maximum, while N20 shows an increment of NBOs for all z > 10.

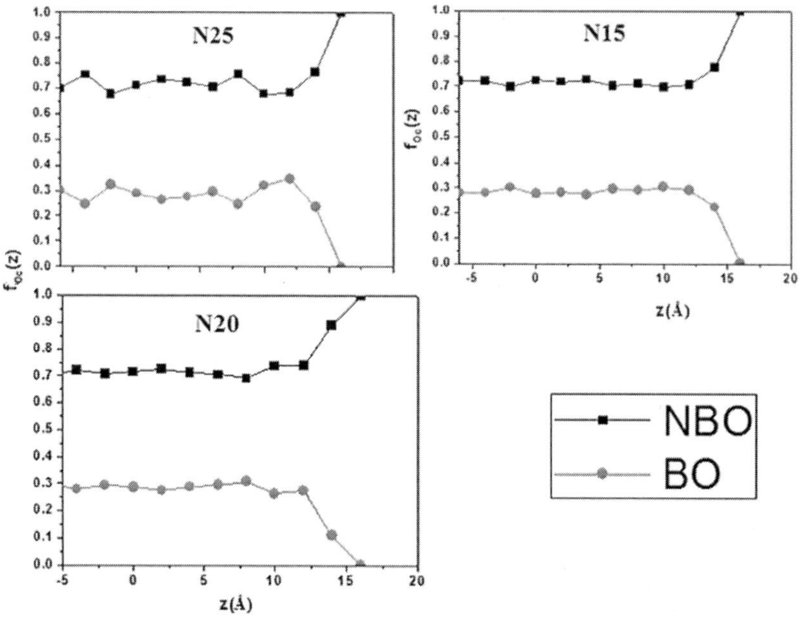

Figure. 5. z-profiles. Fraction of NBO and BO in the slab model, calculated as in Fig. 4.

3.2. Surface network

Other structural attributes of the surface, that are useful to analyse and that contribute to explain the different behaviour of NBOs in N20, are the distribution of the Q^nspecies (where Q^n denotes a phosphorous atom sharing an oxygen with *n* others). The Q^n distributions for the surfaces (using only the top 10 Å of the slab) of the three glasses are shown in Table 4. For all three compositions, at surface level the phosphorus atoms prefer to share two oxygen atoms with two other P atoms, which is similar behaviour to that observed in the bulk in both theoretical and experimental studies [1,10], but there is a significant increase in the cross-linking of just two P tetrahedra. Moreover, there is an evident increase in Q^1 and decrease in Q^2 with respect to previous calculations of the bulk (Table 4), which is partially due to the surface fracture, but is also strongly linked to the composition of the cation modifiers, which are known to alter the phosphate network [1,10,12,14]. We observe the same trends in the surfaces and bulk, i.e. an increase in Q^1 and decrease in Q^2 with the increase in Ca^{2+} content in the glass (N25 to N20), as reported in theoretical studies of the bulk glass [1,10], but there is a small drop in Q^1 from N20 to N15, in agreement with experimental findings in the bulk glasses [1,32].

Table 4. Q^n distribution (%) calculated for the surface region of the slab in comparison with the bulk. The values are averaged over the four surfaces.

System	Q^0	Q^1	Q^2	Q^3	NC
Bulk[1]					
N25	0.0	25.9	70.1	4.0	1.78
N20	0.0	27.7	67.0	5.1	1.78
N15	0.0	28.7	64.6	6.7	1.78
Surface region					
N25	0.5	38.8	55.6	5.2	1.65
N20	0.8	40.8	53.4	5.0	1.63
N15	0.5	40.3	52.8	6.4	1.65

The small variations in the surface Q^n distribution among the different Ca^{2+}compositions are mirrored by the absence of trends as a function of Ca^{2+}/Na^+ ratios in the PO_4^{3-} chain distribution presented in Fig. 6. The average length of the chains is 5.7 (N15), 5.2 (N20) and 5.3 (N25). As is observed in the bulk glass, there is only a small difference in the chain lengths among the compositions. The values are shorter than reported for the bulk, because only the first 10 Å from to the interface are considered where there is an increase in the number of chains with lengths = 2, 3.

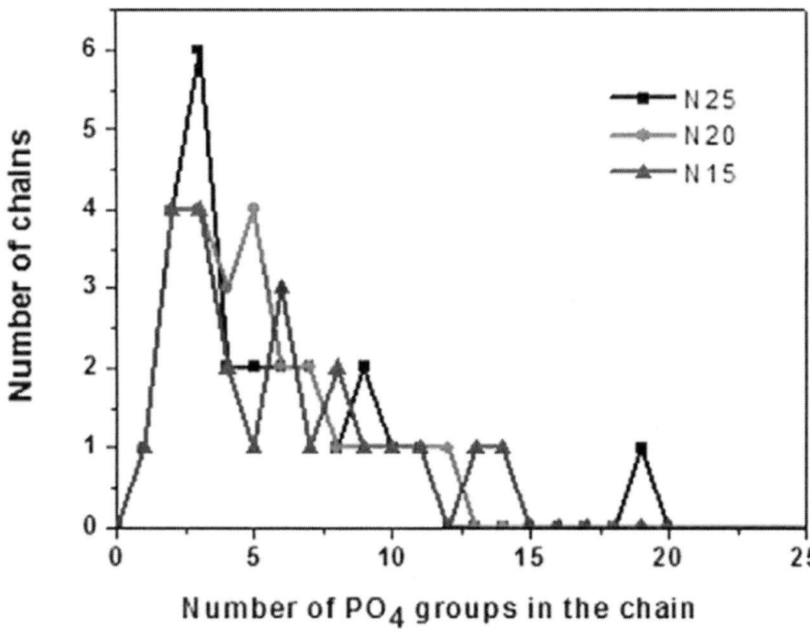

Figure. 6. Distribution of phosphate chains in the surface region of the different glass compositions.

Another distinctive feature in the surfaces is the increase in Q^0 due to the presence of non-connected phosphates, mainly detected as under-coordinated phosphorus (Fig. 3a) and free orthophosphates (Fig. 2b). P_{3C} have been reported before in melt glasses [10] and appear here during relaxation. In the case of N25 there is also a 1% increase of Q3 phosphates caused by the formation of ring structures (Figs. 2a, 3b).

The data in Table 4 show that the surface with composition N20 is the one with the largest shift to lower values of *n* in Q^n, indicating a lower network connectivity (NC) and thus exhibiting more NBOs per total amount of oxygen.

Lower values of network connectivity are closely related to the bioactivity and increase in the solubility of the glass, but, as stated in previous studies [12], the value of NC is only associated with the number of phosphates and not with the Ca^{2+}/Na^{+} ratio in the PBG composition. It is therefore relevant to analyse the interactions of the cation modifiers in the surfaces.

3.3. Network modifiers

3.3.1. Cation clustering at the surfaces

Previous MD simulations of silicate glasses have underlined a link between the aggregation of modifier cations in certain areas of the glass and its stability and bioactivity [14,20].

A statistical measure of the inclination of cations of species x to form clusters is the ratio $R_{x\text{-}x}$ of the total number of other x atoms observed in its coordination sphere (N_{obs}) and the corresponding number N_{hom} which would result from a homogeneous distribution, where all the ions x uniformly occupy the available volume [8].

$$R_{X-X} = \frac{N_{obs}}{N_{hom}} = \frac{CN_{X-X} + 1}{4\pi r_c^3 \dfrac{N_X}{V_S}} \tag{1}$$

where CN_{X-X} is the x-x coordination number calculated for the distance r_c, the minimum value obtained for their radial distribution function ($r_c = 5$ Å). N_x is the number of x ions at the surface and V_s is the volume of the surface section of the slab. Within the definition of Eq. (1), a value of $R > 1$ denotes aggregation of the cations at the interface.

The values of R for the three glass surfaces are reported in Table 5 and clearly show clustering of both modifiers in the region near the interface. As expected from the z-profiles and Table 3, the effect is more pronounced for Na^+ than for Ca^{2+}. In the case of calcium there is no difference in the clustering trends for the N25 and N20 compositions, whereas for sodium the tendency to aggregate varies among all compositions. In Fig. 4, it can also be seen that the amount of sodium increases more rapidly for $z > 10$ Å in N20, although in the N20 slabs the Na^+ remains in a region away from the interface. The difference observed among N25, N20 and N15 is more related to the initial Na^+ composition in the materials.

Table 5. Clustering of the modifiers in the surfaces for the different glass compositions.

Surface composition			
	N25	N20	N15
$\mathbf{R}_{\mathbf{Ca\text{-}Ca}}$	1.52	1.52	1.37
$\mathbf{R}_{\mathbf{Na\text{-}Na}}$	1.80	1.95	1.61

3.3.2. Cation coordination

We have analysed the coordination numbers (CN) of both Ca^{2+} and Na^{+} modifiers in the surface region, as collected in Table 6. Owing to the creation of the surface from the bulk, it is expected that the coordination numbers of the ions closer to the interface are lower than those found in the bulk (Fig. 7). We also found that the decrease in coordination is at the expense of Me-NBO coordination (where Me = Ca^{2+}, Na^{+}) with the two extreme calcium compositions (N25, N15) the most significantly affected for calcium coordination, while the sodium experiences the same drop in NBO coordination compared to the bulk values for the three compositions, showing the lowest coordination number in N25.

Table 6. Coordination numbers for the modifiers in the surface. The values are averaged over the four surfaces.

	Bulk [1]			**Surface**		
	N25	**N20**	**N15**	**N25**	**N20**	**N15**
Ca CN	6.86	6.82	6.87	5.05	5.37	5.22
Ca-BO	0.59	0.55	0.62	0.44	0.46	0.45
Ca-NBO	6.27	6.27	6.25	4.62	4.91	4.78
$\mathbf{CN}^{\mathbf{Ca\text{-}NBO/Ca\text{-}BO}}$	10.63	11.40	10.08	10.5	10.7	11.23
Na CN	6.49	6.59	6.57	4.86	4.95	4.97
Na-BO	1.26	1.30	1.32	0.89	0.92	0.97
Na-NBO	5.23	5.29	5.25	3.97	4.03	4.01
$\mathbf{CN}^{\mathbf{Na\text{-}NBO/Na\text{-}BO}}$	4.15	4.07	3.98	4.46	4.38	4.13

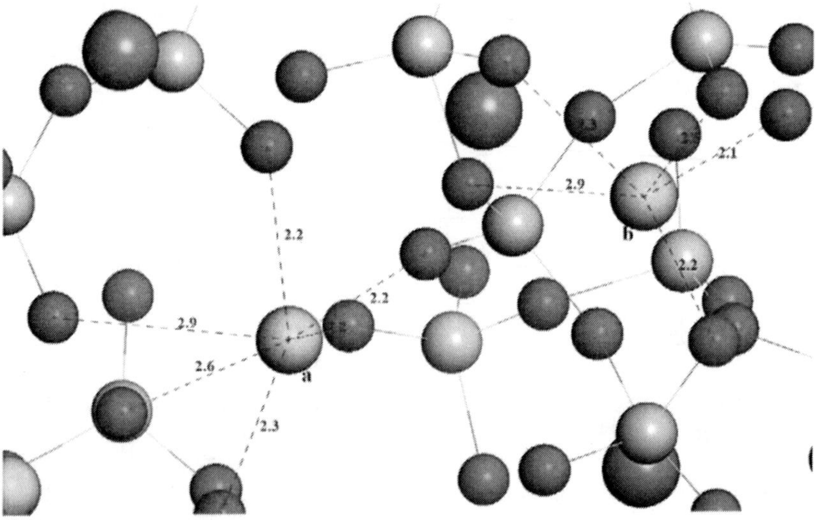

Figure. 7. Coordination of calcium at different values of z in the surface. (a) $z \leq 10$ Å, CN = 6. (b) $z \geq 13$ Å CN = 5. Color code: P (pink), O (red), Ca (green) and Na (violet).

Similarly to the behaviour in the bulk, $[CN^{Ca\text{-}NBO/Ca\text{-}BO}] > [CN^{Na\text{-}NBO/Na\text{-}BO}]$ for surfaces of the three compositions. Calcium has a higher field strength than sodium and will bond to more NBOs, cross-linking more phosphates and strengthening the glass structure [12].

The lowest Ca-NBO and $[CN^{Ca\text{-}NBO/Ca\text{-}BO}]$ for N25 (Table 6) indicates that the chains closer to the interface in these surfaces are structurally weaker in comparison with those of different compositions in this study. In the same way the higher values obtained for calcium coordination in N15 indicate that these surfaces are more resistant to processes like dissolution.

In order to confirm the suggestions obtained through the CN we have also computed the number of chains bonded to each modifier (Table 7).

It is reported for the bulk glasses that, on average, a Ca^{2+} is bonded to 3.9–4.0 chains, while a Na^{+} is only bonded to 3.2–3.3 chains [12]. We note that these values have fallen in the surfaces to 2.8–3.0 chains bonded to Ca^{2+} and 2.5–2.7 chains bonded to Na^{+}, suggesting that a lower calcium content in the PBG will lead to weaker structures and higher solubility rates.

Table 7. Phosphate chains on the surfaces bonded to the modifiers. The values are averaged over the four surfaces and presented as percentages.

Modifier	**Number of chains (%)**								
		0	**1**	**2**	**3**	**4**	**5**	**6**	**Ave**
Ca	N25	0.0	5.5	21.1	39.1	13.9	3.0	1.6	2.84
	N20	0.0	4.7	23.3	46.7	20.7	3.9	0.6	2.98
	N15	0.0	2.8	35.9	37.9	20.1	3.3	0.0	2.85
	Number of chains (%)								
Na		0	1	2	3	4	5	6	Ave
	N25	0.0	6.7	28.9	37.0	11.7	1.8	0.9	2.68
	N20	0.0	10.1	37.6	34.5	12.3	1.1	0.1	2.55
	N15	0.6	8.7	41.0	41.9	7.3	0.5	0.0	2.50

We have calculated additional significant structural features that also reflect the differences in dissolution rates and have been reported before in theoretical studies of the bulk, such as the intra-tetrahedral bonding, a phenomenon describing how many oxygens of the same phosphate tetrahedron are coordinated to a given modifier cation [12] (Fig. 8). We found that an average Ca^{2+} had 0.8 intra-tetrahedral bonds at surface level (0.9–1.1 in the bulk [12]), while an average Na^{+} had 1.5–1.6 (Table 8) compared to 1.2–1.3 in the bulk [12].

Table 8. Intra-tetrahedral M—O—PO_2—O—M bonding of the modifiers in the surfaces. The values are averaged over the four surfaces for each composition.

Surface composition			
Modifier	N25	N20	N15
Ca	0.76	0.83	0.81
Na	1.53	1.55	1.57

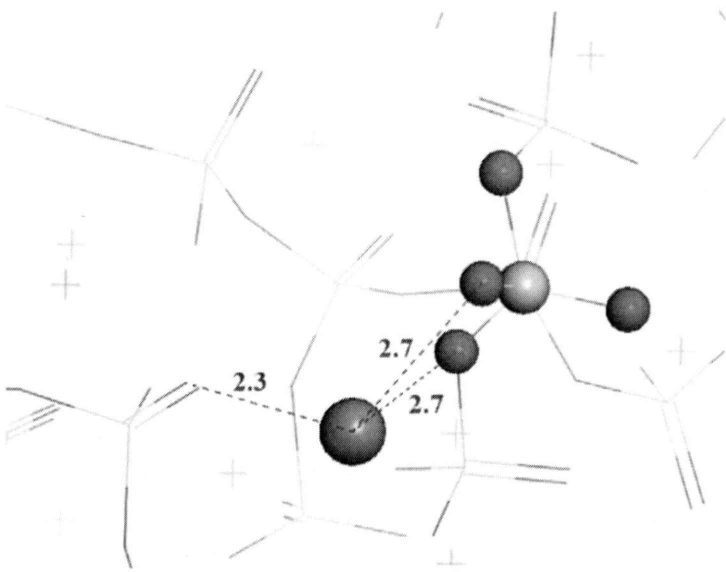

Figure. 8. Na^+ coordinating to two oxygens of the same phosphate (intra-tetrahedral coordination highlighted in balls) Color code: Na violet, Ca green, O red, P pink.

As occurs in the bulk, the lesser tendency of calcium for intra-tetrahedral bonding is likely to contribute strongly to the higher number of phosphates found around a single cation, and hence the strengthening effect on the network. The slight increase in the intra-tetrahedral bonding for sodium at surface level is related to the higher Na^+concentrations observed compared to those in the bulk. As there are more Na^+ near the interface, these ions have to satisfy their need of coordination through using more than one oxygen from the same phosphate group.

4. CONCLUSIONS

The glass bioactivity is closely linked to the ability of its components to disconnect from the surface [18,33], and we thus need to investigate all the different surface features that could be involved in this process. The surfaces of the three compositions are fragmented in a similar fashion, but the slightly higher values of Q^1 and Q^0 in N20 suggest that these surfaces would be more prone to attack by solvent. However, the network modifiers play a key role in the dissolution. All the surfaces of the compositions studied here experience an enrichment in sodium, but it is the higher Na^+ fraction in the bulk that determines the sodium content in the surfaces. Furthermore, in N20 sodium has a preference to aggregate more in a region just below the interface, and there are therefore not as many Na^+ ions exposed at the interface, i.e. closer to the solvent, (see concentrations in Table 3), as in other compositions. Previous studies have shown that Na^+—water interactions are potential initiators of surface degradation [34], whereas it is also known that a quick release of Na^+ will favour Na^+/H^+(solvent) exchange, leading to high alkalinity where the increasing amount of OH^- induces the breaking of P—O—P bonds [18,34,35]. The surfaces with higher concentrations of sodium at the interface will thus dissolve more quickly. The calcium, however, does not show a distinct tendency towards clustering in the studied range of compositions. The Ca^{2+} cations are responsible for strengthening the network, cross-linking various chains through coordination to NBOs and it is in the N25 (lowest Ca^{2+} composition) surfaces, where calcium coordination numbers are lower, that the cations are bonded to fewer phosphate chains. These factors are the strongest contributors to a weaker network in the N25 surfaces and therefore facilitators of the dissolution process. Similar to silicate glasses, the network dissolution has a direct impact on bioactivity. The dissolved species will nucleate others, such as calcium phosphates, thereby encouraging, for example, apatite formation.

Like silicates, immediately after contact with water the top phosphate groups will be rapidly hydroxylated by protonation of most external NBOs. Table 3 suggests that the network formers in the surfaces with N25 and N15 compositions should be hydroxylated to the highest degree. This is the initial step in hydrolysis, but the difference in hydroxylation between these two concentrations may not be enough to override other factors like network strength.

Small 3 M rings are understood to be relevant to the long-term solubility of the glasses, since metal poly-phosphates are highly insoluble [36]. In this study, they are only formed at the interface $z > 14$, with very few cations in the first coordination sphere, and at almost all times disconnected from the main phosphate network. Experimentally it is known that, once in solution, the phosphate rings are very stable [2,37], but their very low concentrations observed in the MD, the lack of solvent in the model employed and no other information from higher levels of theory, make it difficult to predict their impact.

In summary, we have carried out a series of molecular dynamics simulations of different phosphate-based glass surfaces, with three different CaO and Na_2O compositions, in an effort to link their structure to their aptitude towards dissolution and bioactivity. Our results are in a very good agreement with previous predictions based on bulk materials, highlighting N25 as the most soluble material, and they show the critical role of the enrichment of sodium at surface levels. However, following similar work on the dissolution of silicate surfaces [37], further work including explicit solvent and pH effects is needed to assess clearly the diffusion of the ions and more complex phenomena like the formation of the small rings and their role in the solubility of the PBG materials.

ACKNOWLEDGMENTS

We acknowledge the UK Engineering and Physical Sciences Research Council (EP/J008095/1) for funding. This work made use the computational facilities of the Advanced Research Computing@Cardiff (ARCCA) Division, Cardiff University.

REFERENCES

1. R.I. Ainsworth, D. Di Tommaso, J.K. Christie, N.H. de LeeuwJ. Chem. Phys., 137 (2012), pp. 234502-234510
2. E.A. Abou Neel, D.M. Pickup, S.P. Valappil, R.J. Newport, J.C. KnowlesJ. Mater. Chem., 19 (2009), pp. 690-701
3. I.Ahmed, M. Lewis, I. Olsen, J.C. KnowlesBiomaterials, 25 (2004), pp. 501-507
4. W.M. Allen, B.F. Sansom, P.T. Gleed, C.B. Mallinson, C.F. DrakeVet. Rec., 115 (1984), pp. 55-57
5. V. Salih, M. Bitar, V. Mudera, J.C. Knowles, M.P. LewisBiomaterials, 25 (2004), pp. 2283-2292
6. S.P. Valappil, D.M. Pickup, D.L. Carroll, C.K. Hope, J. Pratten, R.J. Newport, M.E. Smith, M. Wilson, J.C. KnowlesAntimicrob. Agents Chemother., 51 (2007), pp. 4453-4461
7. A.M. Mulligan, M. Wilson, J.C. KnowlesBiomaterials, 24 (2003), pp. 1797-1807
8. J.K. Christie, J. Malik, A. TiloccaPhys. Chem. Chem. Phys., 13 (2011), pp. 17749-17755
9. D.M. Pickup, R.J. Newport, J.C. KnowlesJ. Biomater. Appl., 26 (2012), pp. 613-622
10. D. Di Tommaso, R.I. Ainsworth, E. Tang, N.H. de LeeuwJ. Mater. Chem. B., 1 (2013), pp. 5054-5066
11. K. Franks, I. Abrahams, J.C. KnowlesJ. Mater. Sci. Mater. Med., 11 (2000), pp. 609-614
12. J.K. Christie, R.I. Ainsworth, D. Di Tommaso, N.H. De LeeuwJ. Phys. Chem. B, 117 (2013), pp. 10652-10657
13. R.I. Ainsworth, J.K. Christie, N.H. de LeeuwPhys. Chem. Chem. Phys., 16 (2014), pp. 21135-21143
14. J.K. Christie, R.I. Ainsworth, N.H. de LeeuwJ. Non-Cryst. Solids, 432 (2015), pp. 1-4
15. A.TiloccaJ. Mater. Chem., 2 (2010), pp. 6848-6858
16. A.TiloccaJ. Mater. Chem., 21 (2011), pp. 12660-12667
17. A.Tilocca, N.H. De Leeuw, A.N. CormackPhys. Rev. B Condens. Matter Mater. Phys., 73 (2006), pp. 1-14
18. A.Tilocca, A.N. CormackLangmuir, 26 (2010), pp. 545-551
19. A.TiloccaPhys. Chem. Chem. Phys., 16 (2014), pp. 3874-3880
20. A.TiloccaProc. R. Soc. A Math. Phys. Eng. Sci., 465 (2009), pp. 1003-1027
21. M. Born, K. Huang**Dynamical Theory of Crystal Lattices**Oxford University Press, Oxford, Oxford Uni (1954)

22. B.G. Dick, A.W. OverhauserPhys. Rev., 112 (1958), pp. 90-103
23. W. Smith, T.R. ForesterJ. Mol. Graph., 14 (1996), pp. 136-141
24. D.J. Evans, O.P. MorrissComput. Phys. Reports., 1 (1984), pp. 297-343
25. L. Vocadlo, G.A. de Wijs, G. Kresse, M. Gillan, G.D. PriceFaraday Discuss., 106 (1997), pp. 205-218
26. R. Ferneyhough, D. Fincham, J.D. Price, M.J. GillanModel. Simul. Mater. Sci. Eng., 2 (1994), pp. 1101-1110
27. J. Du, A.N. CormackJ. Am. Ceram. Soc., 88 (2005), pp. 2532-2539
28. I.Abrahams, K. Franks, G.E. Hawkes, G. Philippou, J. KnowlesT. Nunes, 7 (1997), pp. 1573-1580
29. A.Tilocca, A.N. CormackWater. (2008), pp. 11936-11945
30. S. Wallace, J.K. West, L.L. HenchJ. Non-Cryst. Solids, 152 (1993), pp. 101-108
31. N. Sahai, M. AnseauBiomaterials, 26 (2005), pp. 5763-5770
32. I.Ahmed, M. Lewis, I. Olsen, J.C. KnowlesBiomaterials, 25 (2004), pp. 491-499
33. D. Arcos, D.C. Greenspan, M. Vallet-Regí, J. BiomedMater. Res. Part A., 65A (2003), pp. 344-351
34. A.Tilocca, A.N. CormackACS Appl. Mater. Interfaces, 1 (2009), pp. 1324-1333
35. R. Pyare, L.J. Lai, V.C. Joshi, V.K. SinghJ. Am. Ceram. Soc., 79 (1996), pp. 1329-1334
36. D.E.C. Corbridge**The Structural Chemistry of Phosphorus**Elsevier Scientific Pub. Co., Amsterdam; New York (1974)
37. B.C. Sales, L.A. Boatner, J.O. RameyJ. Non. Cryst. Solids., 263 (2000), pp. 155-166

CHAPTER 21

To the Question of Validity Grüneisen Solid State Equation

Vladimir Kh. Kozlovskiy

Jewish Scientific Society, Berlin, Germany.

ABSTRACT

The first justified theory of solid state was proposed by Grüneisen in the year 1912 and was based on the virial theorem. The forces of interaction between two atoms were assumed as changing with distance between them according to inverse power laws. But only virial theorem is insufficient to deduce the equation of state, so this author has introduced some relations, which are correct, when the forces linearly depend on displacement of atoms. But with such law of interaction the phase transitions cannot take place. Debye received Grüneisen equation in another way. He deduced the expression for thermocapacity, using Plank formula for energy of harmonic vibrator. Taking into account the dependence of atomic vibration frequency from distance between atoms, when the forces of interaction are anharmonic, he received the equation of state, which in classical limit turns to Grüneisen equation. The question, formulated by Debye is—How can we come to phase transitions, when Plank formula for harmonic vibrator was used? Debye solved this question not perfectly, because he was born to small anharmonicity. In the presented work a chain of atoms is considered, and their movement is analysed by means of relations, equivalent to virial theorem and theorem of Lucas (disappearing of mean force). Both are the results of variation principle of Hamilton. The Grüneisen equation for low temperature (not very low, where quantum expression for energy is essential) was obtained, and a family of isotherms and isobars are drown, which show the existence of spinodals, where phase transitions occur. So, Grüneisen equation is an equation of state for low temperatures.

Keywords: Grüneisen Equation; Phase Transitions; Atomic Chain; Atomic Interactions

1. INTRODUCTION

The molecular-dynamical solid state equation, based on virial theorem, was derived by Sutherland [1], Mie [2], Grüneisen [3,4]. Another way of deriving, based on the Debye theory of heat capacity at low temperatures [5], which used the Plank expression for the energy of harmonic vibrator, was chosen by Ratnowsky [6], Eisenmann [7], Grüneisen [8], Debye [9,10]. In classical limit the quantum equation of state turns to classical Grüneisen equation. But here arises a question, which Debye formulated in such manner: because an expression of harmonic vibrator energy is used, can the Grüneisen equation be regarded as equation of state that describes phase transitions, which are non linear phenomena? In classical deduction of equation of state Grüneisen besides the virial theorem (which alone is insufficient to deduce equation of state) used some equalities that are strictly correct for forces that depends linearly from displacement of atoms. Debye tried to solve the formulated question, but, because he was based on relations correct for small anharmonicity only, the analysis cannot be regarded as complete. We employ another way for equation of state deduction that begins from variation principle of Hamilton, which can be a base for deriving of the virial theorem [11]. The approximations, used in solving of equations, are not burned to small anharmonicity. Here is given a brief account of equations deduction, which in details is published elsewhere [12]. The calculations for atomic chain show that Grüneisen equation is valid as equation of state for no high temperatures. The families of isotherms and isobars illustrate the picture of phase transitions in the chain.

2. THE BASIC EQUATIONS OF SOLID STATE

The deduction of virial theorem from Hamilton variation principle is briefly described below. The variation of mean time Lagrange function L is transformed as follows

$$\delta\overline{L} = \delta\left(\frac{1}{t_2 - t_1}\int_{t_1}^{t_2} L\mathrm{d}t\right) = \frac{1}{t_2 - t_1}\int_{t_1}^{t_2}\sum_{k=1}^{N}\left(\frac{\partial L}{\partial q_k}\delta q_k + \frac{\partial L}{\partial \dot{q}_k}\delta \dot{q}_k\right)\mathrm{d}t$$

$$= \frac{1}{t_2 - t_1}\sum_{k=1}^{N}\int_{t_1}^{t_2}\left[\frac{\partial L}{\partial q_k} - \frac{\partial}{\partial t}\left(\frac{\partial L}{\partial \dot{q}_k}\right)\right]\delta q_k \mathrm{d}t + \frac{1}{t_2 - t_1}\sum_{k=1}^{N}\frac{\partial L}{\partial \dot{q}_k}\delta q_k \tag{1}$$

The first term of last expression is zero as consequence of movements equations, the second—for infinite time interval. The product of a coordinate with variable factor λ represents a varied coordinate: $\tilde{q}_k = \lambda q_k$, so, so $\delta q_k = \delta\lambda \cdot q_k$, $\delta\dot{q}_k = \delta\lambda \cdot \dot{q}_k$ These variations substituted in the left side of last Equation (1) give a relation

$$\sum_{k=1}^{N}\left(\overline{\frac{\partial L}{\partial q_k} q_k}\right) + \sum_{k=1}^{N}\left(\overline{\frac{\partial L}{\partial \dot{q}_k}\dot{q}_k}\right) = 0 \tag{2}$$

If the kinetic energy W does not depend on coordinates (in crystals rectilinear coordinates of atoms are used), it follows the virial theorem

$$\sum_{k=1}^{N}\overline{\frac{\partial U}{\partial q_k} q_k} = \sum_{k=1}^{N}\overline{\frac{\partial W}{\partial \dot{q}_k}\dot{q}_k} = 2\overline{W} = NkT \tag{3}$$

The represented variation procedure may be generalised. The variable coordinates are represented in the form $\tilde{q}_k = \mu_k + \lambda_k q_k$ where μ_k, λ_k are variable parameters (further generalization of variation procedure is described in [12]). The variations of coordinates and velocities are expressed as $\delta q_k = \delta\mu_k + \delta\lambda_k q_k$, $\delta\dot{q}_k = \delta\lambda_k \dot{q}_k$ and after substitution in (1) a relation follows

$$\sum_k\left(\overline{\frac{\partial L}{\partial q_k}}\delta\mu_k + \overline{\frac{\partial L}{\partial q_k} q_k}\delta\lambda_k + \overline{\frac{\partial L}{\partial \dot{q}_k}\dot{q}_k}\delta\lambda_k\right) = 0 \tag{4}$$

So the equations are obtained

$$\overline{\frac{\partial L}{\partial q_k}} = 0 \tag{5}$$

$$-\overline{\frac{\partial L}{\partial q_k} q_k} = \overline{\frac{\partial L}{\partial \dot{q}_k}\dot{q}_k} = \overline{\frac{\partial W}{\partial \dot{q}_k}\dot{q}_k} = kT \tag{6}$$

where the theorem about uniform distribution of kinetic energy is used. Further step in calculations consists in representation of coordinates as functions of time in the form

$$q_k = s_k + u_k \theta_k(t) \tag{7}$$

where S_k are the mean values of coordinates, u_k is root mean square amplitude of vibrations, so the vibration functions satisfy the relations

$$\overline{\theta_k(t)} = 0, \overline{\theta_k^2(t)} = 1 \tag{8}$$

If the kinetic energy depends only on velocities then (5) is represented as

$$\frac{\partial \overline{U}}{\partial s_k} = 0 \tag{9}$$

and

$$-\frac{\partial L}{\partial q_k} q_k = \frac{\partial U}{\partial s_k} s_k + u_k \theta_k \frac{\frac{\partial U}{\partial u_k}}{\frac{\partial q_k}{\partial u_k}} \tag{10}$$

Introducing in (6) taking into account (7) and (9) we obtain

$$u_k \frac{\partial \overline{U}}{\partial u_k} = kT \tag{11}$$

If the system of atoms interacts with external corps, the positions of which are determined by means of coordinates a_i, the mean forces acting on the system are determined by expressions

$$\overline{A_i} = \frac{\partial \overline{U}}{\partial a_i} \tag{12}$$

By shifting of the corps the forces produce a work, so the variation of energy by variation of all parameters is given by the expression

$$\delta H = \delta \overline{W} + \delta \overline{U}$$
$$= \frac{N}{2} k \delta T + \sum_k \left(\frac{\partial \overline{U}}{\partial s_k} \delta s_k + \frac{\partial \overline{U}}{\partial u_k} \delta u_k \right) + \sum_i \overline{A_i} \delta a_i \tag{13}$$

Taking into account (9), (11) a relation will be obtained

$$\frac{\delta H - \sum \overline{A_i} \delta a_i}{T} = k \delta \ln \left(T^{\frac{N}{2}} \prod_k u_k \right) \tag{14}$$

It can be seen that the quantity *T* which till now is only the indication for mean kinetic energy, an integrant divisor for the quantity of heat is, hence proportional to absolute temperature. The expression for entropy is the follows

$$S = k \ln \left(T^{\frac{N}{2}} \prod_k u_k \right) \tag{15}$$

We introduce a function of variables as an analogy of thermodynamic potential

$$\Phi = \overline{U} - kT \ln \left(\prod_k u_k \right) - \sum_i \overline{A_i} a_i \tag{16}$$

(items depending on temperature only are omitted). The condition of minimum Φ by constant temperature leads to Equations (9), (11), (12) and allows to choose stable values of variables. In a crystal the mean values of displacements and amplitudes are equal for atoms belonging to the same sublattice, so the number of unknowns may be not great. Because in crystals the rectilinear coordinates are employed, the kinetic energy will not depend on coordinates.

3. ANHARMONIC VIBRATOR WITH TWO EQUILIBRIUM POSITIONS

Additional information about dynamical variation procedure is obtained through considering of a simple anharmonic vibrator with coordinate x and potential energy of the form

$$U = \frac{c}{2} x^2 + \frac{b}{4} x^4; \quad c < 0,\ b > 0 \tag{17}$$

which is represented on **Figure 1**

It is necessary to determine mean time potential energy, so it will be written

$$x = s + u\theta(t) \tag{18}$$

$$x^2 = s^2 + 2su\theta(t) + u^2\theta^2(t) \tag{19}$$

$$x^4 = s^4 + 4s^3u\theta(t) + 6s^2u^2\theta^2(t) + 4su^3\theta^3(t) + u^4\theta^4(t) \tag{20}$$

$$\overline{x^2} = s^2 + u^2 \,;\, \overline{x^4} = s^4 + 6s^2u^2 + u^4\overline{\theta^4} \tag{21}$$

where is put $\overline{\theta^3} = 0$ because $\theta(t)$ is an accidental function with zero mean value, so with equal probability receives positive and negative values, and in such manner behaves $\theta^3(t)$. The mean value of $\overline{\theta^4}$ is 3/2, if the function $\theta(t)$ is harmonic, but is greater, if on the path is potential barrier, which reduces the speed. Later we show what value of $\overline{\theta^4}$ is worthwhile to put in equations. So

$$\overline{U} = \frac{c}{2}\left(s^2 + u^2\right) + \frac{b}{4}\left(s^4 + 6s^2u^2 + \overline{\theta^4}u^4\right) \tag{22}$$

The equations of state will be written as follows

$$\frac{\partial \overline{U}}{\partial s} = 0 \,;\, s\left(c + bs^2 + 3bu^2\right) = 0 \tag{23}$$

$$u\frac{\partial \overline{U}}{\partial u} = kT \,;\, cu^2 + 3bs^2u^2 + \overline{\theta^4}bu^4 = kT \tag{24}$$

It follows two solutions

$$s = 0 \,;\, cu^2 + \overline{\theta^4}bu^4 = kT \tag{25}$$

The first solution corresponds to vibrations above the potential barrier, the second to vibrations in left or right potential pit. If we put $\overline{\theta^4}$ = 3, then the left sides of second equalities in (25), (26) differ only through multiplier, consequently intersect the abscissa axis in the same point. For this case the Equations (25) and (26) are represented graphically on **Figure 2**.

Parts of the curves that have physical meaning, correspond to positive temperature. Heating or cooling leads to transitions from one curve to the other, which is manifested in jumps of amplitude.

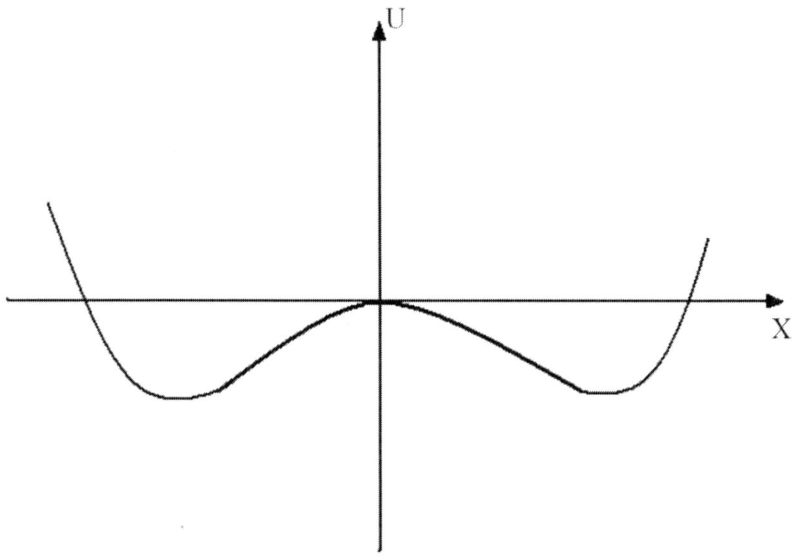

Figure 1. Potential pit with two minima.

Braunbek [13] considered the phenomena of crystal melting, using the equation of sublattice motion. The potential curve was represented by a sinusoidal function of coordinate. The integration could be fulfilled with the help of elliptical functions. By low temperatures the atoms moves in one of minimums of potential curve, by more high temperature—above the maxima. This corresponds to transition from crystalline to non crystalline state. On the **Figure 3** is shown the calculated dependence of the temperature from the energy of vibrations. The amplitude increases with energy, so it exists a possibility to compare the graphs, received with both methods of calculation. The graphs are similar, and the point of intersection of curves, describing finite and infinite movements (in our model also finite, but taking more place in the space) lies on abscissa axis. So, it is reasonable to accept $\theta^4 = 3$

A serious question arises by comparing our calculations with calculations of Syrkin [14]. On the **Figure 4** is shown a form of potential pit: symmetrical double minima rectangular one.

Syrkin calculated the dipole moment in electric field for charged atom by means of formula for statistical distribution, integrating over all space of the pit. It turned out that the dipole moment continuously decreases with

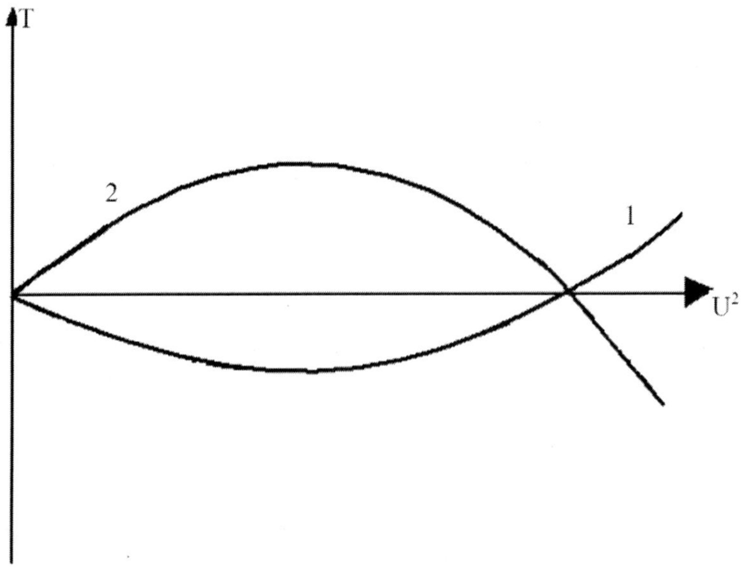

Figure 2. Relations between temperature and amplitude for vibrations; 1—Above the barrier; 2—In the region of minimum.

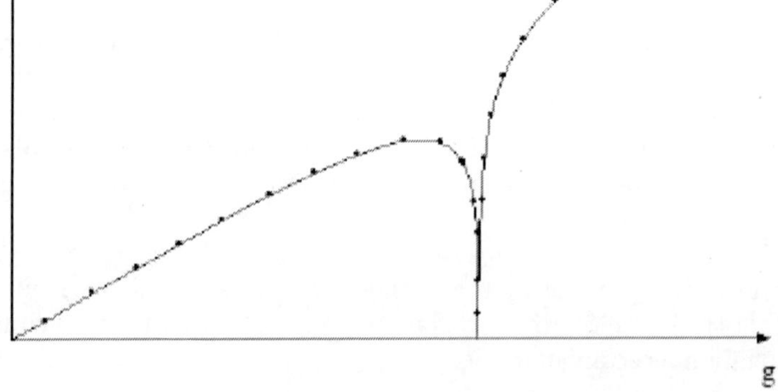

Figure 3. Relation between temperature and energy according [13].

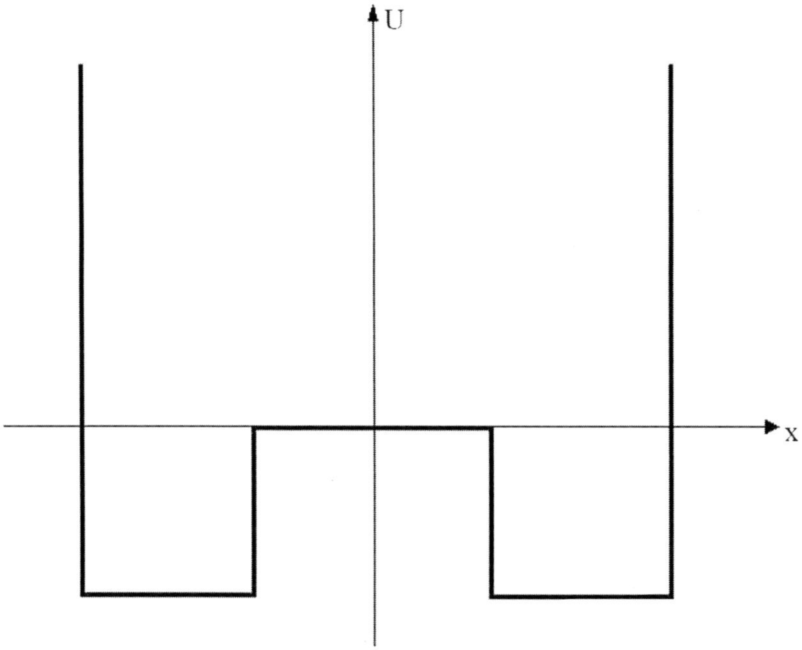

Figure 4. Graph of potential pit according [14].

increasing temperature (as polarisability for thermal polarization according to Debye formula). So, a discrepancy can be marked between dynamical and statistical calculations, but really this discrepancy is illusory, because such question was considered in the literature. In the course of Frenkel [15] is discussed a metastable state with conclusion, that such state corresponds to a region of phase space, where the molecular system exists long time and can be observed, then for determination of phase integral the integration must be extended over this region. But Syrkin integrates over all space, which includes both minima and potential barrier. This means that in no part of the space the atom delays for a long time, but jumps from one minimum to the other. Model of Syrkin corresponds to a hole in a crystal, where the atom is not bound tight to one minimum.

4. EQUATIONS OF STATE OF AN ATOMIC CHAIN

It will be considered a chain of identical atoms situated on x-axis, where atoms interact with nearest neighbours. The extreme left atom with number 0 is fixed in the origin and on the extreme right atom a constant force f directed along the axis is acting (**Figure 5**).

Atom, which has the number *k*, interacts with left and right atom with numbers $k - 1$ and $k + 1$, which are placed on distances l_k, l_{k+1} from the considered.

The energy of interaction for the whole chain is represented by the sum

$$U = \sum_{k=1}^{N} U(l_k) \tag{27}$$

If the mean time distance between neighbouring atoms is *l*—identical for all atoms, then the distance in moment t will be written as

$$l_k = l + u\theta_k(t) \tag{28}$$

where the mean square amplitude is written also as identical for all atoms. The energy of interaction is further expended in power series with respect to amplitude and the order of derivative is marked by means of Roman numerals

$$U(l_k) = U(l) + U^{I}(l)u\theta_k(t) + \frac{1}{2}U^{II}(l)u^2\theta_k^2(t)$$
$$+\frac{1}{6}U^{III}(l)u^3\theta_k^3(t) + \frac{1}{24}U^{IV}(l)u^4\theta_k^4(t) + \cdots \tag{29}$$

For taking the mean time value of (29) the relations (8) must be used and the mean time of (29) has the value

$$\overline{U(l_k)} = U(l) + \frac{1}{2}U^{II}(l)u^2 + \frac{\overline{\theta^4}}{24}U^{IV}(l)u^4 + \cdots \tag{30}$$

For the function Φ determined in (16) and calculated for one atom we receive, taking into account (30), an expression

$$\Phi = U(l) + \frac{1}{2}U^{II}(l)u^2$$
$$+\frac{\overline{\theta^4}}{24}U^{IV}(l)u^4 - kT\ln u - fl \tag{31}$$

where *f* is a constant stretching force (hanging load). The terms depending only on temperature are omitted. The conditions of minimum have the form

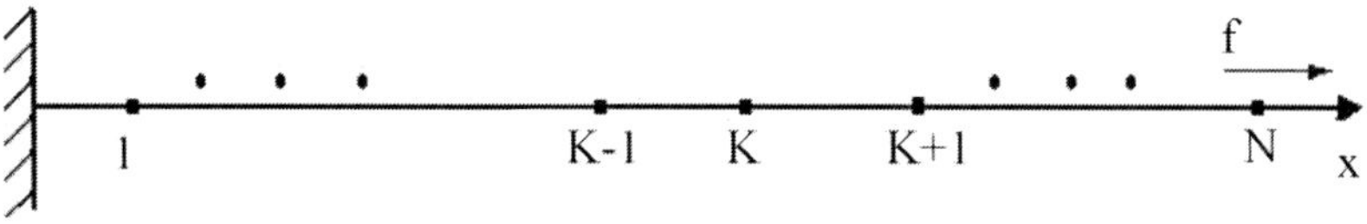

Figure 5. The atomic chain.

$$\frac{\partial \Phi}{\partial l} = U^{I}(l) + \frac{1}{2}U^{III}(l)u^2 + \frac{\overline{\theta^4}}{24}U^{V}(l)u^4 - f = 0 \tag{32}$$

$$u\frac{\partial \Phi}{\partial u} = U^{II}(l)u^2 + \frac{\overline{\theta^4}}{6}U^{IV}(l)u^4 - kT = 0 \tag{33}$$

If we neglect the terms containing fourth power of amplitude, an equation will be received

$$f = U^{I}(l) + \frac{U^{III}(l)}{2U^{II}(l)}kT \tag{34}$$

This is Grüneisen type equation. Now we deduce an equation with regard to fourth power of amplitude. We multiply (32) with U^{IV} and (33) with $\frac{1}{4}U^{V}$ and subtract the second from the first, than it turns to be

$$U^{I}U^{IV} + \frac{1}{2}\left(U^{III}U^{IV} - \frac{1}{2}U^{II}U^{V}\right)u^2$$
$$-U^{IV}f + \frac{1}{4}U^{V}kT = 0 \tag{35}$$

The Equation (33) will be solved with respect to u^2 u^2

$$u^2 = \frac{3}{\overline{\theta^4}}\left(U^{IV}\right)^{-1}\left[-U^{II} + \sqrt{\left(U^{II}\right)^2 + (2/3)\overline{\theta^4}U^{IV}kT}\right] \tag{36}$$

The radical is taken with positive sign, because the amplitude increases with increasing temperature and $U^{IV} > 0$ which will be clear later. Expanding the radical and retaining terms quadratic in temperature, we receive.

$$u^2 = \frac{kT}{U^{II}}\left[1 - \frac{\overline{\theta^4}}{6}U^{IV}\frac{kT}{\left(U^{II}\right)^2}\right] \tag{37}$$

Introducing in (35), we obtain the equation of state

$$f = U^I + \frac{U^{III}}{2U^{II}}kT + \frac{\overline{\theta^4}}{24}\frac{U^{II}U^V - 2U^{III}U^{IV}}{\left(U^{II}\right)^3}(kT)^2 \tag{38}$$

Further calculations concerning the transitions in the chain will be carried out retaining only linear in temperature term.

5. THE EQUATION OF STATE FOR FORCES, CHANGING WITH DISTANCE ACCORDING TO INVERSE POWER LAW

The potential energy of two atoms, one of which is placed in the origin, other on the x-axis on the distance l will be represented by means of a usual function

$$U(l) = \frac{\lambda_n}{l^n} - \frac{\lambda_m}{l^m} \tag{39}$$

which may be written in more convenient form

$$U = \frac{1}{n}\Lambda\left(\frac{l_M}{l}\right)^n - \frac{1}{m}\Lambda\left(\frac{l_M}{l}\right)^m \tag{40}$$

This expression is mathematically equivalent to given in [16].

The minimum energy is displaced on distance l_M, determined through equation

$$\frac{n\lambda_n}{l_M^n} = \frac{m\lambda_m}{l_M^m} = \Lambda \tag{41}$$

The force constants may be expressed through $l_{M,}$ Λ

$$\lambda_n = \frac{1}{n}\Lambda l_M^n, \ \lambda_M = \frac{1}{m}\Lambda l_M^m \tag{42}$$

It follows the expression for the energy in the minimum

$$U_M = -\frac{n-m}{nm}\Lambda \tag{43}$$

Further we find

$$U^I = -\frac{\Lambda}{l_M}\left(\frac{l_M}{l}\right)^{n+1} + \frac{\Lambda}{l_M}\left(\frac{l_M}{l}\right)^{m+1} \tag{44}$$

$$U^{II} = \frac{n+1}{l_M^2}\Lambda\left(\frac{l_M}{l}\right)^{n+2} - \frac{m+1}{l_M^2}\Lambda\left(\frac{l_M}{l}\right)^{m+2} \tag{45}$$

$$U^{III} = -\frac{(n+1)(n+2)}{l_M^3}\Lambda\left(\frac{l_M}{l}\right)^{n+3} + \frac{(m+1)(m+2)}{l_M^3}\Lambda\left(\frac{l_M}{l}\right)^{m+3} \tag{46}$$

As can be seen, in the vicinity of the point *M* it must be$U^{II}>0$, because $n>m$ and the same is true for U^{IV}

These values for derivatives of interaction energy substituted in (34) give the Grüneisen equation for accepted law of interaction. We undertake now a mathematical investigation of this equation for the purpose to receive equations for isotherms and isobars and analyse their behaviour in order to discover transitions. As was noticed [17] the calculations are most simple for values $n = 12$, $m = 6$ that we introduce in above expressions and use the notation

$$\left(\frac{l_M}{l}\right)^6 = q \tag{47}$$

The expressions for energy and its derivatives are the follows

$$U = \frac{1}{12}\Lambda q(q-2) \tag{48}$$

$$U^I = -\frac{\Lambda}{l_M} q^{\frac{1}{6}} q(q-1) \tag{49}$$

$$U^{II} = \frac{\Lambda}{l_M^2} q^{\frac{1}{3}} q(13q-7) \tag{50}$$

$$U^{III} = -\frac{14\Lambda}{l_M^3} q^{\frac{1}{2}} q(13q-4) \tag{51}$$

The Equation (36) now turns to the form

$$f = -\frac{\Lambda}{l_M} q^{\frac{1}{6}} \left[q(q-1) + \frac{13q-4}{13q-7} \frac{7kT}{\Lambda} \right] \tag{52}$$

As the multiplier in front of the bracket varies slowly with the length, we put $q^{\frac{1}{6}} \approx 1$ and accept following approximate equation of state

$$f = -\frac{\Lambda}{l_M} \left[q(q-1) + \frac{13q-4}{13q-7} \frac{7kT}{\Lambda} \right] \tag{53}$$

From this equation, as is usual in physical chemistry, a family of isotherms and isobars will be deduced.

6. ISOTHERMS OF THE ATOMIC CHAIN

The expression in brackets, including the sign in front of it, consists of two items—The first is represented as a parabola, that goes to negative infinity intersecting abscissa axis in points q = 0 and q = 1, the top of which arises above abscissa axis on 0.25. The second item is a fraction where numerator and denominator are linear functions of q and it aspires to constant negative value when q goes to infinity, by diminishing of q also diminishes till the value q = 7/13, where the fraction turns to negative infinity. So the function (53), when the temperature does not exceed definite value, is represented till this point by a curve with roots, lying between values of abscissa 0 and 1, and with top placed above the abscissa axis. By rising of the temperature the roots and the top coincide on the abscissa axis, and farther the chain can exist only under the action of negative (compressing) force. As it was noted in [17], the solid corps must resist to the action of stretching force, otherwise it is not a solid corps. The family of isotherms is depicted on the **Figure 6**. The physical meaning has parts of isotherms to the right side of the top, where condition of stability $\partial f / \partial q \leq 0$ is fulfilled [17]. In the top of the curves the transition in other phase takes place, but the structure of new phase is not described by the present considerations. It may be noticed in passing, that the branch of the curve to the left side of the top approaches the abscissa axis more abruptly, then that to the right side because the third derivative in the top is positive (when it is so, then in the vicinity of the top

$$\left(\frac{\partial f}{\partial q}\right)_{\text{top}-\Delta}+\left(\frac{\partial f}{\partial q}\right)_{\text{top}+\Delta}>0$$

Expanding in series with respect to **Δ**, we arrive at the formulated condition).

Each curve on **Figure 6** is characterised by the position and height of maximum and position of intersection points with abscissa axis. The position of maximum q_m is determined by the equation

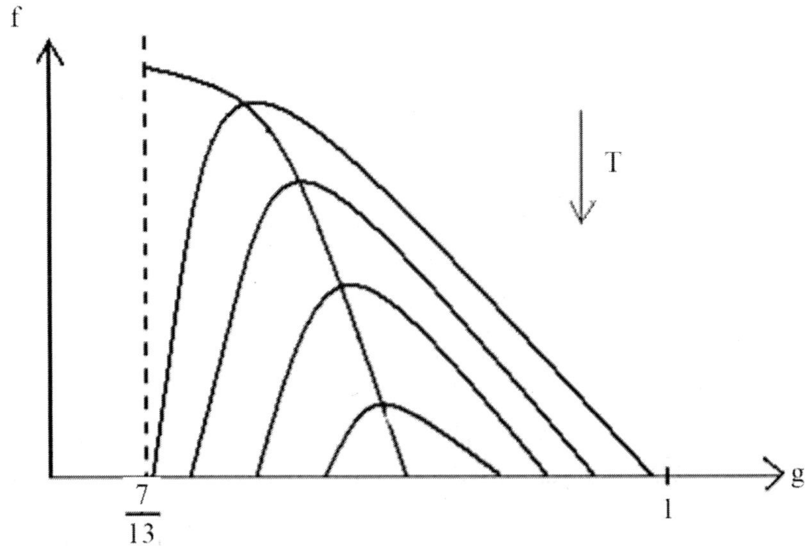

Figure 6. Family of isotherms of chain equation of state.

$$\frac{7kT}{\Lambda}=\frac{1}{39}\left(2q_m-1\right)\left(13q_m-7\right)^2 \tag{54}$$

If this expression of temperature will be substituted in Equation (53), then the equation of the curve, connecting maxima of the isotherms (spinodal) will be obtained

$$\frac{f l_M}{\Lambda} = -q_m (q_m - 1) - \frac{1}{39}(2q_m - 1)(13q_m - 4)(13q_m - 7) \qquad (55)$$

The right side decreases with increasing of the abscissa from the value 7/13=0.538, where the left side has the value 42/169=0.249, till the intersection point with abscissa axis, where the top of the isotherm reaches it. As numerical calculation shows, here

$$q_m = 0.745; \ \frac{kT_m}{\Lambda} = 0.0130 \qquad (56)$$

The diagram (*f*, *T*) is represented in parametric form by expressions (54), (55).

Now we determine the positions of the points, where the isotherm branches intersect the abscissa axis. According to (53) one must put $f = 0$, then solve the equation

$$\frac{7kT}{\Lambda} = -\frac{q(q-1)(13q-7)}{13q-4} \qquad (57)$$

We shall not solve this cubic equation to express the abscissa as a function of temperature, but shall confine us with consideration of solving procedure. The right side is a curve, intersecting the abscissa axis in the points $q = 7/13$ and $q = 1$, between them has a maximum. The left side is a straight line parallel to abscissa axis, that at no high temperatures intersect the curve in two points, corresponding to roots of Equation (57). By rising of the temperature the roots converge till the straight line touches the top of the curve, then the roots and point of maximum coincide. The corresponding values of abscissa and temperature are (56).

7. THE ISOBARS OF ATOMIC CHAIN

From Equation (53) follows the expression for temperature

$$\frac{7kT}{\Lambda} = -\frac{13q-7}{13q-4}\left[\frac{f l_M}{\Lambda} + q(q-1)\right] \qquad (58)$$

The right side is a product of a fraction with negative sign, whose numerator and denominator are linear functions of abscissa, and a quadratic function. The graphs of multipliers are represented on **Figure 7** (the letter f on vertical axis means function).

The temperature is positive, so physical meaning has the interval on abscissa axis, where the ordinates of both curves have the same (negative) sign that means interval between points, where the fraction and parabola intersect the abscissa axis. The first point is fixed; the second depends on the value of the force and may disappear (if parabola is placed above the abscissa axis). So, the family of isobars has the form, shown in **Figure 8**.

Only the part to the right side of the top, where $\frac{\partial T}{\partial q} < 0$ have physical meaning, because here the length of the chain increases with increasing temperature, what is designated as thermal dilatation (notation 47).

The position of the point Q, where parabola intersects the abscissa axis is given by the expression

$$q_Q = \frac{1}{2} + \sqrt{\frac{1}{4} - \frac{fl_M}{\Lambda}} \qquad (59)$$

With increasing force this point moves toward the point with abscissa 7/13 and coincides with it by the value of the force

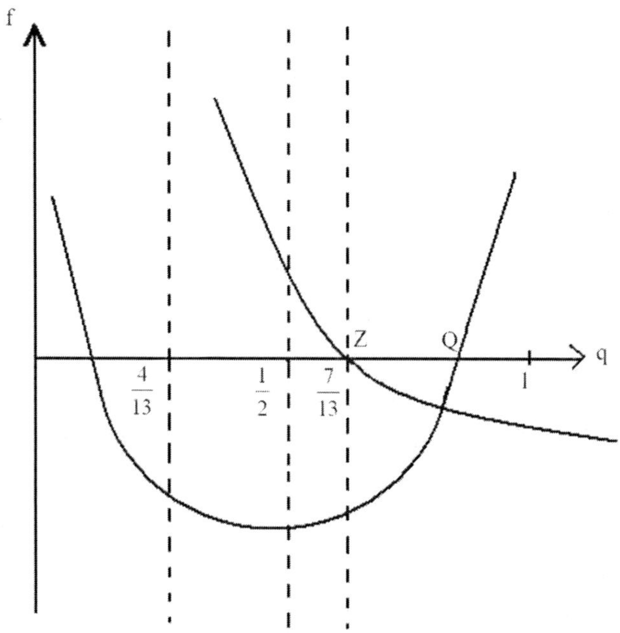

Figure 7. Graphs of multipliers in Equation (58).

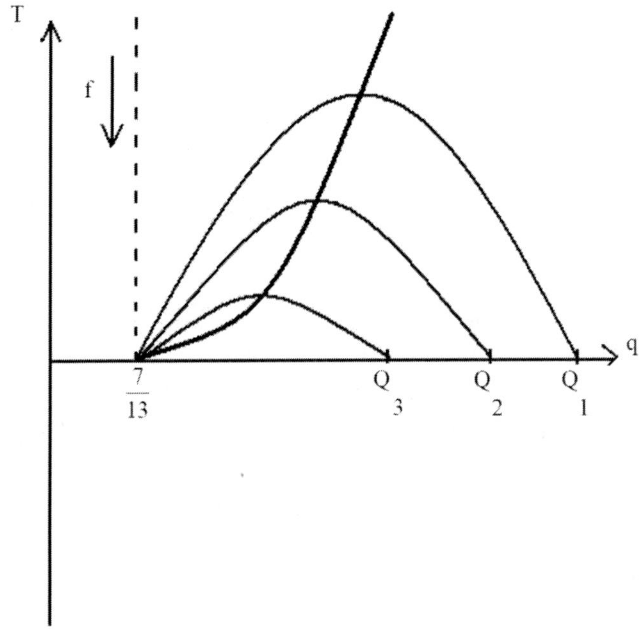

Figure 8. Family of isobars of chain equation of state.

$$\frac{f\,l_M}{\Lambda} = \frac{42}{169}$$

By this value the isobar disappears, and the system of atoms does not exist as regular structure.

For determination of maximum the expression (58) will be differentiated, and the result is

$$\frac{f\,l_M}{\Lambda} = -q_m\left(q_m - 1\right) - \frac{1}{39}\left(13q_m - 7\right)\left(13q_m - 4\right)\left(2q_m - 1\right) \tag{60}$$

This equation has the form of Equation (55). The value of temperature maximum on the curves, shown in **Figure 8**, we receive, if substitute the value of force (60) in (58), where the value of abscissa is q_m The temperature maximum is

$$\frac{7kT}{\Lambda} = \frac{1}{39}\left(2q_m - 1\right)\left(13q_m - 7\right)^2 \tag{61}$$

and has the same mathematical form as (54). It is the equation of spinodale that joints the points of maxima on **Figure 8**.

8. THE ISOBARS TEMPERATURE—AMPLITUDE

The initial equations are (37), (50), (58) and the problem consists in excluding the quantity q. This is made as follows: Excluding the temperature leads to the equation for q (here $q^{\frac{1}{3}} \approx 1$)

$$(91w+1))q^2-(28w+1)q+p=0 \tag{62}$$

Where $w=(u/l_M)^2$. $p=\frac{fl_M}{\Lambda}$ The solution has the form

$$q=\frac{28w+1+R}{2(91w+1)} \tag{63}$$

where

$$R=\sqrt{(28w+1)^2-4(91w+1)p}$$

Introducing (63) in (37), leads to the expression for temperature

$$\frac{kT}{\Lambda}=\frac{w}{4(91w+1)^2}(28w+1+R)\left[-(910w+1)+13R\right] \tag{64}$$

The first multiplier turns to zero, when w = 0, the last multiplier when p = 42/169 turns to zero also. When p = 0, the last multiplier is zero at w = 2/91. It can be proved directly that for p laying between these extreme values the last multiplier is zero at

$$w=\frac{2}{91}\left(1-\frac{169}{42}p\right) \tag{65}$$

So, the temperature has two roots with a maximum between them. The dependence of the temperature from the amplitude is shown in **Figure 9**. Physical meaning has increasing part of the curve from zero to maximum. In the vicinity of the maximum the temperature is nearly constant, so the square of vibration frequency, which is proportional to the fraction, kT/w^2 diminishes with nearing to the maximum. Such conclusion is formulated in Reference [18].

Gilvarry states that root mean square amplitude divided through mean distance between atoms at transition point has the same value for all substances with the same structure [19] and the same assertion is found in the article [20]. Such conclusion follows also from presented calculations. From (37) and (50) follows

$$u^2 q^{\frac{1}{3}} = \frac{kT}{\Lambda} \frac{l_M^2}{q(13q-7)} \tag{66}$$

With using the expression (47) for q this equality turns to

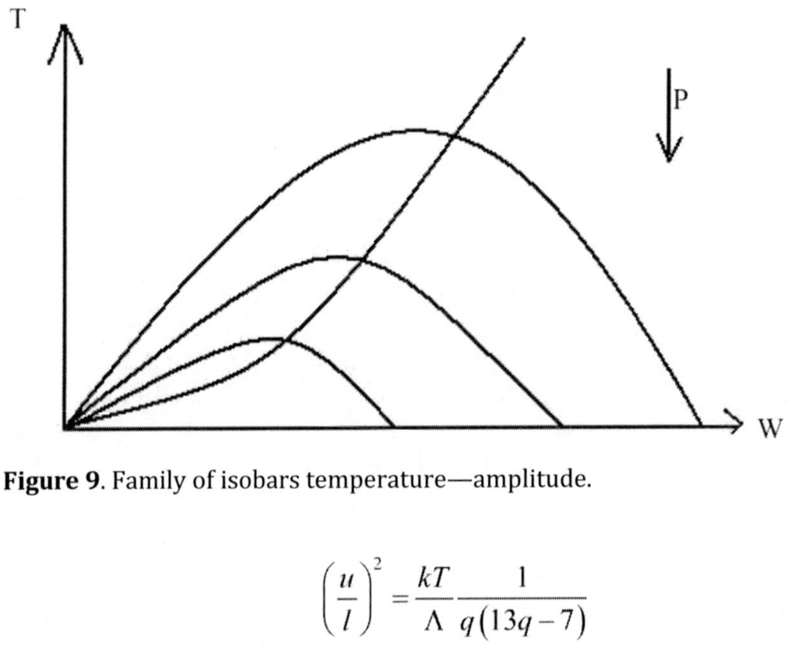

Figure 9. Family of isobars temperature—amplitude.

$$\left(\frac{u}{l}\right)^2 = \frac{kT}{\Lambda} \frac{1}{q(13q-7)} \tag{67}$$

By means of equation of state (58), where is put f = 0, this equation transforms to

$$\left(\frac{u}{l}\right)^2 = \frac{1}{7} \frac{1-q}{13q-4} \tag{68}$$

To receive the value of q in the transition point, one must differentiate (58) and the derivative equate to zero. Under the condition f = 0 definite value of q and consequently definite value of fraction u/l will be obtained.

9. CONCLUDING REMARKS

On the base of dynamic equations, that follow from variation principle of Hamilton, a thermal behavior of linear chain of atoms, interacting with nearest neighbors with forces, obeys inverse power law, and stretched with external force was considered. The mean potential energy is developed in series with respect to the amplitude of atomic vibrations and only second powers were retained. This restricts the temperature from above, and the Grüneisen type equation of state was obtained. The families of isotherms and isobars were drowing, and spinodals are marked. So, Grüneisen equation is an equation of state, which describes phase transitions. This is the answer to the question, formulated by Debye: Is the Grüneisen equation an equation of state? The original calculations contain assertions, which are strictly correct for linear dependence of interacting forces from displacements of atoms, so no transitions can occur. It is a correct reasoning, and if we put in Equations (37), (38) the numerical values (56), we come to result that linear and quadratic in temperature items are nearly equal, so at such temperature quadratic terms cannot be ignored, and Grüneisen equation is an approximation useful for lower temperature. In a discussion about Grüneisen work Nernst expressed his opinion [3]:

"You all have received an impression, that we are so far, that can consider also the solid state from moleculartheoretical point of view, and one may hope that we soon receive a similar perfectly true theory also for solid state, that we for long time possess for gases."

Vladimir Kh. Kozlovskiy The Grüneisen equation in essential features turns into Mie equation, when in the fraction the terms, arises from attraction interaction, will be stroked out. So, the foundation for solid state equation development is the Grü- neisen equation. The equation that takes into account terms with second potent of temperature will be essentially more complicated, than Grüneisen equation.

REFERENCES

1. W. Sutherland, "A Kinetic Theory of Solids, with an Experimental Introduction," Philosophical Magazine Series V, Vol. 32, No. 199, 1891, pp. 524-553. doi:10.1080/14786449108620220
2. G. Mie, "To the Kinetics Theory of One-Atomic Corps," Annals of Physics, Vol. 11, No. 8, 1903, pp. 657-697.
3. E. Grüneisen, "To the Theory of One-Atomic Solid Corps," Physical Magazine, Vol. 12, No. 22-23, 1911, pp. 1023- 1026.
4. E. Grüneisen, "Theory of Solid State One-Atomic Elements," Annals of Physics, Vol. 39, No. 12, 1912, pp. 257- 306.

5. P. Debye, "To the Theory of Heat Capacities," Annals of Physics, Vol. 39, 1912, pp. 789-839.
6. S. Ratnowsky, "One-Atomic Solid Corps Equation of State and the Quantum Theory," Annals of Physics, Vol. 38, 1912, pp. 637-648.
7. K. Eisenmann, "One-atomic Solid Corps Equation of State, according to the Quantum Theory," Annals of Physics, Vol. 39, No. 16, 1912, pp. 1165-1174.
8. E. Grüneisen, "Molecular Theory of Solid Corps," The Structure of the Matter, Conference of Physics, Bruxeles, October 1913.
9. P. Debye, "Equation of State and Quantum Hypothesis," Physical Magazine, Vol. 14, 1913, pp. 259-260.
10. P. Debye, "The Equation of State and Quantum Hypothesis with Supplement of Heat Conduction," Lectures on the Kinetic Theory of Matter and Electricity, Leipzig and Berlin, B. G. Teubner, 1914, pp. 19-46.
11. V. Fock and G. Krutkow, "Remarks to the Virial Theorem of Classical Mechanic," Physical Magazine of USSR, Vol. 1, No. 6, 1932, pp. 756-758.
12. V. Ch. Kozlovskiy, "Dynamical Equations for TimeAveraged Coordinates at Thermal Equilibrium," Soviet Physics Journal, Vol. 34, No. 4, 1991, pp. 353-359.
13. W. Braunbek, "Gratings Dynamics Theory of Melting Phenomenon," Physical Magazine, Vol. 38, No. 6-7, 1926, pp. 549-572.
14. L. N. Syrkin, "To the Question of Temperature Dependence of Thermal Ionic Polarization," Journal of Technical Physics, Vol. 26, No. 6, 1956, pp. 1163-1165.
15. Ja. I. Frenkel, "Statistical Physics, Moscow," Academy of Sciences, Leningrad, 1948.
16. Yu. M. Goryachev, L. L. Moiseenko and E. I. Shvartsman, "Parameters of Elastodynamic State of Dodecaborides of Rare Earth Metals," Soviet Physics Journal, Vol. 34, No. 5, 1991, pp. 455-458.
17. D. S. Lemon and C. M. Lund, "Thermodynamics of High Temperature Mie-Grüneisen Solids," American Journal of Physics, Vol. 67, No. 12, 1999, pp. 1105-1108. doi:10.1119/1.19091
18. F. Lindemann, "About the Calculation of Molecular Own Frequencies," Physical Magazine, Vol. 11, No. 14, 1910, pp. 609-612.
19. J. J. Gilvarry, "The Lindemann and Grüneisen Laws," Physical Review, Vol. 102, No. 2, 1956, pp. 308-316. doi:10.1103/PhysRev.102.308
20. E. Grüneisen, "To the Theory of One-atomic Solid Corps," Conference of German Physical Society, Vol. 13, 1911, pp. 836-847.

CHAPTER 22

Medium Range Structural Order in Amorphous Tantala Spatially Resolved with Changes to Atomic Structure by Thermal Annealing

Martin J. Hart[a,*], Riccardo Bassiri[b], Konstantin B. Borisenko[c], Muriel Véron[d], Edgar F. Rauch[e], IainW.Martin a, Sheila Rowan[a], MartinM. Fejer [b], IanMacLaren[a]

a SUPA, School of Physics and Astronomy, University of Glasgow, G12 8QQ, United Kingdom

b E. L. Ginzton Laboratory, Stanford University, Stanford, CA 94305, USA

c Department of Materials, University of Oxford, Parks Road, Oxford OX1 3PH, United Kingdom

d Univ. Grenoble Alpes, SIMAP, F-38000 Grenoble, France

e CNRS, SIMAP, F-38000 Grenoble, France

ABSTRACT

Amorphous tantala (a-Ta_2O_5) is an important technological material that has wide ranging applications in electronics,optics and the biomedical industry. It is used as the high refractive index layers in the multi-layer dielectric mirror coatings in the latest generation of gravitational wave interferometers, as well as other precision interferometers. One of the current limitations in sensitivity of gravitational wave detectors is Brownian thermal noise that arises from the tantala mirror coatings. Measurements have shown differences in mechanical loss of the mirror coatings, which is directly related to Brownian thermal noise, in response to thermal annealing. We utilise scanning electron diffraction to perform a modified version of Fluctuation Electron Microscopy (FEM) on Ion Beam Sputtered (IBS) amorphous tantala coatings, definitively showing an increase in the medium range order (MRO), as determined fromthe variance between the diffraction patterns in the scan, due to thermal annealing at increasing temperatures. Moreover, we employ Virtual Dark-Field Imaging (VDFi) to spatially resolve the FEM signal, enabling investigation of the persistence of the fragments responsible for the medium range order, as well as the extent of the ordering over nm length scales, and show ordered patches larger than 5 nm in the highest temperature annealed sample. These structural changes directly correlate with the observed changes in mechanical loss.

Keywords: Amorphous materials Fluctuation electron microscopy Nanodiffraction Medium-range order Tantalum-pentoxide Internal friction

1. INTRODUCTION

Ion-beam sputtered amorphous tantala (a-Ta_2O_5) is often the material of choice for the high refractive index layer of highly reflective thin film coatings and find widespread applications that range from optical atomic clocks [1], ring laser gyroscopes [2], frequency comb techniques [3] and high-precision interferometers such as the Laser Interferometer Gravitational-wave Observatory (LIGO) [4]. Amorphous tantala also has applications that include insulating films with high dielectric constant for electronics [5] and corrosion resistant coatings for biomedical applications [6].

However, the performance of the coatings must be improved to make them a viable option for future upgrades to ultra-high precision gravitational wave interferometers, which are currently expected to be limited, at their most sensitive frequencies, by thermal noise arising from the coatings. To do so, it is necessary to understand changes in the atomic structure that occur during manufacturing and post-processing. Previous studies have shown that doping and annealing of the thin films cause considerable changes to the macroscopic properties such as optical absorption, scattering and mechanical loss [7–9]. Mechanical loss is equivalent to internal friction and is defined as the reciprocal of the mechanical Q factor, a quantity that describes the level of damping in an oscillator; a higher Q value indicates a lower rate of energy loss per oscillation with respect to the energy that is stored within the oscillator. This measure of mechanical stability and its relationship to Brownian thermal noise is quantified through the fluctuation dissipation theorem by Callen and Greene [10]. In general, only small changes are observed in the atomic structure of the coatings of the same material prepared by different methods when studied by electron diffraction reduced density function (RDF) analysis (which appear to be only sensitive to short range order, principally around the 1st and 2nd nearest neighbours, and up to a maximum of about 1 nm, and consequently cannot distinguish between atomistic models that contain or do not contain nanoscale order) [11–13]. A previous study using this technique has however demonstrated a correlation between mechanical loss and the concentrations of titania in titania-doped tantala [14]. Extended X-ray Absorption Fine Structure (EXAFS) shows more extensive order to perhaps the 3rd or 4th nearest neighbour and is chemically sensitive to the ordering about specific atoms, although this still does not reveal any order beyond 1 nm. Our recent work has shown that this provides further insights into the behaviour on annealing [15]. The previous electron diffraction studies have shown the atomic structure to be homogeneous at volumes probed with electron beams of 50 nm in lateral extent up to a 600 °C heat treatment, whereas in the present study the atomic structure is shown to be heterogeneous over volumes probed with an electron beam of 2 nm in lateral extent. It has been suggested previously, that this apparent homogeneity of the coating atomic structure is a consequence of the scale at which the structure has been examined; the inherent averaging over volumes containing hundreds of millions to hundreds of billions of atoms averages out any local structural differences in the materials. It is then expected for there to be a peak in the heterogeneity of the atomic structure at a scale determined by the volume of the material probed, coinciding with the maximum variance in structural order. In this work, the volume of material probed contains in the order of thousands of atoms, and proper variation in the ordering of the structure at the medium range can be quantified.

Fluctuation Electron Microscopy (FEM) is a diffraction and/or imaging technique that quantifies medium range order (MRO) in the roughly 1 to 3 nm range. The original formulation of FEM Gibson and Treacy [17] examines the MRO by measuring spatially resolved diffracted intensity fluctuations from nano-volumes in the sample material through the normalised variance,

$$V(k,Q) = \frac{\left\langle I^2(\mathbf{r},k,Q) \right\rangle}{\langle I(\mathbf{r},k,Q)\rangle^2} - 1 \qquad (1)$$

where I ($\mathbf{R}$, k, Q) is the diffracted intensity as a function of position $\mathbf{R}$ on the sample, scattering vector k, probe size Q, and $\langle ... \rangle$ indicates averaging over $\mathbf{R}$. The technique is sensitive to three- and four-body correlations [16], and the fluctuations are maximally sensitive when the electron probe size is of comparable length scale to the MRO structural ordering being probed. So the extent of MRO is quantified through the magnitude of the variance of the diffracted intensity, as a function of scattering vector over a length scale determined by the size of probe used. Originally proposed by Gibson and Treacy [17], the technique was initially carried out using dark-field imaging in the TEM, although an equivalent experiment can be carried out using scanning diffraction [18]. This latter experimental approach has distinct advantages, especially on modern scanning transmission electron microscopes, where probes well below 1 nm in diameter can be routinely produced, provided the diffraction patterns can be acquired reasonably quickly (which is now possible due to advances in imaging detectors).

Atomistic models have shown that the variance displays clear trends as a function of the size and volume fraction of the ordered regions [19, 20] and, to date, the technique has been employed to show variation in the nanoscale order of amorphous silicon [21–26] and amorphous germanium [16, 18, 27] thin films, phase change chalcogenide materials [28–30], and a selection of amorphous metals [31, 32]. In these experiments, qualitative differences in FEM variance were observed and attributed to fundamental physical phenomena such as differences in film deposition condition [23], the existence and thermal ripening of subcritical nuclei that precede crystallisation [29, 30], and the effect of alloying on crystallisation kinetics [33]. Quantitative FEM analysis has thus far proven challenging, but with recent developments such as variable resolution FEM, information about the extent of the nanometre-scale ordering can be extracted [26, 34]. Nevertheless, a number of recent studies have been successful in relating the scattering covariance and angular correlations in FEM data to structural information [13, 35, 36].

In recent work, we used scanning nano-diffraction FEM to collect data, in a similar way to that described by Voyles and Muller [16], and demonstrated the existence of MRO in a-Ta_2O_5[37]. In the version of FEM applied in the present work, we depart from the standard formalism in Eq. (1), and by assuming noise-free kinematic coherent diffraction to be Gaussian distributed, compute the variance of standardised correlation coefficients obtained from a normalised cross-correlation of a Gaussian filter with the raw diffraction data,

$$V(\gamma, k, Q) = \left\{ \langle \gamma^2 (\mathrm{I}(\mathbf{r}, x, y, Q), \mathrm{t}) \rangle - \langle \gamma(\mathrm{I}(\mathbf{r}, x, y, Q), \mathrm{t}) \rangle^2 \right\}_k . \quad (2)$$

where γ, the correlation coefficient is obtained from:

$$\gamma(x,y)=\frac{\sum_{x,y}\left(I(x,y)-\bar{I}_{u,v}\right)\left(t(x-u,y-v)-\bar{t}\right)}{\sqrt{\sum_{x,y}\left(I(x,y)-\bar{I}_{u,v}\right)^{2}\left(t(x-u,y-v)-\bar{t}\right)^{2}}} \tag{3}$$

In Eq. (3), t is the Gaussian filter, $\langle ... \rangle$ indicates averaging over **R**, and I is as in Eq. (1) with the exception that here the variance is computed on a pixel by pixel basis (x, y) through the diffraction pattern stack resulting in a variance map which is thereafter azimuthally averaged (represented by $\{...\}_k$) to obtain the variance as a function of scattering vector k. Eq. (2) is a standard expression for variance which can be found in any statistical reference manual and is easily recognised by the mnemonic "mean of the square minus square of the mean"; it differs in form from Eq. (1) by the change of variable and the normalisation factor in the denominator; we normalise our data through Eq. (2) prior to computing the variance. Whilst the Gaussian filtering was initially intended to mitigate noise, it became apparent that a change of variable in the variance calculations from intensity to a normalised score of the intensities structural significance simultaneously removed noise, background and standardised the datasets. In Eq. (3), $I(x, y)$ denotes the intensity value of the diffraction pattern at the point (x, y),$I_{u,v}$ is the mean value of $I(x, y)$ within the area of the Gaussian filter t at the point (x, y), and t is the mean value of the Gaussian filter. The denominator in Eq. (3) contains the variance of the zero mean diffraction pattern function $I(x,y)-\bar{I}_{x,y}$ and the zero mean Gaussian filter function $\bar{t}-\bar{t}$ at the point (x, y). The 2-D normalised cross correlation is used here as a standardised means to evaluate the significance of the raw diffracted intensity at each point in the diffraction pattern, and is scored upon the similarity of the local distribution of intensity around each point to our model diffraction maxima centred on that point. Our model diffraction maximum is a 7 × 7 pixel rotationally symmetric normalised Gaussian filter with a two pixel standard deviation, obtained by fitting a 2-D Gaussian function to a sharp Bragg spot in a diffraction pattern from the same sample series, which had crystallised after a 800 °C heat treatment. Using this approach, the absolute magnitude of the scattered intensity is irrelevant and instead, it is the shape of the intensity distribution around each pixel that becomes relevant, allowing coherent diffraction with poor SNR to emerge from the background. The resulting normalised correlation map is then a standardised transform of the diffraction pattern into a map of the diffracted intensity's structural significance, where scores range in value between − 1 (maximally anti-correlated), zero (uncorrelated) and 1 (maximally correlated). Only positive scores are deemed structurally significant as we assume coherent diffraction to be approximately Gaussian, whereas the diffuse background and single pixel events are not Gaussian distributed. We thus remove the negatively scored intensity contributions from calculations which anti-correlate with our model Gaussian filter, and we assume that much of the noise and diffuse background in the system will result in this negative range of scored intensities. As a result of this normalisation, $\gamma(x, y)$ is invariant to brightness or contrast variations in the diffraction patterns (including from diffuse inelastic scattering), which are related to the values of the mean and the standard deviation; this has the effect of standardisation of the data-sets and preservation of real diffraction spots deemed structurally significant through positive correlation, whilst rejecting single pixel noise or X-ray events. Intuitively, this approach seems well suited to the study of structural change due to thermal annealing; each diffraction pattern across the sample series is rendered statistically equivalent by the transformation of raw intensity pixel values to a normalised correlation coefficient which itself is determined by the similarity of that pixel and its surrounding area (size of the filter) to the Gaussian filter centred on that pixel. As the Gaussian filter is a representation of diffraction maxima from the specimen that has attained long range order by thermal annealing at 800 °C, it serves as a reference in which the distribution of diffracted intensity in the remaining thermal annealing series, which have not attained that level of order, can be meaningfully compared using the relative values of the assigned correlation coefficients, which describe the degree structural significance that the raw intensity values contribute to the diffraction pattern. Inspired by methods used in image analysis for feature tracking, this approach renders the data invariant to inconsistencies in the illumination conditions during data acquisition [38], and removes the need to normalise after the variance is computed. It has also been shown that reliable extraction of information using the formal FEM technique is highly dependent upon the quality and reproducibility of the experimental data, and as such, steps have been laid out to accurately identify and correct artefacts in affected datasets [39]which we review in relation to our data.

Early work in scanned diffraction used a very small number of diffraction patterns and was very slow to collect [18] due to the limitations of the imaging detectors available at the time. More recently, rapid advances were made in high speed acquisition of scanned diffraction data using an optically coupled camera [40–42]. Whilst these advances were principally made with precession electron diffraction in mind [41], spatial resolution can be increased by deactivating the precession and this is then ideal for application in FEM with a resolution of ~ 1 nm, although the noise in the system requires some special treatment, as described below. Future work will be further aided by the use of a new generation of direct electron detectors that allow high frame rate collection (~ 1 kHz) with almost perfect quantum efficiency [43–45].

In addition to the classic variance-based analysis of the scanned diffraction data [18], it is also possible to image features of the scanned diffraction dataset in real space due to the spatially indexed nature of the diffraction pattern collection which we discuss further in Section 3.2. Using the Virtual Dark-Field Imaging (VDFi) approach of Rauch and Véron, real space images are formed from the indexed scanned diffraction dataset by using the intensity in selected parts of each diffraction pattern [46]. This is a counterpart to classic dark field imaging, where images are formed using only selected scattered electrons instead of the undiffracted central spot, and this is realised by the insertion of a physical aperture in the back focal plane of the objective lens. This has the effect of masking almost all the diffraction pattern, except the one reflection which is visible through the aperture. For the VDFi construction [47], the whole back focal plane is collected with a spatially resolved detector and later numerically reconstructed to rebuild real space representations of the materials, using any desired combination of integrated intensity (unlimited combinations of numerical apertures can be employed) out of the diffraction patterns, in order to create the pixel values of the reconstructed real-space images. This has one clear advantage over conventional dark field imaging: the whole back focal plane is collected allowing any number of different dark field images to be reconstructed afterwards, whereas conventional dark field imaging has just one reflection selected per image. Of course, this comes with a downside that a vast quantity of data needs to be collected to map any reasonable scan area at a sensible spatial resolution, when only a tiny fraction of the data is used in the production of each VDFi image. Thus far the technique has been used to spatially map phases and orientations of polycrystalline materials [46, 48]; here we extend the concept of FEM to map the evolution of medium range order and nano-crystallite nucleation in glasses as a function of thermal annealing, effectively spatially resolving the FEM results. Instead of a two dimensional plot of variance as a function of scattered intensity $I(k)$, we can now visualise the entire dataset simultaneously as a spatially resolved variance map at a given scattering vector (or indeed using any part of the scattered beam). This is similar to FEM using dark-field imaging, in which the intensity is collected in the form of real space images, which have one scattering vector k, one probe size Q, and many spatial samples $\mathbf{R}$, $k$and Q get changed, and the next image is acquired. On the contrary, in our method of STEM nano-diffraction, $I(\mathbf{R},k,Q)$ is obtained in the form of an electron diffraction pattern acquired with a nanometre-sized probe, which has one Q, one $\mathbf{R}$, and many $k$values. That probe is then rastered across the sample to acquire many $\mathbf{R}$[49]. VDFi therefore enables FEM data acquired via scanned diffraction to be quantified in new ways, by combining the real space and reciprocal space representations in one analysis. This combined approach is used in the present work to probe the changes in amorphous Ta_2O_5 films on thermal annealing.

2. EXPERIMENTAL

2.1. Samples/annealing process/specimen preparation

The amorphous IBS tantala coatings were manufactured by the Commonwealth Scientific and Industrial Research Organisation (CSIRO, Materials Science and Engineering Division, West Lindfield, NSW, Australia). The coatings were deposited onto fused silica substrates and were subject to post-deposition annealing at 300 °C, 400 °C and 600 °C for 24 h in air. Specimens were prepared for the scanning electron diffraction studies using a standard cross section method of gluing two sections of film face to face, encapsulation in a brass tube with epoxy resin, slicing, polishing and dimpling. The samples were then thinned to electron transparency using a Gatan Precision Ion Polishing System (PIPS) (Gatan Inc., Pleasanton, CA, (USA), which used Ar + ion irradiation at a relatively low energy, 4 kV beam, and a final 0.5 kV polishing stage, to avoid any changes to the sample material structure.

2.2. Data collection and reduction

The electron diffraction data were collected on a JEOL JEM2100F operating at 200 keV at the University of Grenoble. The microscope was set to nano-beam mode in TEM, encapsulating diffraction from a volume set by the beam diameter, 2 nm FWHM in this work, with a dedicated hardware solution made by NanoMEGAS which controlled the deflection coils of the TEM allowing it to be operated in a ~ 2 mrad low convergence angle STEM mode [40, 42, 50]. Although we did not measure the probe current absolutely, it was kept low enough that there was no speckle movement under the stationary beam, which indicates that there is no modification of the atomic structure of the sample by the beam. The diffraction patterns were recorded using an external video camera imaging the TEM phosphorous screen with a dynamic range of 8 bits, at a rate of 100 patterns per second over the area of interest. All diffraction patterns were linear unsaturated measurements of the diffracted intensity.

At this probe resolution of ~ 2 nm, the resulting diffraction patterns from amorphous tantala take the form of speckly distributions of diffracted intensity as depicted in Fig. 1(a), as is typical of such FEM experiments, as opposed to the diffuse rings that are conventionally associated with diffraction from larger volumes of amorphous materials seen in Fig. 1(b) (recorded with an exposure time an order of magnitude greater than that of Fig. 1(a), from a volume five orders of magnitude larger). Of course, simply adding up many speckle patterns from different regions would produce a pattern like Fig. 1(b). Whilst the authors are aware of methods that explicitly investigate the length scale of MRO using a range of probe sizes [12], these were not explicitly used in this work, and the 2 nm probe size was found to optimal for FEM studies on this sample with this microscope.

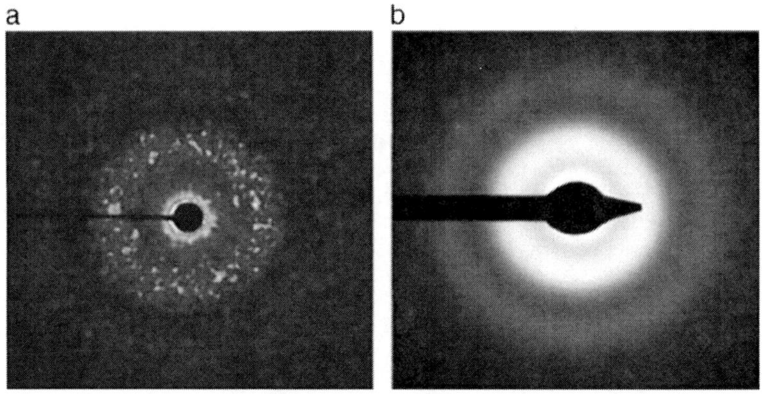

Figure. 1. (a) Typical tantala FEM diffraction pattern using 2 nm probe FWHM, (b) typical tantala diffraction pattern using 50 nm probe FWHM, recorded on a CCD.

The optical coupling in this work, whilst advantageous for rapid data acquisition in low convergence angle STEM mode, has the undesired effect of introducing artefacts and noise into the data. This includes artefacts such as spurious X-rays impinging on the phosphor screen, as well as significant electronic noise in the camera. Methods investigated to reduce these, as well as other sources of noise led to the development of the version of FEM used in this work. Fig. 2 shows an example of a typical diffraction pattern used here for analysis, before and after the data reduction process.

Although the general purpose of the 2D normalised cross correlation is for the determination of the position of a feature within an image using template matching, all of the features of interest in the diffraction pattern bear similarity to our chosen template. Thus the result is not a correlation map of the likeliest position of a feature, but a correlation map of the similarity of regions of the diffraction pattern to the Gaussian filter. Although the variance Eqs. (1) and (2) are not mathematically equivalent,

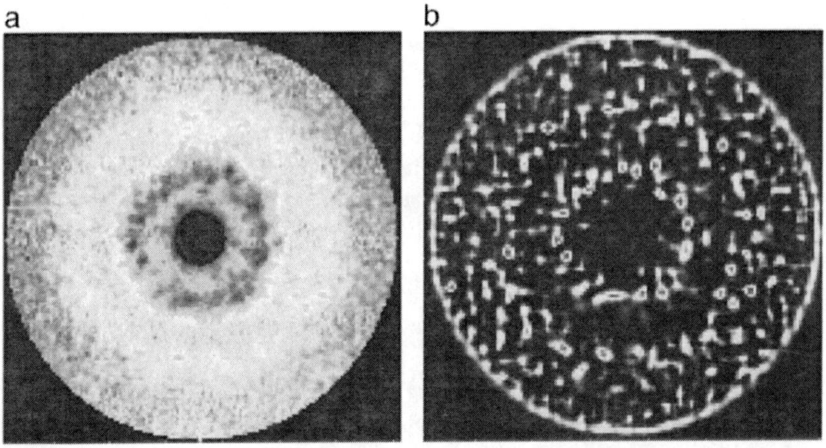

Figure. 2. (a) Raw diffraction pattern, (b) diffraction pattern cross correlated with normalised Gaussian filter, retaining only positive correlations. The central beam has been removed for clearer comparison of features

they both provide a dimensionless value for the variance which are directly related to the structural heterogeneity of the samples. Therefore, we compute the variance of the correlation coefficients of the diffracted intensity with a 2-D Gaussian as opposed to the directly detected intensity values, and use them as a proxy for the scattered intensity by approximating the two dimensional distribution of intensity to be Gaussian in nature for individual diffraction maxima. The use of a constant Gaussian filter throughout the whole data series is key in obtaining meaningful comparison of the standardised data sets of the Ta_2O_5 heat treatment series, and serves as the model which is used to select structurally significant regions of diffracted intensity for calculations.

Li et al. [39] provided a basis upon which FEM can be performed with confidence in the quantitative magnitude of the data, which entails the identification and removal of common artefacts in the dataset. We examine these noise sources here in relation to our data.

The datasets used in this work were from conventionally prepared cross sectional samples which contained some thickness variations, although care was taken to choose areas for analysis which were extremely thin and with very small thickness variations. The consequence was that no saw-tooth pattern was visible in a plot of $I(k)$ versus pattern number at high k before or after data reduction (which would indicate a large thickness change). No voids were identified, neither were any roughness effects. A small amount of carbon contamination was visible in the raw data in a plot of $I(k)$ versus pattern number at high k, although after data reduction the effect vanished. Furthermore, no multiple scattering effects or large nanocrystals were observed.

Additionally, we observed no noticeable dependence of the variance upon the specimen thickness, which was determined by plotting the main variance peaks of 56 FEM signals (100 diffraction patterns each), from parallel lines normal to the small thickness gradient in the dataset, for both the 300 °C and 600 °C data sets. These results confirm that any differences in the volumes of the samples studied were inconsequential to our analysis [39]. The 400 °C data was sampled from areas similar to the 300 °C and 600 °C data sets and underwent identical sample preparation, indicating minimal thickness variation within each sample and from sample to sample. The usual FEM signal is also somewhat sensitive to probe coherence; however using a small semi-convergence angle, and by performing calculations only where scattered intensities have been determined to be structurally significant, as opposed to the total scattered intensity, this sensitivity is reduced.

3. ANALYSIS AND RESULTS

3.1. Fluctuation Electron Microscopy (FEM)

Fig. 3 shows the results of the average variance of five data sets, each using 1000 diffraction patterns, computed for the 300 °C, 400 °C and 600 °C samples, with the error bars showing the standard deviation of the variance computed over the five data sets. These datasets simultaneously confirm the presence of MRO and of a change in short range atomic structure, both as a consequence of thermal annealing.

As mentioned previously, the atomic structure of these coatings annealed up to 600 °C appears homogeneous when probed with a beam of ≥ 50 nm diameter as a function of scattering vector and position on the sample; so unlike the peaks in Fig. 3, a variance plot at that resolution would appear relatively flat and give no hint to any underlying medium range order. Here, probing the structures with a 2 nm diameter probe over thousands of areas of the samples, and obtaining these variance peaks, confirms the heterogeneity of the atomic structure over these length scales. There is

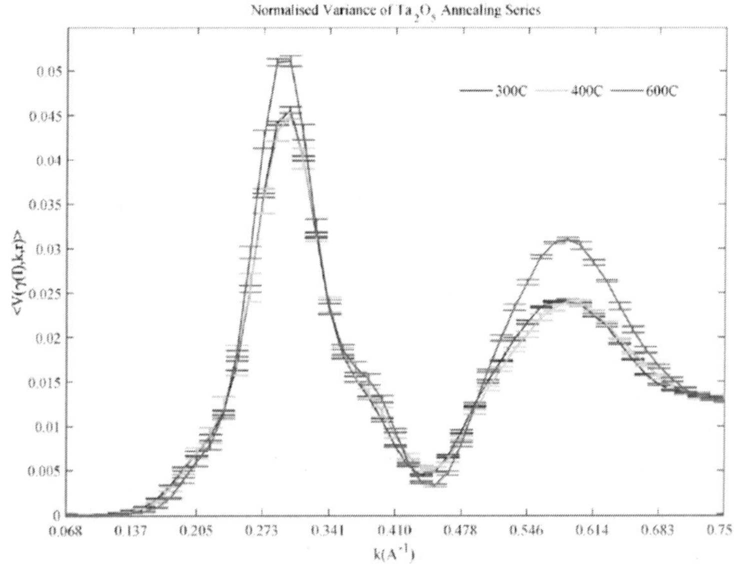

Figure. 3. Comparison of the normalised variance of the tantala thermal annealing series. The variance plot shows that a similar magnitude of change between the first and second peaks is seen as a function of heat-treatment.

always a trade-off between spatial resolution and averaging when selecting a probe size and this technique is effectual only when the probe is sufficiently small that averaging of the volume does not remove resolution in reciprocal space of the diffracted intensity. The 2 nm probe here was selected qualitatively based upon the prominence and contrast of the observed variation in diffracted intensity at different positions on the sample. And whilst a different level of heterogeneity can be observed using different probe sizes, it has been shown that there exists a peak in the heterogeneity of the structure that coincides with the maximum variance, and is observed when the probe size is of comparable dimension to the characteristic length scale of the order within the material [12]. Gibson et al. developed a technique which can quantitatively determine this characteristic length scale using different probe sizes [12] which we plan to use in future studies.

At this scale, the change in atomic structure due to thermal annealing is apparent; the magnitude of the 600 °C variance is significantly larger than that of the 300 °C and 400 °C data, indicating further deviation from homogeneity that can be explained in terms of the structure beginning to organise itself into regions of greater order. This seems to correlate with mechanical loss measurements of the same materials which show no significant difference between the 300 °C and 400 °C coatings, but shows significant changes between those and the 600 °C coating [7].

3.2. Virtual Dark-Field Imaging

The concept behind the Virtual Dark-Field Imaging (VDFi) technique is shown in Fig. 4. Each of the circles represents a diffraction pattern whose resolution Q is dictated by the size of the incident electron probe (2 nm FWHM), rastered across the sample surface. Each of the 1.5 nm × 1.5 nm squares represent a pixel of the reconstructed VDFi image and are the step size in which the electron beam is rastered. The intensity of each pixel is the integrated intensity of any part of the diffraction pattern selected. Fig. 4(a) shows the selection of a line segment containing three Bragg reflections; integrating the resultant intensity from the same location in each diffraction pattern, we obtain an image highlighting the persistence of that structural motif in such a volume over a user defined area. Likewise shown in Fig. 4(b), (c) and (d), any numerical aperture or combination thereof, can be created to enable investigation of the persistence of such nanoscale order over any desired area.

Fig. 5 shows an example of VDFi images of the 300 °C, 400 °C and 600 °C samples created using a virtual annular aperture such as in Fig. 4(c), integrated over the scattering vectors between 0.287 $Å^{-1}$ and 0.300 $Å^{-1}$. The effect of this range selection is to visualise the FEM data in real space where the amorphous structure is furthest from homogeneity, which in this case predominantly describes the heterogeneity of the tantalum–tantalum nearest neighbour environment. Setting the angular range for VDFi at some other point where the variance is less will show less in the way of interesting structure. It is important to note that these VDFi images were constructed using the same data sets that were used in the FEM analysis, and are thus consistent with the FEM results and investigate the spatial persistence of fragments responsible for medium range ordering.

As intimated before, these virtual images can be constructed from the integrated intensity of annuli at all scattering vectors and allow the different changes in structure due to thermal annealing to be charted as a function of such. The results of the Gaussian functions fits to the histograms of intensity distributions of the virtual images shown in Fig. 5 complement that of the FEM study. As heat treatment temperature increases, the mean intensity of the main diffraction peaks increases, with the relative incidence of higher diffracting centres decreasing as would be expected with certain areas becoming more ordered. In this instance we still retain the spatial variance of the diffracted intensity at that scattering vector and in each VDFi image the groupings of red pixels indicate areas of the greatest diffracted intensity, corresponding to the highest degree of structural ordering. In all of these VDFi images, clusters of red pixels are found. Whilst many of the clusters are just 2 pixels wide (about 3 nm), some are as large as 4 pixels across, suggesting patches of order to greater than 5 nm in places.

The results of Fig. 5(a, b, c) (where pixel colours represent the range of integrated intensity and have a one to one correspondence in each of the VDFi images) are quantified in Fig. 5(d), which is a plot of the frequency of occurrence versus mean intensity. Whilst Fig. 5(a) and (b) for the 300 °C and 400 °C annealed materials show a rather similar frequency distribution, as would be expected from their very similar variance plots in Figs. 3, 5(c) shows that the peak of the distribution is moved to higher intensities, although with a lower height. This is fully in accord with the changes in variance observed at the same scattering vector in Fig. 3, and demonstrates an increased ordering after annealing at 600 °C into fewer, larger ordered regions.

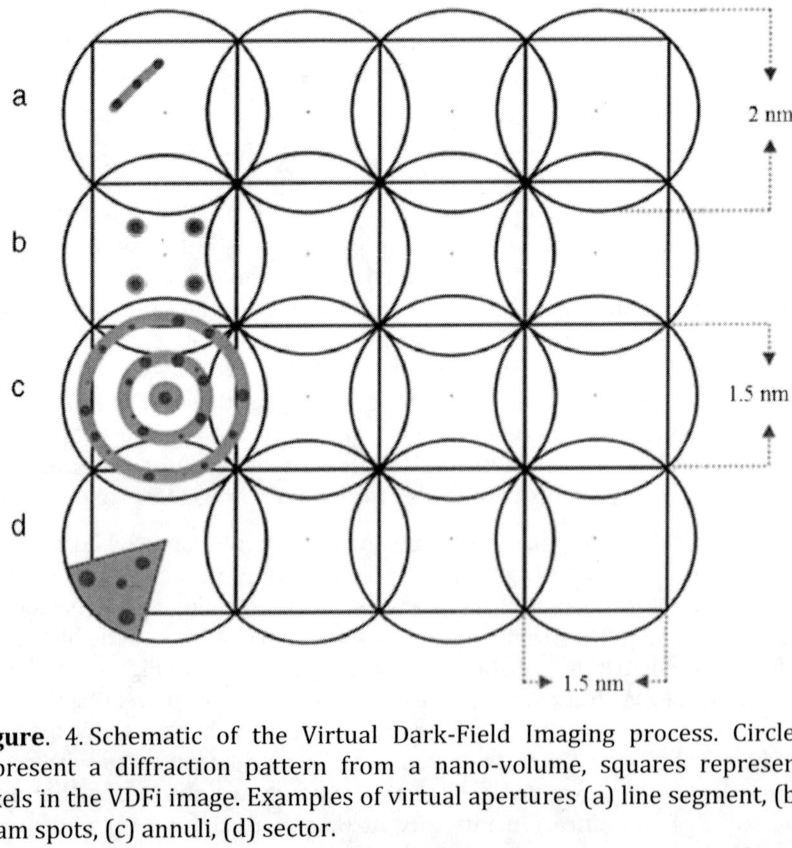

Figure. 4. Schematic of the Virtual Dark-Field Imaging process. Circles represent a diffraction pattern from a nano-volume, squares represent pixels in the VDFi image. Examples of virtual apertures (a) line segment, (b) beam spots, (c) annuli, (d) sector.

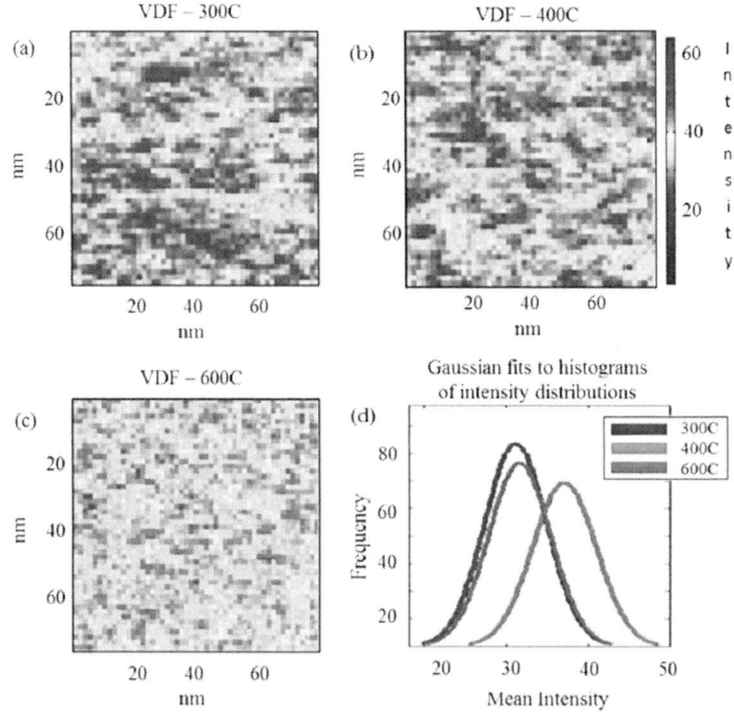

Figure. 5. Virtual Dark-Field Images of (a) 300 °C, (b) 400 °C, (c) 600 °C from the tantala thermal annealing series. (d) Gaussian fits to the histograms of intensity distributions of the VDFi images. The pixel intensities represent integrated intensities of annuli in the range 0.287 $Å^{-1}$ to 0.300 $Å^{-1}$.

4. DISCUSSION

We note that the mechanical loss of these coatings at low temperatures increase with increasing heat treatment [7], and thus it seems plausible that the reason may be related to degree of medium range order. Moreover, the shift of the distributions in Fig. 5(d), due to thermal annealing, very closely resembles the shift of the mechanical loss peaks measured for these same samples [7] and will be investigated further to ascertain if any correlations of the mechanical loss with the presence of nano-crystallites are found. Recent work by Treacy and Borisenko on the local structure of amorphous silicon found para-crystalline structures containing local cubic ordering at the 10 to 20 Å length scale [12]. Similarly, our results suggest that the structure of amorphous tantala can be better described by a phase separated heterogeneous model than the continuously uniform random network model of Zachariasen [51]. Other studies have shown favour to this same nano-crystalline theory of glass [52–54]. Although similar results to the study in [54] initially have been interpreted in terms of Zachariasen's continuous random network model [55], further studies revealed the presence of nanoscale ordering [56].

It is important to consider the connection between the structural trends highlighted as a consequence of annealing in this work, and the trends in mechanical loss with annealing for the same materials. It is found that increased annealing temperature to 600 °C reduces the room-temperature mechanical loss, making the use of 600 °C annealing advantageous for mirrors designed for use at room temperature [7] (such as Advanced LIGO). At the same time, this 600 °C annealing causes the appearance of a significant and well-localised peak in loss at around 20 K, making this annealing possibly less suited for the preparation of coatings for use in cryogenic detectors [7], such as Kagra, currently under construction in Japan [57], and the proposed Einstein Telescope [9]. It may be suggested that the increased ordering affects the number and energetics of defects forming double well potentials at boundaries between paracrystallites. These defects appear to have relatively low energy barrier between two metastable states, reducing the overall losses at elevated temperatures whilst increasing the losses at lower temperatures, similarly to the mechanical loss mechanism proposed for amorphous silica [58].

Future work will further explore the proposed VDFi approach in conjunction with FEM to look for the evolution of structural motifs that correlate with the measured mechanical loss of these coatings resulting from alloying and thermal annealing. It is also planned to perform improved versions of this experiment using the JEOL ARM200F at the University of Glasgow, imaging onto a Medipix III, electron-counting direct electron detector [43–45], in order to study the medium range order in titania-doped tantala.

5. SUMMARY

We have shown using FEM and VDFi that medium range order exists in Ion Beam Sputtered amorphous tantala thin film mirror coatings on the length scale of 1.5–6 nm (from min/max cluster sizes in VDFi data), and that said order increases with increasing levels of thermal annealing, showing definitively for the first time that changes in the medium range ordering of the atomic structure of amorphous tantala result from thermal annealing. It is moreover demonstrated that Virtual Dark-Field Imaging is a valuable spatially resolvable FEM technique for the study of the evolution of local ordering in glasses due to thermal annealing. It seems that the increased medium range order with annealing to 600 °C is correlated with both the reduction in room temperature mechanical loss and the increase in low temperature loss around 20 K, and a tentative model for the mechanism underlying this correlation has been proposed.

ACKNOWLEDGEMENTS

MJH, IM, IWM and SR gratefully acknowledge the support of the STFC(ST/L000946/1 'Investigations in Gravitational Research'), for this work. MJH is grateful to the EPSRC for a PhD studentship. IWM is supported by a Royal Society Research Fellowship. The authors would like to thank Dr. Stavros Nicolopoulos of NanoMegas SPRL for his support for the work in Grenoble, without which this work would not have been possible. RB and MMF gratefully acknowledge the support of the NSF, under award number PHY-1404430.

REFERENCES

1. Y.Y. Jiang, A.D. Ludlow, N.D. Lemke, R.W. Fox, J.A. Sherman, L.-S. Ma, C.W. OatesMaking optical atomic clocks more stable with 10–16-level laser stabilization Nat. Photonics, 5 (3) (2011), pp. 158-161
2. D.A. Andrews, S. Roden, T.A. KingA model for lock-in growth in ring laser gyroscopes IEEE J. Quantum Electron., 31 (9) (1995), pp. 1709-1715
3. K. Numata, A. Kemery, J. CampThermal-noise limit in the frequency stabilization of lasers with rigid cavities Phys. Rev. Lett., 93 (25) (2004), p. 250602
4. G.M. HarryAdvanced LIGO: the next generation of gravitational wave detectors Classical Quantum Gravity, 27 (8) (2010), p. 084006
5. S. Zaima, T. Furuta, Y. Koide, Y. Yasuda, M. LidaConduction mechanism of leakage current in Ta_2O_5 films on Si prepared by Lpcvd J. Electrochem. Soc., 137 (9) (1990), pp. 2876-2879
6. Y.L. Zhou, M. Niinomi, T. Akahori, H. Fukui, H. Toda Corrosion resistance and biocompatibility of Ti–Ta alloys for biomedical applicationsMater. Sci. Eng. A, 398 (1–2) (2005), pp. 28-36
7. I.W. Martin, R. Bassiri, R. Nawrodt, M.M. Fejer, A. Gretarsson, E. Gustafson, G.M. Harry, J. Hough, I. MacLaren, S. Penn, S. Reid, R. Route, S. Rowan, C. Schwarz, P. Seidel, J. Scott, A.L. Woodcraft Effect of heat treatment on mechanical dissipation in Ta_2O_5 coatings Classical Quantum Gravity, 27 (22) (2010), p. 225020
8. G.M. Harry, M.R. Abernathy, A.E. Becerra-Toledo, H. Armandula, E. Black, K. Dooley, M. Eichenfield, C. Nwabugwu, A. Villar, D.R.M. Crooks, G. Cagnoli, J. Hough, C.R. How, I. MacLaren, P. Murray, S. Reid, S. Rowan, P.H. Sneddon, M.M. Fejer, R. Route, S. Penn, P. Ganau, J.M. Mackowski, C. Michel, L. Pinard, A. Remillieux Titania-doped tantala/silica coatings for gravitational-wave detection ClassicalQuantum Gravity, 24 (2) (2007), pp. 405-415
9. ET Design study: The ET Science Team 2011 Einstein gravitational wave telescope conceptual design study. p. ET Technical Document Number ET-0106A-10.

10. H.B. Callen, R.F. GreeneOn a theorem of irreversible thermodynamics Phys. Rev., 86 (5) (1952), pp. 702-710
11. R. Bassiri, K.B. Borisenko, D.J.H. Cockayne, J. Hough, I. MacLaren, S. Rowan Probing the atomic structure of amorphous Ta_2O_5 coatings Appl. Phys. Lett., 98 (2011), p. 031904
12. M.M.J. Treacy, K.B. BorisenkoThe local structure of amorphous siliconScience, 335 (6071) (2012), pp. 950-953
13. A Yan, T. Sun, K.B. Borisenko, D.B. Buchholz, R.P.H. Chang, A.I. Kirkland, V.P. Dravid Multi-scale order in amorphous transparent oxide thin films J. Appl. Phys., 112 (5) (2012), p. 054907
14. R. Bassiri, M.J. Hart, R.L. Byer, K.B. Borisenko, K. Evans, M.M. Fejer, A.C. Lin, I. MacLaren, A.S. Markosyan, I.W. Martin, R.K. Route, S. RowanCorrelations between the mechanical loss and atomic structure of amorphous TiO_2-doped Ta_2O_5 coatingsActa Mater., 61 (4) (2013), pp. 1070-1077
15. R. Bassiri, F. Liou, M.R. Abernathy, A.C. Lin, N. Kim, A. Mehta, B. Shyam, R.L. Byer, E.K. Gustafson, M.J. Hart, I. MacLaren, I.W. Martin, R.K. Route, S. Rowan, J. Stebbins, M.M. Fejer Order within disorder: the atomic structure of ion-beam sputtered amorphous tantala (a-Ta_2O_5)APL Mater., 3 (3) (2015), p. 036103
16. J.M. Gibson, M.M.J. Treacy, P.M. Voyles Atom pair persistence in disordered materials from fluctuation microscopy Ultramicroscopy, 83 (3–4) (2000), pp. 169-178
17. M.M.J. Treacy, J.M. Gibson Variable coherence microscopy: a rich source of structural information from disordered materials Acta Crystallogr. A, 52 (2) (1996), pp. 212-220
18. P.M. Voyles, D.A. Muller Fluctuation microscopy in the STEMUl tramicroscopy, 93 (2) (2002), pp. 147-159
19. S.N. Bogle, P.M. Voyles, S.V. Khare, J.R. Abelson Quantifying nanoscale order in amorphous materials: simulating fluctuation electron microscopy of amorphous silicon J. Phys. Condens. Matter, 19 (45) (2007), p. 455204
20. F. Yi, P.M. Voyles Analytical and computational modeling of fluctuation electron microscopy from a nanocrystal/amorphous composite Ultramicroscopy, 122 (2012), pp. 37-47
21. J.M. Gibson, M.M.J. Treacy, P.M. Voyles, J.R. Abelson Structural disorder induced in hydrogenated amorphous silicon by light soaking Appl. Phys. Lett., 73 (21) (1998), pp. 3093-3095
22. J.-Y. Cheng, J.M. Gibson, D.C. Jacobson Observations of structural order in ion-implanted amorphous silicon J. Mater. Res., 16 (11) (2001), pp. 3030-3033
23. P.M. Voyles, J.E. Gerbi, M.M. Treacy, J.M. Gibson, J.R. AbelsonAbsence of an abrupt phase change from polycrystalline to amorphous in silicon with deposition temperature Phys. Rev. Lett., 86 (24) (2001), pp. 5514-5517
24. P.M. Voyles, J.E. Gerbi, M.M. Treacy, J.M. Gibson, J.R. Abelson Increased medium-range order in amorphous silicon with increased substrate temperature J. Non-Cryst. Solids, 293–295 (0) (2001), pp. 45-52
25. L.N. Nittala, S. Jayaraman, B.A. Sperling, J.R. Abelson Hydrogen-induced modification of the medium-range structural order in amorphous silicon films Appl. Phys. Lett., 87 (24) (2005), p. 241915
26. S.N. Bogle, L.N. Nittala, R.D. Twesten, P.M. Voyles, J.R. AbelsonSize analysis of nanoscale order in amorphous materials by variable-resolution fluctuation electron microscopy Ultramicroscopy, 110 (10) (2010), pp. 1273-1278
27. J.M. Gibson, M.M.J. Treacy Diminished medium-range order observed in annealed amorphous germanium Phys. Rev. Lett., 78 (6) (1997), pp. 1074-1077
28. M.-H. Kwon, B.-S. Lee, S.N. Bogle, L.N. Lakshmi, S.G. Bishop, J.R. Abelson, S. Raoux, B.-K. Cheong, K.-B. Kim Nanometer-scale order in amorphous $Ge_2Sb_2Te_2$ analyzed by fluctuation electron microscopy Appl. Phys. Lett., 90 (2) (2007), p. 021923
29. B.-S. Lee, G.W. Burr, R.M. Shelby, S. Raoux, C.T. Rettner, S.N. Bogle, K. Darmawikarta, S.G. Bishop, J.R. Abelson Observation of the role of subcritical nuclei in crystallization of a glassy solid Science, 326 (5955) (2009), pp. 980-984

30. K. Darmawikarta, B.-S. Lee, R.M. Shelby, S. Raoux, S.G. Bishop, J.R. Abelson Quasi-equilibrium size distribution of subcritical nuclei in amorphous phase change AgIn-Sb_2TeJ. Appl. Phys., 114 (3) (2013), p. 034904
31. Stratton W.G., *et al.*Aluminum nanoscale order in amorphous $Al_{92}Sm_8$ measured by fluctuation electron microscopy Appl. Phys. Lett., 86 (14) (2005), p. 141910
32. W.G. Stratton, J. Hamann, J.H. Perepezko, P.M. Voyles, X. Mao, S.V. Khare Medium-range order in high Al-content amorphous alloys measured by fluctuation electron microscopy Microsc. Microanal., 10 (SupplementS02) (2004), pp. 788-789
33. K. Darmawikarta, S. Raoux, P. Tchoulfian, T.T. Li, J.R. Abelson Evolution of subcritical nuclei in nitrogen-alloyed $Ge_2Sb_2Te_5$ J. Appl. Phys., 112 (12) (2012), p. 124907
34. J. Hwang, P.M. Voyles Variable resolution fluctuation electron microscopy on Cu–Zr metallic glass using a wide range of coherent STEM probe size Microsc. Microanal., 17 (01) (2011), pp. 67-74
35. J.M. Gibson, M.M. Treacy, T. Sun, N.J. Zaluzec Substantial crystalline topology in amorphous silicon Phys. Rev. Lett., 105 (2010), p. 125504
36. T.T. Li, K. Darmawikarta, J.R. Abelson Quantifying nanoscale order in amorphous materials via scattering covariance in fluctuation electron microscopy Ultramicroscopy, 133 (2013), pp. 95-100
37. R. Bassiri, M.J. Hart, R.L. Byer, K.B. Borisenko, K. Evans, M.M. Fejer, A.C. Lin, I. MacLaren, A.S. Markosyan, I.W. Martin, R.K. Route, S. Rowan Investigating the medium range order in amorphous Ta_2O_5coatings J. Phys. Conf. Ser., 522 (2014), p. 012043
38. J.P. Lewis Fast normalized cross-correlationVis. interface, 10 (1) (1995), pp. 120-123
39. T.T. Li, S.N. Bogle, J.R. AbelsonQuantitative fluctuation electron microscopy in the STEM: methods to identify, avoid, and correct for artifacts Microsc. Microanal., 20 (05) (2014), pp. 1605-1618
40. E.F. Rauch, M. Véron, J. Portillo, Y. Maniette, S. Nicolopoulos Automatic crystal orientation and phase mapping in TEM by precession diffraction Microsc. Anal., 93 (2008), pp. S5-S8
41. E.F. Rauch, J. Portillo, S. Nicolopoulos, D. Bultreys, S. Rouvimov, P. MoeckAutomated nanocrystal orientation and phase mapping in the transmission electron microscope on the basis of precession electron diffraction Z. Kristallographie, 225 (2–3) (2010), pp. 103-109
42. J. Portillo, E.F. Rauch, S. Nicolopoulos, M. Gemmi, D. Bultreys Precession electron diffraction assisted orientation mapping in the transmission electron microscope Advanced Electron Microsc. Nanomateri., 644 (2010), pp. 1-7
43. R. Beacham, A. MacRaighne, D. Maneuski, V. O'Shea, S. McVitie, D. McGrouther Medipix2/timepix detector for time resolved transmission electron microscopy J. Instrum., 6 (12) (2011), Article C12052
44. A MacRaighne, G.V. Fernandez, D. Maneuski, D. McGrouther, V. O'Shea Medipix2 as a highly flexible scanning/imaging detector for transmission electron microscopy J. Instrum., 6 (01) (2011), Article C01047
45. G. McMullan, D.M. Cattermole, S. Chen, R. Henderson, X. Llopart, C. Summerfield, L. Tlustos, A.R. Faruqi Electron imaging with Medipix2 hybrid pixel detector Ultramicroscopy, 107 (4–5) (2007), pp. 401-413
46. E.F. Rauch, M. Véron Virtual dark-field images reconstructed from electron diffraction patterns Eur. Phys. J. Appl. Phys., 66 (1) (2014), p. 10701
47. E.F. Rauch, M. Véron Coupled microstructural observations and local texture measurements with an automated crystallographic orientation mapping tool attached to a tem Mater. Werkst., 36 (10) (2005), pp. 552-556
48. A Garner, A. Gholinia, P. Frankel, M. Gass, I. MacLaren, M. PreussThe microstructure and microtexture of zirconium oxide films studied by transmission electron backscatter diffraction and automated crystal orientation mapping with transmission electron microscopy Acta Mater., 80 (2014), pp. 159-171

49. W.G. Stratton, P.M. Voyles Comparison of fluctuation electron microscopy theories and experimental methods J. Phys. Condens. Matter, 19 (45) (2007), p. 455203
50. S. Nicolopoulos, D. Bultreys, Precession electron diffraction and TEM applications www.nanomegas.com.
51. W.H. ZachariasenThe atomic arrangement in glassJ. Am. Chem. Soc., 54 (10) (1932), pp. 3841-3851
52. F.R. Eder, J. Kotakoski, U. Kaiser, J.C. Meyer A journey from order to disorder — atom by atom transformation from graphene to a 2D carbon glass Sci. Rep., 4 (2014)
53. M.R. Ebeid, S.K. Abdelraheem, E.M. Abdel-Minem, K. Abdel-Hady, A.A. Ramadan Verification of crystallite theory of glass modeling using Rietveld method, Egypt J. Sol., 23 (1) (2000)
54. L. Lichtenstein, M. Heyde, H.-J. Freund Atomic arrangement in two-dimensional silica: from crystalline to vitreous structures J. Phys. Chem. C, 116 (38) (2012), pp. 20426-20432
55. P.Y. Huang, S. Kurasch, A. Srivastava, V. Skakalova, J. Kotakoski, A.V. Krasheninnikov, R. Hovden, Q. Mao, J.C. Meyer, J. Smet, D.A. Muller, U. Kaiser Direct imaging of a two-dimensional silica glass on grapheme Nano Lett., 12 (2) (2012), pp. 1081-1086
56. A.C. Wright Crystalline-like ordering in melt-quenched network glasses J. Non-Cryst. Solids, 401 (1) (October 2014), pp. 4-26
57. K. SomiyaDetector configuration of KAGRA—the Japanese cryogenic gravitational-wave detector Classical Quantum Gravity, 29 (12) (2012), p. 124007
58. S. Hunklinger, M. Von Schickfus Amorphous Solids: Low-temperature Properties Springer-Verlag, Berlin (1981), p. 81

Index